Elementary Mathematics Teaching

Elementary Mathematics Teaching

Insights from 30 years of the Symposia on Elementary Mathematics Teaching (SEMT)

Edited by

BRIAN DOIG

Deakin University, Australia

JARMILA NOVOTNÁ

Charles University, Faculty of Education, Prague, Czech Republic and Chercheur titulaire, CeDS, Université Bordeaux, France

BERINDERJEET KAUR

National Institute of Education, Nanyang Technological University, Singapore

DAVID K. PUGALEE

University of North Carolina, Charlotte, USA

OXFORD
UNIVERSITY PRESS

OXFORD
UNIVERSITY PRESS

Great Clarendon Street, Oxford, OX2 6DP,
United Kingdom

Oxford University Press is a department of the University of Oxford.
It furthers the University's objective of excellence in research, scholarship,
and education by publishing worldwide. Oxford is a registered trade mark of
Oxford University Press in the UK and in certain other countries

Published in the United States of America by Oxford University Press
198 Madison Avenue, New York, NY 10016, United States of America

British Library Cataloguing in Publication Data

Data available

Library of Congress Control Number: 2024937887

ISBN 9780192869647

DOI: 10.1093/oso/9780192869647.001.0001

Printed and bound by
CPI Group (UK) Ltd, Croydon, CR0 4YY

This volume is dedicated to the memory of Graham Littler, a strong believer in the value of the Symposium on Elementary Mathematics Teaching (SEMT) from the beginning.

With his assistance and dedication, SEMT blossomed from a relatively small gathering to an international venue for the exchange of a wide range of educational philosophies, ideas, and practices in the teaching and learning of mathematics.

Members of the SEMT family express their debt to his hard work, devotion, and guidance.

Foreword

This book covers input to research in elementary school mathematics education over the last 30 years at the Symposium on Elementary Mathematics Teaching (SEMT) held every two years since 1991 in Prague, the Czech Republic.

The wealth of material contained in the SEMT proceedings has led to the desire to encapsulate the trends from the work presented and how this engages with similar research more generally. Chapter authors were asked to draw on the SEMT material to create chapters which explore and expand from SEMT to the wider mathematics education community.

As is to be expected, with such a range of materials to draw upon, the authors were faced with quite a daunting task. Clearly, trends across time can be exposed and investigated, but in addition, particular aspects of research into elementary mathematics teaching and learning can also be brought to attention.

A feature of the authors is that they span a wide range of backgrounds and experience in research into elementary mathematics teaching and learning. On the other hand, the work of some of the authors is possibly less well known outside their geographical location.

The instructions to potential authors were deliberately left open to interpretation to allow authors the freedom to make their chapter represent their view of the SEMT materials, and also to allow them to make connections to the broader research literature.

Bearing this in mind, the editors have collected chapters under four foci or themes. These are:

Elementary mathematics teaching
This section has a clear focus on the context of learning mathematics: that is to say, its chapters look at the SEMT materials over the years that have dealt with what the student experience has been, or could be. These contexts vary from those that offer an examination of topics like teaching those students with disadvantages, the teaching of problem solving, cultural factors, and teacher professional development.

Elementary mathematics learning
As its name suggests, this section is clearly focused on the SEMT materials that explore how students learn mathematics or, better still, how we can make learning more effective and efficient. Thus, topics in the chapters of this section raise some issues for students with special needs, as well as the needs of students in the early years of school. These two topics are well supported by two chapters, developing from SEMT materials, research about problem solving, problem posing, and problem solving with understanding.

Enhancing the teaching of mathematics
The chapters in this section have a clear focus on mathematics teachers and change. In particular, there is a focus on adaptation of foreign notions of teacher professional development, and changes in teaching ideas during the 30 years of SEMT. Other chapters address some particular teaching strategies, such as the use of visualization, and the competencies of pre-service teachers. SEMT materials can inform the reader about what surrounds the teacher in their practice: that is to say, not only how they teach, but also what affects their role in the classroom.

Enhancing the learning of mathematics
This section draws on SEMT materials that have presented aspects of the classroom that can be enhanced through a focus on students and their learning. Two of the chapters examine the usefulness of Japanese Lesson Study for improving the participation of students in the mathematics classroom, with less 'teaching as telling' and more mathematical discussion by the students. Other chapters deal with mathematical problem solving and problem posing, and how SEMT presentations demonstrate the critical role of these in the mathematical learning of elementary students.

What is in these chapters for the reader?

The relevance to the reader is to be found in chapters that feature worked examples, case studies, and activities, as well as a wealth of references on all topics canvassed by the authors. Thus, while providing some historical background, there is also current thinking on a large range of topics. Given that the authors come from a large range of educational cultures, there are many insights to be had into how things are thought about, what options are considered, and how mathematics education is implemented, in many different educational cultures.

Why another book?

To be precise, this book is unique. The outstanding list of authors and the distinctive use of SEMT presentations to ground the links to current practices worldwide are two unique features of this book. Further, the fact that the chapters are not simply records of current practice, but are embedded within an historical setting, makes this book an outstanding contribution to the literature on elementary mathematics education.

About the SEMT

It should be noted, too, that SEMT is not all work! Within SEMT there is time for attendees to enjoy aspects of Czech life. The SEMT dinner provides Czech food and drink, and social opportunities for attendees to meet with others, and discuss topics of common interest. In addition, SEMT excursions provide a chance for attendees to learn about Czech history and visit some highlights of Prague or the surrounding countryside. Every night there are concerts and recitals in many venues in Prague, and all within easy walking distance from the University. Unfortunately, these delights of the Czech Republic are not mentioned in the following chapters.

In order to simplify the author's work, and the readers' reading, we have adopted a non-standard form for referencing of SEMT presentations:

a) All SEMT symposia proceedings are published by the Pedagogical Faculty of Charles University in Prague, and this remains constant.
b) Further, the date of the reference tells the reader the year of the SEMT, all of which are odd years.
c) Finally, as the editors have changed over the years, these are retained in the referencing.

Thus, the reference below would normally be referenced as, for example:

Askew, M., Brown, M., Denvir, H., and Rhodes, V. (1999). Numeracy in early years: Building on children's understanding or delivering the curriculum? In M. Hejný and J. Novotná (Eds), SEMT (pp. 120–124). Prague, The Czech Republic: Charles University.

But in this volume it appears as:

Askew, M., Brown, M., Denvir, H., and Rhodes, V. (1999). Numeracy in early years: Building on children's understanding or delivering the curriculum? In M. Hejný and J. Novotná (Eds) SEMT 120–124.

Here, the year of citation is the year of the SEMT symposium, as per (b) above, the place and publisher are omitted, see (a), as they are constant, and the editors are included, as per (c). The page numbers complete the reference.

If you cite SEMT materials elsewhere, you will need to restore the missing details, which are: Prague, The Czech Republic (place) and Charles University (publisher).

We hope that this arrangement makes reading the references simpler and clearer.

The future

Finally, we hope that the chapters in this volume stimulate you to consider attending SEMT, presenting your work, and thus, becoming part of the ever-growing SEMT family.

The SEMT family will welcome you to a city, culture, and research that may change your life!

Brian Doig
Jarmila Novotná
Berinderjeet Kaur
David Pugalee

Contents

SECTION 3 ENHANCING THE TEACHING OF MATHEMATICS

Brian Doig

SECTION 4 ASPECTS OF LEARNING MATHEMATICS

David Pugalee

A History of SEMT

Jarmila Novotná

Can we do more to support pre-service and in-service primary school teachers and their educators in their struggle for more effective, interesting, and challenging mathematics teaching? Are 5- to 12-year-old pupils' attitudes towards mathematics and their needs and abilities different? Should more attention be paid to the teaching methods, contents, and activities used in teaching these pupils? Does this group of pupils have any special needs? Are there any cognitive and psychological limits and boundaries that must be respected? What is their prior experience with mathematics with which they enter primary school, and which can be built upon? Are there any special tools and aids that can be successfully used for this specific target group?

These are just some of the questions that made Czech primary school mathematics educators and researchers realize that the platform for the discussion of these issues was not easily available, as international scientific events did not specialize in this age range, as a consequence of which there was a lack of space for discussing the questions specific to this age group. This was the stimulus to organize an international conference focusing on teaching and learning mathematics for children up to the age of 12 years, and teacher education for them. The answer to this need was the Symposium on Elementary Mathematics Teaching (SEMT) that focuses on the teaching of mathematics to children within the age range of 5 to 12 years.

A child, when conceived, has two parents. Our child, SEMT, also has two parents—two colleagues from Charles University in Prague: Michaela Kaslová and Jarmila Novotná. Conceived in 1990, SEMT was born in August 1991 as the only conference focusing on the teaching and learning of elementary mathematics.

Thus, the first SEMT symposium took place in 1991, with the participation of some 39 educators and researchers, and since then, SEMT has occurred every two years. It was organized to be in Prague, at Charles University, in the Faculty of Education, by the Department of Mathematics and Mathematical Education staff. This remains true to this day.

Since its inception, there have been some changes to SEMT. For example, the language of SEMT has changed, so that while in 1991 French and English were used, since 1993 the language has been English only.

Similarly, over time, the programme of events has changed. Whereas in 1991, research reports were the only form of presentation, in 1993, plenary addresses were added, with posters added to the programme in 1995. Finally, in 1999, workshops were introduced and have become a much-appreciated feature of the symposia.

Originally, one committee took care of both programme details and organizational issues, but for SEMT 1995, the first International Programme Committee (IPC) was established, and in 2013, the first Advisory Board was established.

SEMT has also had a number of 'aunts, uncles, and friends' supporting its development. The 'child' has been gradually growing up, developing and assimilating new ideas, meeting new people, both from the Czech Republic and abroad, who helped its development. The SEMT community can now boast a relatively stable community of regular participants while attracting the attention of newcomers from very distant corners of the world. The community involves colleagues from most countries in Europe, the Middle East, Japan, Australia, and America. The mix of nationalities and the return of many old friends contribute considerably to the warmth and friendliness, which epitomizes the SEMT symposia.

Each SEMT focuses on one important central theme in elementary mathematics teaching, and these themes, which are set out below, are a strong indication of the concerns of the SEMT family at the time of the symposium. The themes are:

- 1991: The teaching of mathematics to elementary mathematics pupils
- 1993: The changing face of elementary mathematics
- 1995: Geometry and word problems for elementary mathematics
- 1997: Assessment and evaluation
- 1999: How the world of mathematics emerges from everyday experiences of children
- 2001: What is meant by the competence and confidence of the people involved in the teaching of elementary mathematics?
- 2003: Knowledge starts with pre-conceptions
- 2005: Understanding the environment of the classroom
- 2007: Approaches to teaching mathematics at the elementary level
- 2009: The development of mathematical understanding
- 2011: The mathematical knowledge needed for teaching in elementary schools
- 2013: Tasks and tools in elementary mathematics
- 2015: Developing mathematical language and reasoning
- 2017: Equity and diversity in elementary mathematics education
- 2019: Opportunities in learning and teaching elementary mathematics
- 2021: Broadening experiences in elementary school mathematics

The multicultural background of the participants adds to the vibrancy of the symposium and the vigour of the interchanges between participants. This sharing of different perspectives on mathematics teaching has always been a strong and most attractive aspect of SEMT, as we gain a greater understanding of our own practice through learning about the practice of others.

This multi-national basis of SEMT makes the participants realize that no problem or issue is a problem of just one institution or country. Discussions with fellow participants quickly make them realize that many of the most pressing matters are common for many countries. The chance to discuss these issues in this colourful community helps, and sometimes even provides, a solution.

The range of new ideas for helping teachers to make mathematics both a meaningful and an enjoyable subject, as presented and exchanged in the SEMT history, is enormous. The number of formats for presentation, including plenary addresses, discussion groups, research reports, short oral presentations, and posters, provides enough space for both experienced and new researchers to share their ideas, research results, and concerns. It has also helped new colleagues and post-graduate students to present their first papers to a discerning but appreciative international audience.

As in all international conferences, the contributions of the participants are published in the conference proceedings. There is a strict reviewing process in which all the contributions are reviewed by an international board. For those who attend SEMT, the proceedings act not only as a reminder of their SEMT experience, but also, as the proceedings are available at registration, the participants are given the chance to select from among different parallel sessions according to their interests. Moreover, the proceedings help those who do not feel very confident about their English to follow the oral aspect of presentations with textual support.

The number of copies of the proceedings has been gradually growing, not only in response to the growing number of symposia participants but also to answer the demand of the wider academic and research community. Clearly, SEMT has become an important international event with a high scientific standard.

There is no doubt that the biannual nature of the *Symposium on Elementary Mathematics Teaching* is now a well-established tradition, as attested by its already 30-year history. The growing importance of contributions to the SEMT programme has been documented by the publication of a special issue of the *Mediterranean Journal for Research in Mathematics Education* in 2009 (Volume 8, No. 1), which contains augmented texts of selected plenary lectures from SEMT 2005 and 2007. The full list of plenary addresses follows, and many of these are also cited in the chapters that follow.

1991

No Plenary addresses

1993

- Cannizzaro: The changing faces of numbers
- Kuřina: Constructive dimension of mathematics
- Littler: There is more to mathematics than sums!
- Szetela: North American goals and standards for school mathematics in Grades K-4

1995

- Cannizzaro: On insight in problem solving activity
- Gray: Tackling the problems: an explanation for success and failure
- Hejný: Development of geometrical concepts
- Littler: Geometry for the teacher education student
- Meissner: Geometry: learning by doing
- Semadeni: Developing children's understanding of verbal arithmetical problems

1997

- Arzarello: Assessing long-term processes in the class of mathematics: the role of the teacher as a participant observer

- Berenson, Vidakovic, & Carter: North American perspectives on assessment
- Kuřina, Tichá, & Hošpesová: How to evaluate and utilize our children's geometrical experience
- Pitta & Gray: Evaluating children's mental structures in elementary arithmetic
- Wiliam: High-stakes assessment of open-ended work in mathematics: problems and possibilities

1999

- Groves: Calculators, budgerigars, and milk cartons: linking school mathematics to young children's reality
- Hošpesová: Addition and subtraction concepts in school education
- Lanciano: Look at the sky and the earth through the eyes of geometry
- Pehkonen: On the use of open approach to promote an educational change in mathematics

2001

- Clarke: Understanding, assessing, and developing young children's mathematical thinking: the experience of the Early Numeracy research project
- Gruzczyk-Kolczynska: Why the solving of mathematical problems is so difficult for children? Intellectual and emotional conditions of mathematics education
- Tall: What mathematics is needed by teachers of young children?
- Wittmann: The O-Script/A-Script method

2003

- Mason: Structure of attention in the learning of mathematics
- Tichá: Following the path of discovering fractions
- Pessia Tsamir & Tirosh: Errors in an in-service mathematics teacher classroom: what do we know about errors in the classroom?
- Wollring: Linking pre-service and in-service teacher training: co-operative design and Dissemination of working environments for primary mathematics

2005

- Bednarz: A mathematics teaching approach in "weak classes": a passage from elementary to secondary level rooted in meaning making construal
- van den Heuvel-Panhuizen: Children's perspectives of the mathematics classroom
- Sarrazy & Novotná: Didactical contract: theoretical frame for the analysis of phenomena of teaching mathematics
- Steinbring: Children's ways of mathematical argumentation in the classroom Environment

2007

- Ruthven: Towards a calculator-aware number curriculum
- Scherer: Investigating children learning mathematics/investigating mathematics
- Watson: The development of statistical understanding at the elementary school level
- Zazkis: Number theory in mathematics education: queen and servant

2009

- Cheeseman: Challenging children to think: teacher behaviours that stimulate young children to probe their mathematical understanding
- Hart: Why do we expect so much?
- Hershkovitz: Intuition, schema, and problem solving
- Kilpatrick: Conceptual understanding as a strand of mathematical proficiency
- 2011
- Bartolini-Bussi, Canalini, & Ferri: Towards cultural analysis of content: problems with variation in primary school
- Spangler: Facilitating knowledge development and refinement in elementary school teachers
- Vinner: What should we expect from somebody who teaches mathematics in elementary schools?
- Wittmann: Early mathematical education: a plea for mathematically founded conceptions

2013

- Chapman: Engaging children in learner-focused mathematical tasks
- Griffiths: Working with children in public care who have difficulties in mathematics
- Krauthausen: Digital media in elementary mathematics education
- Mulligan: Inspiring young children's mathematical thinking through pattern and structure
- Young-Loveridge & Bicknell: Using multiplication and division tasks to support young children's part–whole thinking in mathematics

2015

- Fujii: Designing and adapting tasks in the Japanese lesson study: focusing on the role of the quasi-variable
- Nührenbörger: Mathematical argumentation processes of children between calculation and conversion
- Peled: Modelling tasks: what develops?
- Poirier & Bisaillon: The role of language and communication in the learning of maths in primary school: some examples

2017

- Ginsburg & Uscianowski: Stories, stories, and more math stories
- Graven: Blending elementary mathematics education research with development for equity—an ethical imperative enabling qualitatively richer work
- Janík: From content to meaning: semantics of teaching in the tradition of Bildung-centred didactics
- Levenson: Promoting mathematical creativity in heterogeneous classes

2019

- Kaur: Opportunity to learn: Reason and communicate
- Savard: How teaching mathematics in elementary school can support financial literacy education?
- Tzur: Elementary conceptual progressions (trajectories): reality check + implications

2021

- Civil & Hunter: Reflections on participation in elementary school mathematics: views from two different contexts
- Goos: Numeracy across the curriculum as a stimulus for interdisciplinary inquiry in elementary school mathematics
- Straková: Mathematical education in the context of educational inequalities in Czech primary schools

With the symposium venue in the very centre of Prague, SEMT offers its participants unforgettable social and cultural experiences. The symposium organizers are well aware of the fact that it is often during informal meetings that ideas are born, and so they pay careful attention to the organization of the symposium dinner, a welcome party, and excursions. Good food, live music, and historical surroundings all foster an atmosphere that prepares the ground for making new friends and starting co-operative research. A testimony to this is the increasing number of members of the SEMT family over the years, and continuing today.

SECTION 1
ELEMENTARY MATHEMATICS TEACHING

Jarmila Novotná
Charles University, Faculty of Education, Prague, Czech Republic;
Chercheur titulaire, CeDS, Université Bordeaux, France

Introduction

This section has a clear focus on the teacher of mathematics, since the context in which mathematics is taught has a major impact on a teacher's work and their professional training. The context of the successful learning of mathematics concerns teacher education and the effects it may have on students' mathematical working and attitude. It covers a wide range of topics, from classroom teaching to teacher education. Materials of all types from the Symposium on Elementary Mathematics Teaching (SEMT) can be a source of information about what surrounds the teacher in their practices—that is to say, not only how they teach, but also what affects their role in the classroom.

The content of this section focuses on the following issues, each of which is the concern of one chapter, and based on SEMT presentations:

- What interventions are there to create equity in the under-resourced and underserved contexts of teaching and learning, and what is known about them?
- What are good practices, and what knowledge is necessary among teachers, in the domain of mathematical problem solving?
- What developments have been seen in the research focusing on cultural and linguistic heterogeneity in mathematics classrooms, and what are the trends in this domain?
- What is appropriate for pre-primary and primary teachers' professional development, leading to the development of teachers as agents of change?

Thus, the first chapter has a focus on interventions whose objective is to strengthen teaching mathematics in contexts of disadvantage. Graven and Venkat give an overview of 14 SEMT presentations, focusing on interventions that aim for equity for learners in contexts of disadvantage. They focus on different ways of strengthening teaching and learning in such conditions, and deal with the situation in nine countries, both developed and developing. The variety of environments in which the issue is studied offers the reader interesting comparisons between approaches in different national settings, which may well stimulate emulation or assist in local adaptation.

The learning of mathematics at all levels is largely based on problem solving. Brousseau (1997, p. 43) states that:

> The more the student has been drilled in formal exercises, the more it is difficult for her, later, to restore a fruitful functioning of concepts so acquired. 'Application' of learned, ready-made knowledge goes badly because the logic of the articulation of the acquisitions that compose it is exclusively that of the knowledge itself and because the role of situations has been excluded *a priori.* (p. 43)

Thus, Piñeiro and Castro-Rodríguez, in Chapter 1.2, review SEMT presentations, focusing on those addressing teachers' knowledge of mathematical problem solving. Besides this, they describe good practices identified in pre-service education of primary teachers and their professional development. This chapter examines and analyses practices from 14 different countries.

Chapter 1.3, by Novotná and Moraová, is focused on the topic of linguistic and cultural heterogeneity, which has become important in recent decades. This topic is very complex, and can be perceived and studied from many different perspectives, from theoretical aspects of language and communication in classrooms to practical aspects such as the materials suitable for these classrooms, and their production, evaluation, and implementation. Novotná and Moraová's analysis clearly shows that there is a growing research interest in this topic. This theme is not isolated in the educational domain, and is, among others, very closely linked to the issues discussed by Graven and Venkat in their earlier chapter.

In the final chapter, Hošpesová and Kaur present a survey of SEMT presentations focusing on the professional development (PD) of pre-primary and primary teachers. They emphasize the importance of this topic for these teachers, as they are usually generalists, and only a minority of them are specialized in teaching mathematics. The authors focus on three broad themes: *Development of teachers' mathematics knowledge for teaching; Context, content, and structure of PD programmes; and Impact of PD programmes on participants's instruction.* This topic has gained increasing research attention in recent decades. In different countries it may have different forms, be organized in different ways, but it is always about teachers learning and transforming their knowledge into practice with the objective of supporting students' growth. However, there are differences in the impact of PD programmes on individual teachers and groups of teachers. Therefore, it is of utmost importance to study and discuss these programmes and their outcomes. This importance is mirrored in the number of presentations analysed from 21 countries.

Reference

Brousseau, G. (1997). *Theory of Didactical Situations in Mathematics 1970–1990.* Dordrecht, The Netherlands: Kluwer Academic Publishers.

1.1

Interventions to strengthen teaching and learning in contexts of disadvantage

Mellony Graven
Rhodes University, South Africa
Hansa Venkat
University of the Witwatersrand, Johannesburg, South Africa

Introduction

In this chapter, we look at the different ways in which intervention studies directed at strengthening teaching and learning in contexts of disadvantage featured in SEMT presentations. This focus was motivated by the twin drivers of increasing globalization of research activity in mathematics education over the last two decades and evidence of increasing inequality, both globally and within several specific nations. SEMT, like other symposia in the mathematics education field, has been both witness and party to these global tides. Given our work on research-based interventions in South Africa, we were particularly interested in understanding how these tides have featured in the work of SEMT, and whether the emphases have changed over time.

In the global context, two interlinked currents continue to pull in opposing directions. Firstly, there is the push for access to universal primary education (UPE), where, while significant gains have been achieved in some countries in the last two decades, global initiatives are described as having 'stalled' in terms of progress (UNESCO Institute for Statistics and UNICEF 2015). Alongside this, growing concerns have been raised about a lack of epistemic access for many children in the context of increasing physical access to schooling (Pritchett 2013), with this push encompassing countries with established mass access to education and those where mass access is relatively recent. Across developed and developing country systems, strong associations continue to be seen between lack of epistemic access and poverty and disadvantage. Intervention research seeking to address this problem and work towards greater equity is the focus of this chapter.

One distinction between developed and developing country systems proved particularly salient to our reading of the ways in which this problem has featured over time at SEMT. This distinction relates to questions of scale. Specifically, a key difference across developed and developing countries is that, in the latter, the 'marginalized' striving for access are more likely to be the majority rather than a minority. This distinction renders imperatives for scaling-up interventions showing promise as a more urgent and broad-based need in developing countries. This is not to say that

Mellony Graven and Hansa Venkat, *Interventions to strengthen teaching and learning in contexts of disadvantage*. In: *Elementary Mathematics Teaching*. Edited by: Brian Doig, Jarmila Novotná, Berinderjeet Kaur and David K. Pugalee, Oxford University Press. © Oxford University Press (2024). DOI: 10.1093/oso/9780192869647.003.0001 smaller-scale intervention studies are not prevalent in the developing world—our review of SEMT presentations indeed found such studies in Mexico and South Africa. But it does point to some differences in the scale (and scalability) of the studies that we overviewed for this chapter in ways that reflected the majority, and minority, marginalized contextual background. Fundamentally, what this translates into is a more urgent imperative to find ways of working with the system that provides schooling for the poor—that is, the general education system for public schools—rather than minority sectors within the general education. This distinction therefore informs the way we have structured this chapter. We begin with an outline of our approach to identifying SEMT presentations related to the chapter theme of intervention studies, focused on strengthening teaching and learning in contexts of disadvantage, and present a tabular summary of the presentations falling within this remit. We then go on to offer brief summaries of these, sorted according to the developed or developing country, and minority or majority marginalized, distinction. We follow this with an analysis of emergent sub-themes within the two categories and of the ways in which the primary distinction of scale plays into the methodologies, approaches, and implications of presentations in the two overarching categories. This analysis leads into a brief outline and discussion of key features of an intervention project that we worked on collaboratively, and implemented in South Africa. This illustration highlights several of the distinctions and sub-themes identified in the first part of the chapter.

Approach taken to the review of SEMT

Our review includes 14 SEMT presentations that we identified that focused on interventions aimed at equity for learners in contexts of disadvantage. In order to identify presentations on this topic, we scanned the titles, abstracts, and keywords of all plenaries and research presentations in the proceedings, from when the first SEMT proceedings were published to the most recent from SEMT 2019. In order to focus our review on the topic, and to find a manageable and meaningful set of presentations for the topic, we chose to only review those presentations that explicitly indicated in their titles or abstracts that their focus was on *interventions* for more *equitable* mathematical access in contexts of *disadvantage*. We thus do not include presentations in which research and, or, development work relates more tangentially to the topic, or that relate to only one or two of these three highlighted aspects. Thus, we wish to acknowledge that other SEMT presentations (not included in our review) considered issues of equity and, or, spoke to the implications of their research for interventions. For example, especially in SEMT 2017, where equity was built into the theme of the symposium, several presentations that address the issue of equity have been excluded from our review because they were not focused on an intervention. For example, Kasmer, Harrison, and Billings (2017) explore the accessibility of mathematics textbooks written in English for multilingual learners in Tanzania, where the medium of instruction in elementary schools is English. The presentation highlights implications for learning and equity, but we did not include it in our review, as we have focused on interventions. On the other hand, presentations that focused on interventions for poorly performing learners, or learners 'at risk', within a general education context (rather than within a context of disadvantage in particular) were also excluded from our review (e.g. Pearn 1999).

Table 1.1.1 below provides a summary of the 14 presentations we found on our topic. The table indicates the year the presentation appeared at SEMT, the country in which the intervention occurred, the focal age or grade indicated, and the equity focus or specific context of disadvantage, along with the participants in the intervention. Using UNESCO statistics for the proportions of children of primary age out of school in 1991. and a range of background economic data from 1991 (https://countryeconomy.com/gdp?year=1991) as proxy indicators for developing and developed mass education systems, we were also able to classify the presentations in the review based on their location in one of these two categories.

In terms of regional distribution, the 14 presentations come from nine countries, namely: the USA (1), Canada (1), Slovakia (1), Germany (1), Sweden (1), New Zealand (1), Mexico (1), the UK (3), and South Africa (4). Most countries have thus contributed single presentations, and, while the UK contributes three presentations, these are from a single contributor. South Africa has contributed the highest number of presentations (4).

A summary of SEMT presentations on intervention studies in contexts of disadvantage

Given that the first SEMT symposium was held in 1991, it is of interest that the first presentation on our focus topic appeared in 2005. This was a presentation by Bednarz. While there were mostly single presentations on the topic in subsequent SEMTs, the two most recent symposia have seen much expansion on the topic. That is, of the 14 presentations covered in our review, more than half of them (9) have been presented at the most recent SEMTs in 2017 and 2019. While from 2005 to 2015 there were only single presentations on the topic of interventions for equity in under-resourced or under-served learning and teaching contexts, in 2017, in coherence with the theme of the symposium—namely 'Equity and diversity in elementary mathematics education'—the number of contributions jumped to six. Looking through the SEMT themes, this was the first symposium with a theme focused on equity, or equity-related issues. It should be noted, however, that symposia proceedings prior to 2007 do not indicate themes.

The six 2017 SEMT presentations, focused on interventions for equity, span developing countries (Mexico and South Africa) and developed countries (Slovakia, New Zealand, and Germany). While the 2019 SEMT theme was not explicitly focused on equity, it focused on *Opportunities in Learning and Teaching Elementary Mathematics*. Three 2019 presentations focused on providing equitable opportunities in learning and teaching in relation to contexts of disadvantage or disadvantaged groups of learners. The three presentations spanned both the Northern and Southern Hemispheres, and developed and developing countries (the USA, Sweden, and South Africa).

While we see that the presentations span nine countries, several continents, the Northern and Southern Hemispheres, and developed and developing countries, the majority of presentations (9) have emerged from developed countries in the Northern

Hemisphere. In these presentations the equity focus is on marginalized minority groups who are under-resourced, or under-served, in terms of their educational needs within wealthier functioning education systems. This said, since 2015, when the first South African presentation appears, there has been an even balance of presentations focused on interventions in wealthier and poorer countries.

Looking across the presentations, a distinction worth noting between presentations emerging in wealthier and poorer countries is that the interventions in wealthier countries focus on minority groups within the education system (Bednarz 2005; Griffiths 2007; Hunter and Miller 2017; Žilková, Kopáčová, and Rybanský 2017; Larsson, Andrews, and Ridderlind 2019; Haydar 2019), while the interventions in poorer countries focus on low socio-economic status (SES) groups representative of the majority of learners in the country (Stott and Graven 2015; Graven 2017; Višňovská and Cortina 2017). Below we provide brief summaries of the presentations across the categories of developed country intervention presentations and developing country intervention presentations. At the end of each section, we pull together important ways in which these presentations, though comprising small numbers, derive from, and contribute to, broader currents in the mathematics education research literature, addressing issues of access and equity for marginalized learners.

Developed country presentations: equity-focused interventions for marginalized minority groups

At SEMT 2005, the Bednarz presentation focused on a collaborative research intervention in one classroom in an underprivileged secondary school in Canada. The classroom intervention involved co-teaching by the researcher and the teacher, and the intervention itself was jointly constructed. A key focus of the teaching involved the use of games 'in an attempt to engage students differently in mathematical learning' (Bednarz 2005, p. 16). Furthermore activities and concepts were embedded in meaningful contexts, with an emphasis on the use of visual representations to support conceptual development.

After Bednarz' SEMT 2005 presentation, at the next two SEMT symposia (2007, 2009) and then SEMT 2013, Griffiths presented her work on the Letterbox Club, which is focused on children aged between 7 and 11 in foster care in the UK. In her 2007 presentation, Griffiths presented the findings of the pilot Letterbox Club from 2003 to 2006 across urban and rural areas. Twenty children were enrolled in the club in the first year, and another 37 children in the second year. The intervention involved providing foster-care children with membership of the club, which provided a package of mathematics books and activities tailored to their particular reading and numeracy level. These included 'fill-in practice worksheets; fill-in puzzle sheets; card games; board games; and bingo games' (p. 122) to be completed in five-to-ten-minute sessions. Data were obtained from 23 of the 57 children, and showed positive responses in terms of enjoyment of activities and positive feedback from carers. Griffiths (2007) emphasized the importance of sending the parcels to the child rather than the adult, noting how this enabled personal involvement. She emphasized the importance of the activities being accessible and at the appropriate level, and ensuring

Table 1.1.1 Summary of presentations focused explicitly on interventions for equity in contexts of disadvantage

Year	Author/s	Country (Developed or Developing)	Focus age or Grade	Equity focus, context of disadvantage, and participants
2005	Bednarz	Canada Developed	12–15	Underprivileged school, poorly performing learners 1 teacher 1 class of 28 students
2007	Griffiths	UK Developed	7–11	Children in foster care—pilot involved 37 children Later expanded to hundreds of children
2009	Griffiths	UK Developed	7–11	Children in foster care—focus of presentation on design of assessment items used with > 1400 children
2013	Griffiths	UK Developed	7–10	Children in foster care Case study of 5 children
2015	Stott and Graven	South Africa Developing	8–11	Low SES Under-performing 1 teacher 1 class for pilot, then > 26 teachers and classes
2017	Višňovská and Cortina	Mexico Developing	Grade 5	Under-performing and low SES 1 teacher 1 class
2017	Hunter and Miller	New Zealand Developed	6 Grade 2	Pāsifika and Maori learners, under-performing, low SES, high poverty 1 teacher 1 class of 27 learners
2017	Žilková, Kopáčová, and Rybanský	Slovakia	Grade 4	Low socio-economic status, under-performing
2017	Stott	South Africa Developed	9–12	Low SES, under-performing Research on 2 clubs (20 children), presentation focuses on 2 children

Continued

Table 1.1.1 *Continued*

Year	Author/s	Country (Developed or Developing)	Focus age or Grade	Equity focus, context of disadvantage, and participants
2017	Götze	Germany Developed	Grades 3–4	Low SES, including refugee learners, German-language learners 4 classes, 65 children
2017	Graven	South Africa Developing	5–11 Grades R-6	Low SES, under-performing >100 teachers and district officials involved in PD, multiple clubs of learners, multiple family mathematics programmes and events, multiple researchers involved in development programmes
2019	Larsen, Andrews, and Ridderland	Sweden Developed	7–11	Children in foster care and in families receiving economic support Low SES Pilot 77 learners (and families)
2019	Haydar	USA Developed	Grades 1–3	Emergent bilinguals, under-performing in inner-city schools Prospective and beginning elementary teachers with their emergent bilingual students Research included recordings of 39 interviews, focused on two case studies
2019	Venkat	South Africa Developing	Grades 1–3	Low SES Under-performance >100 teachers and district officials involved in PD and class interventions

that all resources needed were provided and that the games were adaptable. Her 2009 presentation expanded on this intervention and shared an approach to developing assessment items for use with a larger scaled-up national pilot involving over 1400 children in foster care. Her 2013 presentation again focused on supporting children in public care, this time featuring case study research with two children aged seven and nine, and shared pedagogical approaches that proved successful in working with these children over a year. The approaches shared that were found to be useful with the children included using the clinical interview, using counting equipment, video recordings, and stimulated recall, using multiple problem-solving strategies, having learners pose problems, and providing opportunity for learning number facts.

Moving to Europe, Götze, in 2017, shared her language-and-mathematics-integrated intervention aimed at supporting learner understanding of division and divisibility, with 45 Grade 3 and 4 learners in German schools with low-SES catchment areas. Götze (2017) argued that there is a lack of equal opportunity in the German school system for children from migrant backgrounds (including refugees) who do not speak German (the language of instruction). Aiming to support these students' learning opportunities Götze introduced a design experiment focused on a rich learning task, namely the partition tree. The task involved a starting natural number progressively partitioned into its natural divisors. Four pre-service teachers were trained to be resource teachers for four classes in four low-SES schools. Their training involved an explicit focus on providing deliberate and explicit focus on 'teaching language, teaching through language, and teaching about language (Hammond & Gibbons, 2005, p. 9)' (Götze 2017, p. 180). The empirical data derived from the design experiment lessons highlighted that it was fruitful to deliberately connect multiplication and division, and that technical vocabulary support was centrally important for successful learning because 'misconceptions can only be rendered verbally diagnosable by the use of technical terms and phrases in the first place' (p. 185). Thus, a focus on minority groups of children, who are marginalized in terms of access to mathematics learning because they are not fluent in the language of instruction in German schools, is at the forefront across this body of work.

Also in Europe, Žilková, Kopáčová, and Rybanský's (2017) presentation focuses on exploring problems that Grade 4 Slovakian pupils have with the identification of rectangles. The research explored similarities and differences in pupil performance on tasks from different socio-economic backgrounds. While their presentation focused on the (N=345) pupils' results on a pencil and presentation test, the presentation linked its work and findings to key intervention programmes in Slovakia that aimed to 'mitigate the impact of socially disadvantaged backgrounds for the educational results of children, including mathematics' (p. 467). They referred to projects such as the Project of Inclusive Education (PRINED), which included a focus on improving mathematical performance for low-SES groups of students, particularly Roma children. Such pilot programmes were intended to inform the national system of education on ways to support equity for marginalized minority groups of learners who under-perform within the system. A key intention of Žilková, Kopáčová, and Rybanský's research was that they would be in a position to use their results to 'modify educational approaches in mathematics' (p. 467) for these target groups.

Moving east to the New Zealand context, at SEMT 2017, Hunter and Miller shared their study on implementing contextual Pāsifika and Maori patterning tasks aimed at supporting young learners to develop understanding of algebraic growing patterns. The intervention involved a ten-lesson teaching experiment in a low-SES, high-poverty urban school with 27 six-year-old students: 'The students were predominantly of Pāsifika descent (n=22) with three students from an indigenous New Zealand Maori background, and two students from South East Asia ... All lessons drew on a cultural context relevant to the students' background and patterns were drawn from Pāsifika and Maori culture' (p. 218). The presentation highlighted the way in which drawing on Pāsifika and Maori patterns used in craft and dance provided a positive, and successful, experience for students to explore mathematical patterns. Furthermore, the activities enabled students 'to provide factual, conceptual and symbolic generalisations' (p. 222). They advocated for the use of such cultural patterns in the teaching of young learners, in the New Zealand context, to support more equitable access to mathematical learning.

At SEMT 2019, a presentation by Larsen, Andrews, and Ridderland in a Swedish context was focused on supporting children in public care and children of families receiving long-term economic support from the government. Drawing on the aforementioned work of Griffiths in the UK, they piloted the implementation of the Letterbox Club project with 77 children from the ages of 7 to 11. Swedish adaptations were made to the materials and games. The piloted intervention was accompanied by pre- and post-tests with 48 Letterbox Club members to ascertain mathematical progress, and to compare this progress with that of their peers. The findings indicated significant improvements in the performance of Letterbox Club members between pre- to post-tests compared with that of their peers. These promising findings indicated that there was value in expanding the Letterbox Club concept as a viable intervention for increased equity in other countries and contexts, for groups of learners who are under-performing due to their being disadvantaged by personal and socio-economic circumstances.

Moving across to the USA context, in SEMT 2019, Haydar (2019) shared the results of an intervention that focused on prospective and beginning elementary school teachers, using clinical interviews with emergent bilingual students to enhance teachers own learning about the language needs of emergent bilinguals, in order to provide increased access to mathematical understanding for these learners. Noting that schools in the USA are becoming increasingly diverse 'with the emergent bilingual population representing the fastest growing group' (p. 193), Haydar argued that more must be done to develop equitable quality mathematics instruction for emergent bilinguals so as 'to narrow the persistent mathematics achievement gap' (p. 193). His intervention prepared prospective and beginner elementary teachers to conduct clinical interviews focused on creating a safe space for teachers and emergent bilinguals to 'try out' various mathematical ideas. The research included video recordings of 39 interviews conducted with elementary-grade learners in urban schools in New York. The research findings pointed to powerful teacher learning, illuminated by the case studies of two teachers, through the safe space of one-to-one interviews with emergent bilingual learners. This learning enabled prospective and beginning teachers to explicitly plan for meeting the language needs of emergent bilingual learners when teaching mathematics and thus increased learners' access to mathematical understanding.

Emerging themes from the developed country presentations

In looking across this set of presentations, several key aspects of writing in the fields of access and equity are represented. Across the Götze and Haydar presentations, the need for attention to language as a resource for learning, rather viewing it as a barrier to learning, is in focus. This strand represents an area of work that now has a long history in mathematics education, with writing that has taken in the need to look at mechanisms for supporting access, as seen in the two presentations in this review (e.g. see Moschkovich 2010), as well as the socio-cultural and socio-political issues surrounding language choice and language use in multilingual contexts (Setati 2008; Planas and Setati-Phakeng 2014).

Hunter and Miller's presentation makes reference to the body of work, pointing to the ways in which attention to culture provides an important resource base for supporting learning and access. This idea has been pulled in 'directly' in the creation of tasks and resources. This connects with earlier ethno-mathematical approaches to identifying mathematics in cultural artefacts, using these to improve access to mathematics for cultural groups whose mathematical experiences and knowledge have been marginalized and devalued within the school curriculum (Gerdes 1988). Their work also connects, in more indirect ways, with a range of theoretical approaches to supporting mathematical access and cultural affirmation of 'funds of knowledge'-based studies (González et al. 2001).

In the presentations by Griffiths and Larsson et al., interventions directed at particular student communities are in focus—that is, learners in foster care in these studies. This approach echoes the attention across all of the presentations in this category to student communities, recognized in a variety of ways, as marginalized within a country's population.

Within the clear identification of the target group—that is, the focus of the equity intervention—across all the presentations in developed countries, there is also a clear demarcation between minority student groupings and the broader population. Furthermore, and of interest here, there is also a common reference to the issues in focus as growing concerns increasingly needing attention and intervention. For example, in Haydar's note that emergent bilingual learners are a fast-growing group in the USA context, there is recognition that learning to teach mathematics effectively with bi- and multilingual learners in classrooms is a skill that more and more teachers in the USA (and elsewhere) are going to need in the context of international patterns of migration.

Developing country presentations: equity-focused interventions for a 'marginalized' majority

In SEMT 2015, moving south and from developed to developing world contexts, Stott and Graven contributed a presentation based on their work in under-resourced and low-SES schools in the South African context. While the broader project in which their work emerged focused on a range of teacher and learner interventions, including professional development (PD) programmes and after-school learner clubs, this presentation focused on the implementation of collaborative teaching of 'number talks' (after Boaler 2015a; 2015b, among others) in multiple classes. The focus of the presentation was on the experiences of the pilot, which took place with one teacher in her Grade 4 class in a fee-free (low-SES) school. Three number talks were provided, one by the teacher, one by Graven, and another by Stott. The medium of instruction at the school was isiXhosa up to the end of Grade 3, and learners had only begun learning mathematics and other subjects in English from Grade 4: this is typical of many schools in South Africa, which change to English as the medium of instruction from Grade 4 onwards. Following positive experiences by the teacher and learners, and researchers and learners, of this pilot, 'number talks' were introduced, with all teachers participating in the PD programme provided by the South African Numeracy Chair Project. The presentation shared the 26 teacher responses regarding their experiences of the number talks initiative. The most frequent response from teachers related to the way in which the talks enabled conceptual development and numeracy progression for learners.

At SEMT 2017, Graven's presentation, in line with the symposium theme, focused on mathematics education research with development for equity. In this presentation, Graven shared the multiple intervention projects that her South African Numeracy Chair Project had implemented in underprivileged South African communities, particularly in the Eastern Cape. She focused on the way in which various interventionist project spaces created opportunities for collaboration between researchers, teachers, departmental advisors, families, and learners. The presentation indicated that a key mandate of the Chair Project is that the research and implementation of interventions must speak to opportunities for national scalability. Since, in the South African context, it is the majority of learners who are living in contexts of social, economic, and educational disadvantage, the majority of learners are marginalized from successful mathematical learning. The imperative to scale up researched interventions is discussed further in the final section as we discuss our collaborative work on the Numeracy Chair, which is currently being expanded to a national intervention run by the South African Department of Basic Education (DBE).

Stott's SEMT 2017 presentation focused on researching interactive gesturing, with two young South African learners involved in a task-based interview. The presentation draws on her research and development work within the South African Numeracy Chair Project's after-school club intervention programme. The club intervention focused on developing number sense and productive learning dispositions through a range of fun problem-solving and game-based activities. Stott's research focused on two after-school clubs, which she ran with Grade 3 and 4 learners over a two-year period. The two learners in the presentation were first-language isi-Xhosa speakers, although the medium of instruction at school was English. Stott argued, from her analysis of the task-based interviews, that interactive gesturing is particularly useful for young learners where the language of learning and teaching is not the first language of the learners.

Višňovská and Cortina's SEMT 2017 presentation focuses on their 'research collaboration with Irene, a Mexican teacher who undertook to trial a research-based instructional sequence on Fractions as Measures, in her under-performing fifth grade classroom' (p. 431). The instructional sequence was guided by the instructional design of Realistic Mathematics Education, and their wish to provide all students with access to mathematical activity that is personally meaningful to them. The sequence they worked with involved 'fraction as comparer' rather than 'fraction as fracturer' (after Freudenthal's (1983) distinction). In the presentation they shared the significant fraction-learning gains of students that occurred, as well as how the instructional sequence, and the collaboration, supported Irene's learning and her transition towards more inclusive teaching.

In her SEMT 2019 presentation, 'Improving early number learning in contexts of disadvantage', Venkat described the interventions of the Wits Maths Connect Project, which she established as the South African Numeracy Chair at Witwatersrand University, with the mandate to implement research-based interventions in South African contexts of disadvantage. In this presentation Venkat shared the way in which context and conditions informed the intervention project's focus on tasks, teaching, and resources that emphasized the structure of number, and teaching for connection and coherence. The result of working over time (2011–2018) with teachers and students in ten partner schools, the presentation details the combination of research-based PD and resources that contributed to substantial improvements in learning outcomes related to more efficient working with early number structure. This contribution is discussed further as we discuss our joint Numeracy Chair work across our two institutions and provinces.

Emerging themes: developing countries

Across all five presentations in this set, we see reference to broad evidence of systemic low performance in mathematics. Further, all presentations report on learning gains of different types that point to interventions with the potential to address some of the lags in learning outcomes. In Višňovská and Cortina's (2017) work, a small-scale intervention study is reported that produced exciting learning gains. The case study scope is small here—working with just one teacher; however, the point we want to emphasize is that the problem the authors sought to address had impacts on the majority of Mexican learners and was therefore much more systemic.

In contrast, what we see in the South African work in this small set is attention, not just to the goals and outcomes of local projects, but also to the forward scaling-up work emanating from these projects. In the case of both these projects, this attention was a built-in requirement of the terms of their work. The themes that we identified from the analysis of interventions in developed country contexts—specifically language and culture as resources for teaching and learning—form part of the 'air' that is breathed in these latter studies. Graven and Venkat's presentations mention that their intervention activities took multilingual settings into account, and supported teachers' work in students' home languages, in classrooms. In working with the system in mind, Venkat, Askew, and Morrison (in press) have also detailed their rationales for developing and adapting interventions in ways that take classroom cultures on the ground into account.

Discussion of the presentations reviewed: Distinctions, sub-themes, and changes over time

All presentations pointed to positive experiences and, or, results in relation to the interventions shared and all indicated advocacy for supporting mathematical learning in such under-resourced and/or under-served contexts in order to close the performance gaps which currently have them lagging behind.

From the summary table and the descriptions of the presentations therein, we see that interventions in contexts of disadvantage did not feature with any prominence for an extended period in SEMTs. This has, however, changed more recently, with a remarkable increase in the number of contributions on the topic over the past two SEMTs. Internationalization of SEMT probably plays a role in this growth, given that the early SEMTs were more confined to Eastern Europe and key international partners, whereas more recent SEMTs have drawn on a more diverse base of international participants. The increase in engagement on this topic also coincides with the introduction of the theme of 'Equity and diversity in elementary mathematics education' at SEMT 2017.

The range of student backgrounds that formed the equity focus, across both developed and developing countries, included (and we draw on the phrases used in the presentations here): emergent bilinguals, disadvantaged learners, under-performing learners, children living in foster care, students from low-SES or poverty backgrounds, and indigenous learners. Recurrent themes relating to these equity foci are to provide access to:

- mathematical learning in the context of language diversity and multilingualism
- quality education resources in schools and homes of learners from low-SES backgrounds
- mathematical activity that is personally and culturally meaningful to learners.

A key difference between developed and developing world presentations is that, in the latter, the 'marginalized' striving for access tends to be the majority rather than a minority.

While the focal ages of learners in presentations ranged from 5 to 15 years, most presentations focused on learners between the ages of 7 and 11. While some presentations focused on teacher development interventions, several focused on working with both learners and teachers, and others focused on working directly with learners in and out of school contexts. Of note is the fact that none of the presentations dealt specifically with interventions for equity in the preschool years.

The interventions shared range from small-scale to medium- and large-scale research and development studies. The small-scale intervention studies were often the form of researcher–teacher collaborations, drawing on the design research tradition. While small-scale localized studies make up the vast majority of the presentations, several presentations focus on expanded interventions across different settings (e.g. to include rural and urban and to include multiple provinces or districts) with multiple learners and/or teachers (e.g. Žilková, Kopáčová, and Rybanský 2017; Griffiths 2009; Graven 2017; Venkat 2019). Of interest, in the developing country group, was the fact that four of the five presentations explicitly addressed the issue of scaling up, either from small-scale initial studies, or in mid-level studies (those relating to ten or more schools). What this suggests is that the resources and PD packages created in these studies are geared towards understanding and supporting the possibility of scaled-up interventions rather than focusing only on the results of the intervention locally.

An important part of this distinction may be about failures within functional systems to ensure quality. 'Maths for all' in the developed world contrasted with systemic education failures that have an impact on the majority of learners. The latter context points to the need to investigate initiatives that show possibilities for building more functional systems in the developing world. In this respect, the assumption is that what exists in the education systems in developed and developing countries is very different.

For example, the South African cluster of presentations, while similarly focused on equity for poorer learners and schools who are economically and educationally under-resourced, consider contexts where schools and learners represent a majority rather than a minority. Since this work occurs within an education system, widely considered to be failing to meet the educational needs of the majority of its learners, the work in the South African context must simultaneously attend to the building of more functional systems that serve the majority through:

- scaling up initial small-scale interventions to larger provincial and national interventions
- working simultaneously on teacher PD and materials development
- working closely with provincial and national governments.

In the following section, we discuss our joint South African Numeracy Chair work. Rather than describing the breadth of our work and the various research and development projects therein, we focus on, and exemplify, the theme of the need for scalability of interventions for equity in developing countries. Our Chair mandate is to build research and develop local partnerships with schools, teachers, communities, and district officials that speak to national possibilities for supporting the majority of learners who are under-performing mathematically in our country.

An intervention study in South Africa: an illustrative case

While in post-apartheid South Africa access to education has improved dramatically to the point where almost all children of school-going age attend classes, the quality of that education is significantly varied. Fleisch (2008) and Spaull (2013) both capture the way in which South Africa has a dual education system that leads to a bimodal distribution of mathematical achievement, whereby mathematical learning and reading performance in the wealthiest 20–25% of schools is quite strong, while for the majority of learners in the 75–80% of poorer schools performance is dismal. The majority of South African learners are thus marginalized from effective mathematical learning opportunities and thus competent mathematical participation and performance (Spaull and Kotze 2015; Graven 2014; Venkat and Spaull 2015). The conditions of teaching and learning in the proportionally small percentage of 'schools for the rich' are vastly different from those in the vast majority of 'schools for the poor' (Shalem and Hoadley 2009). This situation of massive inequality has seen South African learners performing towards the bottom of both regional and international comparative studies (Reddy et al. 2015; Spaull 2011, 2013).

This backdrop of the extreme nature of the numeracy challenges faced, and the fact that it is the majority, rather than a minority, of learners who are under-performing, gave rise to an innovative model of research and development of university-based Chairs jointly funded by private institutions, the Department of Science and Technology, and the National Research Foundation. In relation to the dual education system that Fleisch (2008) and Spaull (2013) refer to, whereby the vast majority (75–80%) of schools exist within the disadvantaged and dysfunctional system while a minority (20–25%) operate within an advantaged and functional one, our work is firmly located in the contexts, conditions, and needs of the 80% of learners, teachers, and schools. Our Chairs are mandated to: develop a range of intervention projects that have a positive impact on the schools, teachers, learners, and communities with whom we partner; grow the local research field of early mathematics teaching and learning, and researcher capacity within the field; and speak to sustainable ways forward to the national challenges of mathematics education in South Africa and beyond (see Graven and Venkat 2017).

Our addresses covered some of the intervention projects that we have both implemented since 2011, when our Chairs began. In recent years there is growing departmental interest in expanding our PD programmes (for teachers and district advisors) and the after-school mathematics club programmes to a national scale, in partnership with the South African DBE. In this chapter, however, we focus on only one of our joint Chair programmes that is already undergoing this process, following several years of localized trials with our partner schools, and then more broadly across three provinces. This focus enables us to exemplify how smaller-scale research and development programmes within our remits build towards systemic opportunities, which partner with national structures, to support equity for the majority of learners who are under-performing mathematically and learning in contexts of disadvantage.

The mental starters diagnostic assessment project began in 2015 as a collaboration initiated and led by the two South African Numeracy Chairs (the present authors), between various stakeholder representatives (professional and research organizations, departmental education districts, and non-governmental organizations (NGOs)), international experts (Professors Askew and Wright), and the DBE. Following extensive engagement, lesson starter and assessment design, and planning, the first of six identified strategies was piloted with local partner schools of the two Chairs. Using the Curriculum and Assessment Policy (CAPS) document for the Foundation Phase (Grades R, 1, 2, and 3) (DBE 2011) as a basis, the six strategies we identified as important for all learners to have developed by the end of Grade 3 are: bridging through ten; jump strategy; doubling and halving; understanding the relationship between addition and subtraction; re-ordering; and compensation. These were chosen because the research literature points to them being particularly effective in supporting learners to move on from calculation by counting.

The format for the 'intervention' to promote the teaching and learning of each strategy was a two- to three-week cycle of 10 minutes of lesson starter strategy teaching, beginning and ending with a pre- and post-assessment format. 'Guided' lesson starter outlines were provided to teachers that indicated the use of number lines and part–whole bar diagrams as essential models and tools for understanding and using the strategies. Furthermore, it was suggested that each lesson begin with a one-to-two-minute quick question session, asking learners to recall linked basic facts that were identified as needed for using the strategies. For example, to use the bridging through ten strategy, learners should be able to, among other things, be able to instantly add a single-digit number to a multiple of ten, e.g. 30 + 4 = 34; 50 + 6 = 56.

The pilot results, which included the results of the pre-and post-assessments of the 'bridging through ten' strategy in seven classes across one of the poorest provinces and one of the wealthiest province, indicated strong positive gains between pre- and post-test (see Graven and Venkat 2019). Furthermore, teachers gave positive feedback regarding their experience of implementing the two-to-three-week assessment-lesson starter cycle. This led to a broader and more nationally representative DBE-led trial of the model with the strategy. In these trials, three provinces were included and a much larger and more representative sample of schools and learners were selected to enable statistical analysis of the results. Summaries of the results are provided below (see Table 1.1.1 and Table 1.1.2). These results were presented at the 2020 conference of the Southern African Association for Research in Mathematics, Science and Technology Education (Venkat et al. 2020).

The national pilot data were important for understanding and informing the possibilities for scaling up such an intervention that had previously only shown promise when run with our local partner teachers and schools. For the national pilot, our respective Chair teams ran a half-day training session for Foundation Phase mathematics subject advisors, and other district personnel from the district offices in the provinces. This training provided support for the subject advisors to run similar training sessions on the intervention model for the sampled Grade 3 teachers from nine schools in their districts. District office personnel were responsible for administering, marking, and capturing the data of the pre- and post-tests. The scripts and spreadsheets were submitted to us for checking (we randomly selected one in five scripts). The low prevalence of errors indicated the likelihood that the assessments could be administered effectively and departmentally on a national scale. The national-level sampling is summarized in Table 1.1.2 below, while the overall results of our pre-and post-assessments are summarized in Table 1.1.3. Note the poorest schools within the DBE system are in Quintile 1 (Q1), while the 'least poor' are in Quintile 5 (Department of Education 1998). As Adler and Pillay (2017) note, it is only those in Q5 schools who tend to perform at the required mathematical level generally.

Table 1.1.2 shows a pleasing 17 percentage point increase overall between the pre- and the post-test. While the overall results show smaller gains on the strategic thinking and strategic calculating sections than on the rapid recall section (a mark increase of over 4/20), the results are promising. We are now working with the South African DBE to expand this intervention project to all nine provinces across the country and to the implementation of all six strategies. The DBE is currently arranging translation of the materials into all South African official languages. To this end, our team of collaborators have worked hard to complete the lesson starter guides and the pre- and post-assessments for all six strategies. These were handed to the DBE in September 2020.

Table 1.1.2 2019 national pilot sample—bridging through ten

		District 1					District 2				
		Q1	Q2	Q3	Q4	Q5	Q1	Q2	Q3	Q4	Q5
Prov 1	Learners	111	74	28	103		99	42	75	75	31
N=638	Schools	3	2	1	3		3	1	2	2	1
Prov 2	Learners	127	85	40	42	58	129		106	118	
N=705	Schools	3	2	1	1	2	3		3	3	
Prov 3	Learners			132	39	95			196	33	51
N=546	Schools			4	1	4			6	1	2

Table 1.1.3 2019 national pilot outcomes—bridging through ten

	Rapid recall Mean/20		Strategic calculating Mean/5		Strategic thinking Mean/5		Overall mean %	
n	Pre	Post	Pre	Post	Pre	Post	Pre	Post
1889	8.38	12.63	1.15	1.34	0.26	0.93	32.63	49.67

Conclusion

The growing internationalization of SEMT over the past decade has increasingly highlighted the need for interventions for equity in mathematics education, both in wealthier developed countries, where minority groups of learners are often marginalized from access to effective learning opportunities, and in poorer developing countries, where the majority of learners in the country are performing poorly and learning in challenging contexts.

In this chapter we have highlighted the many innovative contributions of authors from developing and developed countries regarding interventions for equity in contexts of disadvantage. We have highlighted the need in developing contexts, where the majority of learners are marginalized from effective mathematical learning, to have localized research and interventions that pay attention to opportunities for scalability. The increasing proportion of presentations on the topic, and the way in which these build on previous SEMT presentations, points to the power of SEMT as a platform for sharing our experiences and strengthening our interventions for equity in contexts of disadvantage.

Acknowledgements

Our thanks to our author team, Mike Askew, Lynn Bowie, Samantha Morrison, and Pamela Vale, who worked with us on the pre- and post-assessments and lesson starters. We also thank our broader team, who grappled with the author team to conceptualize these: Lawan Abdulhamid, Noluntu Baart, Mark Chetty, Busi Goba, Corin Mathews, Zanele Mofu, Debbie Stott, Thulelah Takane, Herman Tshesane, Lyn Webb, Marie Weitz, Lise Westaway, and Bob Wright. Thanks to the DBE and First Rand Foundation (FRF) for their support of this work.

This project is based on research supported by the South African Research Chairs Initiative of the Department of Science and Technology and National Research Foundation (NRF) (Grant Nos 74658 and 74703). Any opinion, finding, and conclusion or recommendation expressed in this material is that of the author(s) and the NRF does not accept any liability in this regard.

References

Adler, J., and Pillay, V. (2017). Mathematics education in South Africa. In J. Adler and A. Sfard (Eds), *Research for Educational Change: Transforming Researchers' Insights into Improvement in Mathematics Teaching and Learning*, 9–24. New York: Routledge.

Bednarz, N. (2005). A mathematics teaching approach in 'weak classes': A passage from elementary to secondary level rooted in meaning making construal. In J. Novotná (Ed.) SEMT 9–23.

Boaler, J. (2015a). Foreword: The wonder of number talks. In C. Humphreys and R. Parker, *Making Number Talks Matter: Developing Mathematical Practices and Deepening Understanding, Grades 3–10*. Portland, ME: Stenhouse Publishers.

Boaler, J. (2015b). Number talks. You cubed. Retrieved 27 March 2015 from http://www.youcubed.org/from-stanford-onlines-how-to-learn-math-for-teachers-and-parents-number-talks/

Department of Education (1998). National norms and standards for school funding in terms of the South African Schools Act, 1996 (Act No. 84, 1996). Pretoria: Government Printer.

Fleisch, B. (2008). *Primary Education in Crisis: Why South African School Children Underachieve in Reading and Mathematics*. Cape Town, South Africa: Juta.

Freudenthal, H. (1983). *Didactical Phenomenology of Mathematical Structures*. Dordrecht, The Netherlands: Kluwer.

Gerdes, P. (1988). On possible uses of traditional Angolan sand drawings in the mathematics classroom. *Educational Studies in Mathematics*, 19: 3–22. doi.org/10.1007/BF00428382

Goetze, D. (2017). Language- and mathematics-integrated intervention for understanding division and divisibility. In J. Novotná and H. Moraová (Eds) SEMT 177–187.

González, N., Andrade, R., Civil, M., and Moll, L. (2001). Bridging funds of distributed knowledge: Creating zones of practices in mathematics. *Journal of Education for Students Placed at Risk (JESPAR)*, 6(1–2): 115–132. doi.org/10.1207/S15327671ESPR0601-2_7

Graven, M. (2014). Poverty, inequality and mathematics performance: The case of South Africa's post-apartheid context. *ZDM—Mathematics Education*, 46: 1039–1049.

Graven, M. (2017). Blending elementary mathematics education research with development for equity: An ethical imperative enabling qualitatively richer work. In J. Novotná and H. Moraová (Eds) SEMT 20–31.

Graven, M., and Venkat, H. (2017). Advocating linked research and development in the primary mathematics education landscape in contexts of poverty. In M. Graven and H. Venkat (Eds), *Improving Primary Mathematics Education, Teaching and Learning*. 11–13. London: Palgrave Macmillan.

Graven, M., and Venkat, H. (2019). Piloting national diagnostic assessment for strategic calculation. *Mathematics Education Research Journal*, 33: 23–42. doi.org/10.1007/s13394-019-00291-0

Graven, M., Venkat, H., Westaway, L., and Tshesane, H. (2013). Place value without number sense: Exploring the need for mental mathematical skills assessment within the Annual National Assessments. *South African Journal of Childhood Education*, 3(2): 131–143.

Griffiths, R. (2007). The Letterbox Club: The pilot stage of a project aiming to raise achievement in number for children aged 7 to 11 who are in foster care. In J. Novotná and H. Moraová (Eds) SEMT 118–125.
Griffiths, R. (2009). Evaluating the Letterbox Club: Developing assessment items to examine the progress in number of children aged 7 to 11 in public care. In J. Novotná and H. Moraová (Eds) SEMT 126–134.
Griffiths, R. (2013). Working with children in public care who have difficulties in mathematics. In J. Novotná and H. Moraová (Eds) SEMT 21–33.
Haydar, H. (2019). Learning to enhance emergent bilinguals' access to mathematics: Elementary teachers experimenting with the clinical interview. In J. Novotná and H. Moraová (Eds) SEMT 192–201.
Hunter, J., and Miller, J. (2017). Using cultural tasks to develop growing pattern generalisations for young culturally diverse students. In J. Novotná and H. Moraová (Eds) SEMT 215–225.
Larsson, K., Andrews, P., and Ridderland, I. (2019). Mathematical games can make a difference: An intervention for children at risk. In J. Novotná and H. Moraová (Eds) SEMT 241–250.
Kasmer, L., Harrison, D., and Billings, E. (2017). Equitable learning opportunities: Textbook language accessibility in English medium classes. In J. Novotná and H. Moraová (Eds) SEMT 265–272.
Moschkovich, J. (2010). Language(s) and learning mathematics: Resources, challenges and issues for research. In J. Moschkovich (Ed.), *Language and Mathematics: Multiple Perspectives and Directions for Research*, 1–28. Charlotte, NC: Information Age Publishing.
Pearn, C. (1999). Procedures or concepts: Confusion in young children's development of mathematical understanding. In M. Hejný and J. Novotná (Eds) SEMT 80–84.
Planas, N., and Setati-Phakeng, M. (2014). On the process of gaining language as a resource in mathematics education. *ZDM—Mathematics Education*, 46(6): 883–893. doi.org/10.1007/s11858-014-0610-2
Pritchett, L. (2013). *The Rebirth of Education: Schooling Ain't Learning*. Baltimore, MD: Brookings Institution Press.
Reddy, V., Zuze, T., Visser, M., Winnaar, L., Juan, A., Prinsloo, C., Arends, F., and Rogers, S. (2015). *Beyond Benchmarks: What Twenty Years of TIMSS Data Tell Us About South African Education*. Cape Town, South Africa: HSRC Press.
Setati, M. (2008). Access to mathematics versus access to the language of power: The struggle in multilingual mathematics classrooms. *South African Journal of Education*, 28 (1): 103–116.
Shalem, Y., and Hoadley, U. (2009). The dual economy of schooling and teacher morale in South Africa. *International Studies in Sociology of Education*, 19(2): 119–134. doi.org/10.1080/09620210903257224
Spaull, N. (2011). A preliminary analysis of SACMEQ III South Africa. *Stellenbosch Economic Working Papers: 11/11.*
Spaull, N. (2013). Poverty & privilege: Primary school inequality in South Africa. *International Journal of Educational Development*, 33(5): 436–447. doi.org/10.1016/j.ijedudev.2012.09.009
Spaull, N., and Kotze, J. (2015). Starting behind and staying behind in South Africa. *International Journal of Educational Development*, 41: 13–24. doi.org/10.1016/j.ijedudev.2015.01.002
Stott, D. (2017). Interactive gesturing: Deepening understanding of mathematical activity with young learners in a South African context. In J. Novotná and H. Moraová (Eds) SEMT 393–403.
Stott, D., and Graven, M. (2015). Adapting number talks to foreground mathematical progression in South African classrooms. In J. Novotná and H. Moraová (Eds) SEMT 311–322.
UNESCO Institute for Statistics (UIS) and UNICEF (2015). Fixing the broken promise of education for all: Findings from the Global Initiative on Out-of-School Children. Montreal: UIS. doi.org/10.15220/978-92-9189-161-0-en
Venkat, H. (2019). Improving early number learning in contexts of disadvantage. In J. Novotná and H. Moraová (Eds) SEMT 40–51.
Venkat, H., and Spaull, N. (2015). What do we know about primary teachers' mathematical content knowledge in South Africa? An analysis of SACMEQ 2007. *International Journal of Educational Development*, 41: 121–130.

Venkat, H., Graven, M., Askew, M., and Vale, P. (2020). Foundation Phase diagnostic assessments for developing mental number sense: National pilot outcomes 2019. A short presentation presented at the 28th Annual Conference of the Southern African Association for Mathematics Science and Technology Education. Port Elizabeth: SAARMSTE.

Venkat, H., Askew, M., and Morrison, S. (2020). Shape-shifting Davydov's ideas for early number leaning in South Africa. *Educational Studies in Mathematics*, 106: 397–412.

Višňovská, J., and Cortina, J. (2017). Learning to support all students' fraction understanding. In J. Novotná and H. Moraová (Eds) SEMT 430–440.

Žilková, K., Kopáčová, J., and Rybanský, L. (2017). Similarities and differences in identification of rectangles of fourth-grade pupils. In J. Novotná and H. Moraová (Eds) SEMT 466–477.

1.2
Primary teachers' knowledge for teaching mathematical problem solving

Juan Luis Piñeiro
Universidad Metropolitana de Ciencias de la Educación, Chile
Elena Castro-Rodríguez
Universidad de Granada, Spain

Introduction

Teaching mathematical problem solving calls for a series of specific skills that constitute knowledge essential for teachers looking to provide effective instruction geared to improving students' learning (Chapman 2015). Research in the area has, nonetheless, tended to overlook the teachers' role, focusing rather on that of the problem solver (Chapman 2008; Lester 2013).

A number of initiatives have been undertaken on the assumption that higher-quality teaching improves student learning (da Ponte and Chapman 2006). One indicator of such quality is teachers' knowledge about classroom activities, which depends largely on their professional knowledge (Ball, Lubienski, and Mewborn 2001). In other words, teachers' knowledge has a significant impact on their classroom delivery and ultimately on what students learn (Fennema and Franke 1992). Nonetheless, as research on problem solving has assigned the teachers' role to only be of secondary importance, and knowledge models are designed and developed on the grounds of concepts (Foster, Wake, and Swan 2014), the field needs studies identifying the type of knowledge required to teach children how to solve problems (Lester 2013). With a view to filling that gap, this chapter poses two objectives. The first is to contribute to integrating the existing research, and the second to highlight SEMT-driven progress in the analysis of the skills and knowledge needed to teach problem solving. This review, consequently, aims to explore problem solving from the perspective of the part played by primary mathematics teachers. An initial synthesis of what the authors understand as problem-solving knowledge in primary education is supplemented by a discussion of how the dimensions of that knowledge have informed the review and the details of its design. Those sections are followed by a description of the results and prospects for future studies.

Juan Luis Piñeiro and Elena Castro-Rodríguez, *Primary teachers' knowledge for teaching mathematical problem solving*. In: *Elementary Mathematics Teaching*. Edited by: Brian Doig, Jarmila Novotná, Berinderjeet Kaur and David K. Pugalee, Oxford University Press. © Oxford University Press (2024). DOI: 10.1093/oso/9780192869647.003.0002

Professional knowledge for teaching problem solving

Given the slow progress made in the research to date, mathematics teachers' knowledge around problem solving has been identified as an area in need of attention (Weber and Leikin 2016). The studies found in the literature focus primarily on teachers *as* problem solvers, which has led to a paucity of presentations on problem solving from the perspective of their knowledge (Lester 2013). Earlier research along these lines shows that pre-service trainees', and in-service primary teachers', limited knowledge of problem solving (e.g. Boote and Boote 2018) has an impact on their students' ability to solve problems (Depaepe, De Corte, and Verschaffel 2010).

While no consensus has been reached on how to describe teachers' knowledge, its multi-dimensionality has been recognized. Shulman (1986), in a pioneering presentation, defined seven such dimensions. One, which has had a substantial impact on research, is a special type of knowledge that enables teachers to do their job: pedagogical content knowledge (PCK). The limitations to Shulman's description, addressed in the literature (e.g. Depaepe, Verschaffel, and Kelchtermans 2013), have induced researchers to re-interpret, specify, or design new models (e.g. Ball, Thames, and Phelps 2008) '[f]ramed predominantly around mathematical concepts' (Foster, Wake, and Swan 2014, p. 98); however, those knowledge models tend to omit the role of processes such as problem solving. Analysing problem solving as part of teachers' knowledge entails assuming that it is associated with a series of ideas arising from empirical research and the respective theoretical discourse. That premise underlies the distinction drawn here between concepts (knowing) and processes (doing) (NCTM 2000). Ideas around processes, however, do not necessarily constitute mathematical knowledge about some specific concept. To use an arithmetical word problem to its full potential, teachers must not only have a command of the implicit mathematics—i.e. which algorithms to use, or how to do so efficiently (operating 'in the head' or using one or several types of representations); they must also acknowledge when it is a problem for each student.

Foster, Wake, and Swan (2014) re-interpreted the Ball, Thames, and Phelps (2008) study by differentiating between concept and process as components of the contents. The authors noted that teachers' knowledge has generally been interpreted in terms of concepts, omitting the particularities of processes such as problem solving. For Foster, and Swan (2014), the model should conceive of content as arching over concepts and processes. From that perspective, the model deems the knowledge of subject matter to comprise common, specialized, and horizontal content knowledge, which should address concepts and processes. Pedagogical knowledge, in turn, is divided into three areas. The first is an understanding of content in terms of students, defined as the combination of concepts and processes as generally viewed by students, the contexts that encourage them to learn concepts and processes, and the difficulties they may encounter. The second is an understanding of concepts and processes in terms of teaching as related to knowledge, and the ability to use effective strategies to teach processes such as problem solving. The third is the knowledge of concepts and processes in terms of the curriculum, in order to select, and organize sequentially, the most suitable tasks that favour the development of students' skills. This model is regarded as conservative, however (Lin and Rowland 2016). Following Carrillo et al. (2018), the re-write of Ball, Thames, and Phelps' (2008) model made it more difficult to differentiate between common and specialized knowledge of concepts and processes. For instance, which way to solve a problem constitutes common content knowledge, and which specialized content knowledge? Knowing one may plausibly be thought to be tantamount to having common content knowledge and being aware of the existence of the specialized content knowledge. Nonetheless, the decision on which approach to problem solving is deemed common, and which specialized content knowledge, is not straightforward.

Chapman (2015) advanced, along those lines, by focusing specifically on what she called mathematical problem-solving knowledge for teaching (MPSKT). That approach was the result of a review that:

> began with current literature involving 'problem solving' and evolved into considering much earlier studies of 'problem solving' resulting in a sample of studies from 1922 to 2013. This sample was obtained by a search of articles on PS (problem solving) in international mathematics education journals and books involving theory/research on PS in mathematics education. (p. 20)

As the author noted, the knowledge required to teach problem solving is not organized around the categories proposed by Ball, Thames, and Phelps (2008). She contended that teaching problem solving is based on a complex network of interdependent knowledge, with components including teachers' own problem-solving proficiency, problem-solving content knowledge (problems, solving process, and problem posing) and pedagogical problem-solving knowledge (students as problem solvers and problem-solving teaching methods), and affective factors and beliefs.

Taking Chapman's (2015) approach as a starting point, and analysing curricular documents to identify the problem-solving knowledge required by teachers, a recent study identified the components that should be deemed requisite to teachers' problem-solving knowledge (Piñeiro 2019). Piñeiro refined Chapman's focus and, as in the earlier approach, identified such components as teachers' own problem-solving proficiency and their affective inclinations and beliefs. The difference lay in the degree of description provided by the dimensions associated with professional knowledge, problem-solving knowledge, and pedagogical problem-solving knowledge. Three elements were identified in connection with problem-solving knowledge: problem characterization, problem-solving process, and non-cognitive factors, and, in particular, the willingness to confront the problem-solving challenge posed. Of the four elements of pedagogical problem-solving knowledge identified, three relate to learning and one to teaching: students as problem solvers, problem solving as a worthwhile task, non-cognitive factors that affect problem solving, and the pedagogical orchestration of problem-solving instruction.

Focus of the review

In this review, which inter-relates problem solving and teachers' knowledge, certain decisions were adopted based on the following considerations. Earlier reviews of the area have shown that problem-solving research is hard to systematize (e.g. Lesh and Zawojewski 2007). Particularly, Lester (2013) stressed that much of the research conducted has ignored the teachers' role and their knowledge regarding teaching problem solving. Goos (2018) noted that this line of study 'distinguishes between research in which problem solving is the object of study and that which uses problem solving as a research tool to investigate other aspects of mathematics learning' (p. 5).

In that context, the preliminary review of the SEMT presentations on problem solving and teachers brought to light studies on teaching problem solving with no reference to teachers (e.g. Leppäaho and Ahtee 2005). Others addressed teaching problem solving but with no explicit discussion of teachers' knowledge (e.g. Novákova and Novotná 2011). Yet others touch tangentially on problem solving as related to teachers (e.g. Levenson 2017). A fourth series of presentations takes knowledge into consideration, but as the conceit is woven into a broader idea, it cannot be distinguished (e.g. Jirotková and Slezáková 2017). Consequently, given that the intention here is to highlight the contribution made by the SEMT symposia between 1991 and 2019 to primary mathematics teachers' knowledge of problem solving, the decision made was to focus the review on studies explicitly addressing problem solving, in which the participants were primary teachers. That decision was likewise influenced by the present authors' views of the particulars that comprise teachers' knowledge, as set out in the preceding section.

In light of the diversity of theoretical approaches and methodological information identified in the analysed SEMT presentations, the review focuses exclusively on the results reported. Consequently, it does not seek to generalize or compile the knowledge set out in the presentations at the SEMTs. Rather, the aim is to identify the variety characterizing the SEMTs, in the certainty that most of that diversity, while inconclusive, is useful for defining lines of study meriting future research, or information that may help teacher trainers.

The review was organized around the conceptualization of knowledge for teaching problem solving defined in the preceding section—that is, using the categories listed in Table 1.2.1 below. It should be noted that the categories were not deemed mutually exclusive, in the understanding that the content of some presentations lies under more than one heading.

Table 1.2.1 Categories defined for the review

Category	Description
Mathematical problem-solving proficiency	Teachers' own proficiency in solving mathematical problems.
Problem-solving knowledge	Knowledge of problem characterization, problem types, characteristics, and willingness to solve problems. Knowledge of stages and (heuristic) strategies, the role of meta-cognition and of the affective factors and beliefs in problem solving, and the acceptance of the problem-solving challenge.
Pedagogical problem-solving knowledge	Knowledge of students as problem solvers, problem solving as a worthwhile task, problem-solving teaching as a non-cognitive factor influencing problem-solving instruction.
Affective factors and beliefs	Teachers' attitudes and beliefs about problem solving.

Mathematical problem-solving proficiency

Lester (2013) pointed out that 'teachers should be experienced problem solvers and should have a firm grasp of what successful problem solving involves, but care should be taken not to confuse proficiency in teaching students to solve problems with expertise as problem solvers' (p. 250). In that vein, the presentations at SEMT reveal a number of factors around how future primary teachers solve problems, and how they can develop that skill.

A first group of presentations dealt with future teachers' problem-solving proficiency. A study by Koman (1995) noted that future primary school teachers solved problems in three steps: heuristic probing, with more or less random attempts to gain experience with, and a few initial ideas around, the situation posed; heuristic searching, through which they accumulated and categorized the experience, discovering new facts, formulating hypotheses, and choosing knowledge useful for solving the problem; and an aimed heuristic approach, in which they attempted to establish the action, or actions, leading to the final solution. Koman found shortcomings in mathematical knowledge to be common, however, which affect the subjects' problem-solving skills. Anastasiadou (2011) assessed the approach deployed by pre-service teachers to perform simple statistical tasks involving central tendency (mean, median, mode). The findings of that study revealed a lack of flexibility in the application of knowledge, with the subjects undertaking actions that separated each process by type of task, which prevented their use of a given concept in more than one context, or of more than one type of representation. Similarly, Cifarelli et al. (2011) compared the approach adopted by one future primary and one future secondary teacher to solve two problems that required them to apply their knowledge to compare fractions. The findings showed that, due to a limited or superficial understanding of concepts, the future primary teacher's attempts were unsuccessful, despite the deployment of several strategies.

A second group of studies focuses on the characteristics of (pre- or in-service) teacher training designed to develop problem-solving proficiency. One of the findings common to both groups of subjects was related to the use of challenging tasks to engage teachers and encourage them to enhance their proficiency. It showed that, when ranked by difficulty, non-routine sequences of problems drawn from the Mathematical Olympics constituted a useful resource by providing a significant context in which future teachers could develop their problem-solving skills (Schütz 1993). Challenging problems also induced the development of mathematical knowledge and self-confidence (Albert 2001). Albert cautioned, however, that the first problems posed must be accessible and readily solved, after which the difficulty should be gradually increased, and questions asked about the mathematics. As verbalization and personal reflection are elements crucial to problem solving, teachers should be encouraged to work in co-operative teams (Albert 2001).

A study by Tržilová (1995) showed that spreadsheets can be used as a tool to develop problem-solving proficiency. Tržilová recommended that, in the first stage, future teachers should solve simple, familiar problems, designed to enter words, numbers, and formulae on the spreadsheet. In a second stage, they should use exploration and investigation problems that drive discussion of the features of the problem, and the process used to find the solution with the spreadsheet. The rationale was that the spreadsheet enabled participants to discover patterns or prove hypotheses. In the third stage the recommendation was to return to familiar problems to afford students a sufficient command of spreadsheet use to enable them to focus on the problem. In the last stage, the problems posed should be geared to modelling the situation and predicting results.

In all the foregoing studies, tasks constituted a pivotal element. Samková (2017) stressed the importance of identifying the topics in the syllabi of the course which can be explored and investigated through problem solving—that is, those that are unfamiliar, or can be viewed from a new perspective. This author proposed four types of problems that should be assigned to future teachers: tasks requiring the use of recently acquired knowledge, tasks calling for knowledge acquired some time ago but applied from a different perspective, tasks that afford future teachers the opportunity to become acquainted with new topics, and tasks that students find particularly difficult because there is no direct or familiar pathway with which to solve them. In terms of internal structure, Samková divided tasks into three groups. The first group constituted hierarchical tasks involving two or more sub-tasks in which the final solution to some of the sub-tasks may be used as part of the initial situation in others; the second group were tasks with a dynamic starting point, comprising two or more sub-tasks posing the same question, with each sub-task adding new information or new conditions to the initial task; and the third were tasks with a dynamic outcome comprising two or more sub-tasks with the same initial situation in which each adds a new question to the initial task.

Trch and Zapotilová (2001), in turn, pointed out that classroom atmosphere also played an important role in developing problem-solving proficiency. They described three concepts for generating a propitious atmosphere: motivating situations for a teacher, motivating atmosphere, and motivating climates. They nonetheless indicated that such measures take time and that the person delivering the course should use a wide range of simple, open problems that are both appealing and stimulating, along with challenging questions. They should also propose tasks that entail student experimentation as the ultimate aim, and ones in which future teachers are free to choose the topics for experimentation; finally, they should enlarge the set of problems used each year and design new and appealing problems with which future teachers are unfamiliar.

Problem-solving knowledge

The studies under this heading dealt with the knowledge of the stages and strategies, as well as the meta-cognitive and affective factors, which have an impact on problem solving, and willingness, in particular. One of the first such studies was by Koman (1995), who stressed that knowing and thinking about such stages is a natural and suitable way to train future teachers, for it was observed that this enables them to understand problems and see how they work, while narrowing knowledge gaps in certain types of mathematical knowledge, thanks to the inter-relationships that could be defined. Tržilová (1995) described a teacher-training course in which spreadsheets constituted a tool for discussing the process involved in finding solutions and thinking through each stage. Piñeiro et al. (2019), in turn, proposed that teachers should be in possession of knowledge of problem characterization (i.e. whether a task is a problem), and a disposition (i.e. the involvement that is generated when realizing a truly problematic task). These authors also described a tool for assessing future teachers' problem-solving knowledge, suggesting its utility in pre-service training.

Pedagogical problem-solving knowledge

Lester (2013) showed that one of the elements that need to be studied to deepen the understanding of how to improve school children's problem-solving skills is the teaching and learning of that process. SEMT presentations on that topic constitute the majority of the studies reviewed hereunder. For example, the contribution by Piñeiro et al. (2019) contended that, in addition to a knowledge of problem solving, teachers also need pedagogical problem-solving knowledge, including the following elements: students as problem solvers, problem solving as a worthwhile task, any affective attitudes held by teachers and students that may have an impact on teaching and learning, and pedagogical orchestration in problem-solving instruction.

The presentations classified under the pedagogical problem-solving knowledge heading included: a set addressing an understanding of students as problem solvers, a second on problem solving as a worthwhile task, a third on pedagogical orchestration in problem-solving instruction, and a fourth dealing with (pre- and in-service) teacher-training courses on how to teach students to solve problems.

Knowledge of students as problem solvers

One of the matters highlighted in these presentations is the potential that analysing students' replies and behaviour when solving problems has for future teachers. Hošpesová (1995) found that two groups of pre-service trainees, only one of whom had had classroom experience, analysed the mistakes in students' answers to arithmetical word problems very similarly. Both groups identified the reason for the error, defining and detecting students' strategies. The school's culture on how problem solving was taught was observed to have a greater impact than training courses, however, by inducing participants to believe that discussing problem solving takes too much time; instead, it encouraged them to attach more importance to teaching students how to perform the respective operations.

A study by Kratochvílová (2001) revealed that even when future teachers had classroom experience, they were scarcely aware of both students' and their own problem-solving processes. One possible solution proposed was to broach the issue in university courses, using actual student responses and showing how such analysis can contribute to more effective learning. In that vein, Trch and Zapotilová (2001) noted that observing students as they solved problems enabled future teachers to find understandable and suitable ways to communicate with them, to discover and respect their problem-solving proficiency, to identify the difficulties and seek the underlying reasons for them, to give thought to, and accept, problem complexity, and to acknowledge the need for a better understanding of the mathematics involved.

The SEMT presentations reviewed also defined the characteristics of teachers' knowledge of students as problem solvers. Wallach and Even (2001) studied future teachers' interpretations of students' conversations, actions, and behaviour when solving problems. Teachers taking a training course were asked first to solve non-routine problems and then to watch their students as they in turn solved them, but without intervening. The aim was for the subjects to focus on understanding their students' problem-solving process. Wallach and Even reported that, depending on student responses, the future teachers misheard—i.e. heard things students did not say (e.g. students referred to specific cases which the teachers heard as generalizations); did not hear—i.e. ignored some of the things that were said (omitting attempts to find solutions, taking note only of when they were found); misconstrued suitable and unsuitable solutions—i.e. distinguished them on the grounds of their own understanding rather than the solutions accommodated by the problem; or expressed affect—i.e. joy or disappointment. Wallach and Even noted that these elements were based on a number of factors, such as affection for the students, which would explain the future teachers' misheard comments. The future teachers' solving process, and an understanding of how to obtain the result, were constraints that induced mishearing some remarks and not hearing others. The outcome was that they identified problem-solving processes only when students replied with what they deemed suitable answers, and not when they took other routes.

Chopin (2009) explored how, and why, in-service teachers helped their students solve arithmetical problems. She identified four types of feedback. In the first type, phatic regulations, the teacher detected the student's error but feedback was fairly superficial, with strategies such as suggesting the right answer by posing increasingly more explicit questions. The second type, a reminder of a rule or method associated with the relevant knowledge, entailed directly teaching or showing the knowledge needed to find the right answer. In changes in teaching approach, the third type, the didactic strategies involved were often superficial, not referring to the structure of a problem, for instance, but only to its wording or asking several students to work together to tackle the difficulty. With the fourth type, organization of a classroom-wide or team-wide debate, the teacher essentially assumed the role of discussion leader, in which instruction consisted in merely exploring the problem to seek ways to correct errors. Chopin also noted that teachers furnished feedback when watching the work performed by individual students or small teams, when monitoring their work, and when the exercise was corrected collectively.

Knowledge of problem solving as a worthwhile task

This pedagogical knowledge refers to the characteristics of problem solving as a worthwhile task, an activity geared towards learning (Piñeiro 2019). It covers the features that can make it a task with high mathematical potential, how such features can be changed through problem posing, and solving methods that may prove to be helpful for students.

A first group of SEMT presentations dealt with the selection of problems as crucial knowledge for teaching problem solving (Lester 2013). One presentation, by Novák and Uhlířová (1997), showed that future teachers overestimated the difficulty of the tasks they proposed. Priolet (2013), in turn, found that teachers took sociocultural and mathematical contexts into consideration when choosing problems for their students. The subjects also took the language used and the ways problems could be solved (in terms of the concepts and possible strategies involved) into consideration, although exactly which mathematical concepts they had in mind and whether they were aware of the mathematical structure of the problem chosen was unclear.

A second group of presentations addressed problem posing. Ruppeldtová (2009) asked future and in-service primary teachers to pose one or more questions based on given situations, such as 'Alan is 7 cm shorter than David. David is 12 cm taller than Karol. Karol is 9 cm shorter than Roman.' They were then asked to answer the questions as a student would. Ruppeldtová also observed subjects posing questions about someone's height before or after including further information, or asking students to add up all the values, operate with data pairs, find the minimum or maximum heights, or rank the dataset by height. In other words, most of the questions had a simple additive structure.

Hošpesová et al. (2015) explored the nature of the problems posed by pre- and in-service teachers in a course on inquiry-based learning. At the end of the course, participants were asked, among other things, to pose a problem that in their opinion could be applied in a classroom. They then had to put the problem posed into practice with a small group of students and write up the activity. Analysing the problems posed, Hošpesová et al. found the tasks to be insufficiently open, conceived from the perspective of a pastime, or drawn from real-life contexts. As to the degree of difficulty, some of the problems posed were 'sufficiently open' to allow for investigation, but the participants' discourse in their own write-ups revealed that they lowered the cognitive demands of the solution as they guided, or even directed, students to the right answer. The authors concluded that posing suitable problems does not guarantee effective classroom application.

Jančařík (2017) analysed the problems posed by future teachers taking an optional chess course, exploring the general characteristics, problem type, difficulty, and area of mathematics involved. He noted that the problem type most frequently posed was what he called paths, which consisted in describing the movements of a given piece on the chessboard (which promotes visualization and logical reasoning), while problems based on the movements of different pieces, and on how these threaten or do not threaten each other (which promotes guess and check, and visualization), were used by fewer participants.

Knowledge of pedagogical orchestration in problem-solving instruction

A third group of presentations explored the orchestration of problem-solving lessons. Analysing a lesson's didactic structure, forms of social organization, and variability of the arithmetical problem assigned, Sarrazy and Novotná (2005) sought to explain the effects of the environment on mathematical learning. They detected, and compared, three teaching practices. The first, labelled devolving or active pedagogy, was characterized by wide variability in organizing and handling situations, which included encouraging teamwork, using complex problems, engaging in interactive classroom work, and diversifying institutionalized problem-solving strategies. Institutionalizing, the second style, was characterized by a sparse introduction and narrow variety of situations. It was associated with conventional teaching in which the rule seemed to be to deploy the 'show-remember-apply' teaching strategy. Teachers who used this style quickly institutionalized a solving model and then presented students with increasingly difficult exercises. Each student's work was subsequently corrected individually, after which all corrections were shown on the blackboard. In the process, the teachers furnished details about the solution, and, depending on the time available, on occasion, had one of the students come to the board, either to ensure they were paying attention or to review previously acquired knowledge. The third delivery style was labelled intermediary. It was closer to institutionalizing, although sometimes the teachers introduced the situations, affording students the opportunity to tackle investigative problems and discuss their problem-solving processes. According to Sarrazy and Novotná, the style used affected how students interpreted the didactical contract. Students whose teachers' delivery was devolving were much more liable to take risks and ignore didactical contract rules. That did not, however, translate into more effective progress. Students taught by institutionalizing teachers progressed more than the others, a finding attributed by the authors to the *paradox of sensibility* (Sarrazy 1996).

Priolet and Novotná (2007) made a SEMT presentation that provided insight into this paradox. They compared ten teachers' delivery styles to identify practices that would explain the differences observed between student performances. Four of the teachers had taken a course on prior knowledge, conversion of registers of representations, and problem modelling. Four others received no training, and two had been participating for two years in a project to implement problem solving in classrooms. The findings showed that changing teachers' practices takes a long time, for although students learnt more with the teachers who had taken the course than with those who had not, the students taught by the two project participants exhibited the best performance. Those two teachers took an explicit approach that favoured perception of the connection between what was being learnt and other knowledge and problems. They provided only vague hints to encourage investigation and exploration. They also assigned tasks consisting in a series of sub-problems in which the solution to one sub-problem was the grounds for the solution to the next. And they encouraged reasoning and justification, discussion, and the formulation of ideas. All the foregoing was applied in both individual and teamwork situations, and specific assistance was provided for those struggling to identify connections and changes in the referential schemes, guiding them in the use of strategies such as breaking problems down into several simpler sub-problems.

Bacon et al. (2017), in turn, sought to understand the difficulties involved in teaching problem solving, and how teachers confronted them. They reported that part of the complexity was due to students' many different responses, the different ways problems were broached, and the strategies used. They also detected diversity in the mathematical notions applied to solving problems, either from the outset, during the process, or during classroom discussion. They consequently concluded that all those factors should be taken into consideration when preparing a lesson.

In a case study, Fritzlar (2007) asked future teachers to solve a problem that could potentially favour learning, and later assign it to their students. The author used the findings to discuss the difficulties encountered by teachers assigning a good problem. One was that lessons using mathematically complex problems were characterized by students' differing, dynamic, and simultaneous solving processes. As such processes were predictable in some cases only, they often caught teachers by surprise. In addition, students' problem exploration process was affected, at different levels, by a series of interactive pedagogical conditions and decisions. Those circumstances, which made it extremely difficult for teachers to help students avoid error, generated feelings of insecurity. Other studies yielded more optimistic findings, however, in which teachers handled such situations satisfactorily. Cifarelli et al. (2017), for instance, showed that identifying a series of concepts underlying problems helped one teacher, suggesting that her teaching orchestration improved when she understood the problem structure in advance.

Shabtay (2019) explored the links between teaching very cognitively demanding tasks or problems, classroom mathematical discourse, and the opportunity afforded students to perform mathematical tasks. The findings revealed that teacher mediation (suggesting problem breakdown into smaller parts) in students' processes lowered cognitive demand, confining classroom discourse to the initiation-response-evaluation (IRE) pattern, and consequently reducing students' opportunity to engage in mathematics.

In a similar vein, Doig and Groves (2015) sought to explain the details that make structured problem solving, and support, in the form of Japanese Lesson Study, an effective approach. The authors premised that a general structure should contain five essential and inter-related elements: the lesson planning, the problem or task selection, blackboard display practice, whole-class discussion, and teacher review and denouement. The authors noted that the inter-relationships among the elements constituted both the framework and the essence of the structured problem-solving lesson and its orchestration. They added that lessons are based on a symbiotic relationship between objectives, task chosen, and expected and actual student responses. Each item in that relationship was deemed necessary for an overall plan to ensure effective learning. Nonetheless, as Bureš and Clanché (2009) pointed out, teachers with an instrumental conception of problem solving, and the problems *per se* constrained their classroom use by teachers.

Learning to teach problem solving

A final group of presentations on pedagogical knowledge dealt with learning to teach problem solving. For example, Gaulin (1995) described a course that taught pre-service teachers to teach problem solving with approaches *for*, *about*, and *through* problem solving. He proposed that such courses should be based on the following principles: participants should experience genuine problem solving and reflect on the problems encountered to be more sensitive to their students' experience when solving problems and discuss how to perform the task with them; they should be encouraged to understand that, in teaching, problem solving fulfils a number of purposes—for example, one for each teaching approach (for, about, through)—and they should learn to integrate a problem-solving strategy through classroom experiments and implementation. Along those lines, Bartolini Bussi, Canalini, and Ferri (2011) found that when introduced to *biànshì*, a Chinese problem-solving method, teachers who already had knowledge of types of arithmetical word problems and degrees of difficulty acquired new perspectives. The authors stressed that the Chinese approach favoured comprehensive problem instruction.

In that same vein, Cifarelli et al. (2011) showed that analysing students' hypothetical answers enabled teachers to appreciate students' struggles with mathematics, and that in turn revealed the importance of asking students how they solved problems. It also required teachers to think in depth about the mathematics involved in the problem, and assess students' understanding of those concepts. Despite this, Jirotková and Zemanová (2013) noted that pre-service teachers were not very receptive to peer commentary. Approximately one-third of the participants in their study were unaffected by their peers' comments or suggestions about their own reflections on a student's response. Jirotková and Zemanová consequently concluded that teachers had a lack of ability to consider other people's ideas and greater effort will be needed to influence these teachers.

Affective factors and beliefs

Teachers' classroom delivery is greatly influenced by a system of conceptions and beliefs about how problem solving should be taught and learnt (Carrillo et al. 2018). This section reviews SEMT presentations on teachers' own conception of, and beliefs about, problem solving. Two trends were observed in this group of presentations, one assessing and another attempting to encourage beliefs compatible with genuine problem solving.

A study presented at SEMT by Dofková (2019), exemplifying the former trend, explored the degree of mathematical anxiety in pre-service teachers when solving arithmetical word problems (in everyday situations such as summing the price of various articles, making change, or adding up course credits) and non-routine problems. Participants were found to feel least anxious when registering for a mathematics course, checking the ticket after retail purchases, watching a primary school teacher solve a mathematics problem on the blackboard, purchasing mathematics course notes and other materials, listening to other students explain mathematical terms, entering a mathematics class, and opening up materials to begin to perform a task. The level of anxiety was highest during a mathematics course final class one hour prior to a mathematics examination, when the teacher assigned an impromptu examination during a mathematics lesson, or when doing the mathematical part of an entrance examination. Participants suffered further anxiety when they were expected to solve many difficult mathematical problems for a forthcoming class, one day prior to a mathematics test, or while taking a mathematics test, and when receiving their mathematics mark or seeing a page full of mathematical problems. A comparison of types of problems to be solved showed that students' mathematical anxiety was significantly higher when solving non-routine than when solving routine arithmetical word problems.

Evered (1995) explored the degree of subjects' skill and self-confidence through their discourse when engaging in teamwork sessions. He analysed the interactions between skill (based on the type of cognitive participation: facts remembered, examples, or comparisons), self-confidence (assuming more active participation to denote more self-confidence), and performance (stressing whether the problem was answered correctly, focusing on how it was broached). The analysis suggested that individual self-confidence, or skill, or both, affected team achievement, while skill predicted performance, and the combination of self-confidence and skill were closely related to team achievement. Further, team achievement could not be predicted in terms of the skill of the most self-confident members.

A study on the development of beliefs that favoured primary school teachers' problem-solving proficiency was presented by Littler and Taylor (1995). They showed that a change in teaching strategy, introduced in a teacher-training course, had an impact on subjects' attitudes towards mathematics. That strategic change involved the use of mathematical problems and investigation. Koman (1995) found that analysing challenging problem-solving processes as enigmas furthered self-confidence and ultimately contributed to the development of positive attitudes towards mathematics. Albert (2001) reported genuine problem solving (involving exploring and investigating) to improve self-confidence in that endeavour, and encouraged teachers to use that approach with their students. A study by Kamen et al. (2015) suggested that when teachers worked on this type of problem, involving exploring and investigating, their beliefs about teaching methods were more closely aligned with a genuine problem-solving approach. As Trch and Zapotilová (2001) noted, this type of problem may also serve as a basis for classroom atmospheres favourable to pre-service teacher learning.

Conclusion

Teaching problem solving is a complex endeavour. One of the factors contributing to that complexity is the lack of systematized research in the area. Most of the studies reviewed here were based on different theoretical grounds, rendering in-depth systematization very difficult, and requiring the present review to focus exclusively on the research findings. The exercise also revealed the need for comprehensive exploration of problem-solving instruction. A specific example is to be found in the presentations on task sub-division. The review shows that in some cases it furthered learning, while in others it lowered the cognitive demand involved in performing tasks. Such differences could not be fully explained in terms of task sub-division alone, however.

As a rule, according to the literature, research addressing teachers' problem-solving knowledge is conducted from the perspective of the problem solver (e.g. Chapman 2008; Lester 2013). Notwithstanding that, this review of presentations at SEMT found pedagogical knowledge to be the area dealt with more commonly than any other. That may be the result of the specific nature of this review and the interests of researchers attending SEMT symposia. However, it may also denote a difference in the issues addressed in primary, as opposed to secondary, or post-secondary, education research.

Conceptualization of what constitutes a problem as defined in the studies reviewed is a factor not to be overlooked. One set of presentations identified challenging problems as a useful tool for developing problem-solving knowledge. We interpret that to mean tasks truly problematic for participants, with no direct, familiar pathway to follow to solve them. The question that arises, then, is how do the studies not explicitly qualifying problems as challenging, or similar, conceptualize that notion? We believe that to be a key question, for one of the reasons underlying pre-service teachers' difficulties with problem solving is their failure to keep the problem solver in mind when determining when a task is, or is not, a problem (Piñeiro 2019).

We believe this review to be of use for mathematics education researchers, for it identifies the various lines of research dealing with problem solving and teacher knowledge in primary education, along with the findings published to date, and the questions that remain unanswered. The findings highlighted in this review also put forward ideas, and provide good practice, for teacher training.

References

Albert, J. (2001). Working with elementary teachers on challenge problems in mathematics. In J. Novotná and M. Hejný (Eds) SEMT 41–43.

Anastasiadou, S. D. (2011). Greek pre-service teachers' cognitive abilities in understanding central tendency measures: A multilevel statistical analysis. In J. Novotná and H. Moraová (Eds) SEMT 55–63.

Bacon, L., Bednarz, N., Lajoie, C., Maheux, J.-F., and Saboya, M. (2017). Two perspectives on diversity based on the pedagogical consultant's work on problem solving in a teaching context. In J. Novotná and H. Moraová (Eds) SEMT 53–62.

Ball, D. L., Lubienski, S. T., and Mewborn, D. S. (2001). Research on teaching mathematics: The unsolved problem of teachers' mathematical knowledge. In V. Richardson (Ed.), *Handbook of Research on Teaching*, 4th ed., 433–456. Washington, DC: AERA.

Ball, D. L., Thames, M. H., and Phelps, G. (2008). Content knowledge for teaching: What makes it special? *Journal of Teacher Education*, 59(5): 389–407. doi.org/10.1177/0022487108324554

Bartolini Bussi, M. G., Canalini, R., and Ferri, F. (2011). Towards cultural analysis of content: Problems with variation in primary school. In J. Novotná and H. Moraová (Eds) SEMT 9–21.

Boote, S. K., and Boote, D. N. (2018). ABC problem in elementary mathematics education: Arithmetic before comprehension. *Journal of Mathematics Teacher Education*, 21(2): 99–122. doi.org/10.1007/s10857-016-9350-2

Bureš, J., and Clanché, P. (2009). Problem posing as a mathematical activity: The young teachers' perspective. In J. Novotná and H. Moraová (Eds) SEMT 81–88.

Carrillo, J., Climent, N., Montes, M., Contreras, L. C., Flores-Medrano, E., Escudero-Ávila, D., ... Muñoz-Catalán, M. C. (2018). The mathematics teacher's specialised knowledge (MTSK) model. *Research in Mathematics Education*, 20(3): 236–253. doi.org/10.1080/14794802.2018.1479981

Chapman, O. (2008). Instructional practices to facilitate prospective mathematics teachers' learning of problem solving for teaching. In M. Santos-Trigo and Y. Schimizu (Eds), *ICME-11. Topic Study Group 19. Research and Development in Problem Solving in Mathematics Education*, 158–167. Monterrey, México: ICME.

Chapman, O. (2015). Mathematics teachers' knowledge for teaching problem solving. *LUMAT*, 3(1): 19–36.

Chopin, M.-P. (2009). Regulations of pupils' mistakes in teaching: What purpose do they serve? An analysis of the teaching practice in 8 classes during lessons of arithmetic. In J. Novotná and H. Moraová (Eds) SEMT 103–110.
Cifarelli, V. V., Morge, S. P., Goodson-Espy, T., and Pugalee, D. K. (2011). Examining the math content and pedagogical knowledge of pre-service teachers: How Katelyn and Matthew made sense of students' problems. In J. Novotná and H. Moraová (Eds) SEMT 111–117.
Cifarelli, V. V., Stephan, M., Pugalee, D. K., and Wang, C. (2017). Implementing an instructional sequence for solving net worth problems. In J. Novotná and H. Moraová (Eds) SEMT 148–158.
da Ponte, J. P., and Chapman, O. (2006). Mathematics teachers' knowledge and practices. In A. Gutierrez and P. Boero (Eds), *Handbook of Research on the Psychology of Mathematics Education: Past, Present and Future*, 461–494. Rotterdam, The Netherlands: Sense.
Depaepe, F., De Corte, E., and Verschaffel, L. (2010). Teachers' approaches towards word problem solving: Elaborating or restricting the problem context. *Teaching and Teacher Education*, 26(2): 152–160. doi.org/10.1016/j.tate.2009.03.016
Depaepe, F., Verschaffel, L., and Kelchtermans, G. (2013). Pedagogical content knowledge: A systematic review of the way in which the concept has pervaded mathematics educational research. *Teaching and Teacher Education*, 34: 12–25. doi.org/10.1016/J.TATE.2013.03.001
Dofková, R. (2019). Perception of mathematics anxiety among prospective teachers while solving word and non-standard problems in mathematics. In J. Novotná and H. Moraová (Eds) SEMT 148–157.
Doig, B., and Groves, S. (2015). Reaching for the heavenly bead of teaching through structured problem solving. In J. Novotná and H. Moraová (Eds) SEMT 117–127.
Evered, L. J. (1995). Problem solving and collaborative learning: A Vyigotskian approach. In M. Hejný and J. Novotná (Eds) SEMT 73–76.
Fennema, E., and Franke, M. L. (1992). Teachers' knowledge and its impact. In D. A. Grows (Ed.), *Handbook of Research on Mathematics Teaching and Learning*, 147–164. New York: Macmillan.
Foster, C., Wake, G., and Swan, M. (2014). Mathematical knowledge for teaching problem solving: Lessons from lesson study. In S. Oesterle, P. Liljedahl, C. Nicol, and D. Allan (Eds), *Proceedings of the Joint Meeting of PME 38 and PME-NA 36*, Vol. 3, 97–104. Vancouver, Canada: PME.
Fritzlar, T. (2007). Why is it so difficult to use a 'good problem' in a 'good way' for teaching? In J. Novotná and H. Moraová (Eds) SEMT 109–117.
Gaulin, C. (1995). In-service courses for elementary teachers about problem solving in mathematics. In M. Hejný and J. Novotná (Eds) SEMT 39.
Goos, M. (2018). A review of research on understanding and promoting students' mathematical thinking published in Educational Studies in Mathematics 2014–2018. In D. M. Gómez (Ed.), *Proceedings of the First PME Regional Conference: South America*, 1–16. Rancagua, Chile: PME.
Hošpesová, A. (1995). The analysis of children's solutions of word problems. In M. Hejný and J. Novotná (Eds), SEMT 85–86.
Hošpesová, A., Samková, L., Tichá, M., and Roubíček, F. (2015). How primary school teachers posed problem for inquiry based mathematics education. In J. Novotná and H. Moraová (Eds) SEMT 156–165.
Jančařík, A. (2017). The potential of the chess environment in mathematics education – pre-service teachers' perspective. In J. Novotná and H. Moraová (Eds) SEMT 225–235.
Jirotková, D. and Slezáková, J. (2017). Do teachers understand their pupils? In J. Novotná and H. Moraová (Eds) SEMT 235–245.
Jirotková, D. and Zemanová, R. (2013). Student teachers' empathy for pupils thinking process when solving process. In J. Novotná and H. Moraová (Eds) SEMT 155–162.
Kamen, M., Taylor, B., Dutcher, K., Galloway, R., and Wagui, A. (2015). Becoming a problem-based elementary mathematics teacher. In J. Novotná and H. Moraová (Eds) SEMT 381–382.
Koman, M. (1995). Word puzzles—future teachers— pupil relay in teacher education. In M. Hejný and J. Novotná (Eds) SEMT 96–99.
Kratochvílová, J. (2001). The analysis of one undergraduate student's project. In J. Novotná and M. Hejný (Eds) SEMT 101–104.
Leppäaho, H., and Ahtee, M. (2005). Teaching mathematical problem solving with the solving map method at primary school. In J. Novotná (Ed.) SEMT 194–202.

Lesh, R., and Zawojewski, J. (2007). Problem solving and modeling. In F. K. Lester (Ed.), *Second Handbook of Research on Mathematics Teaching and Learning*, Vol. 2, 763–804. Charlotte, NC: NCTM.

Lester, F. K. (2013). Thoughts about research on mathematical problem-solving instruction. *The Mathematics Enthusiast*, 10(1&2): 245–278.

Levenson, E. S. (2017). Promoting mathematical creativity in heterogeneous classes. In J. Novotná and H. Moraová (Eds) SEMT 42–52.

Lin, F.-L., and Rowland, T. (2016). Pre-service and in-service mathematics teachers' knowledge and professional development. In Á. Gutiérrez, G. C. Leder, and P. Boero (Eds), *The Second Handbook of Research on the Psychology of Mathematics Education*, 483–520. Rotterdam, The Netherlands: Sense.

Littler, G. H., and Taylor, V. (1995). Teaching strategies in mathematical education courses for student teachers. In M. Hejný and J. Novotná (Eds) SEMT 133–137.

National Council of Teachers of Mathematics (2000). *Principles and standards for school mathematics*. Reston, VA: Author.

Novák, B., and Uhlířová, M. (1997). Prospective teacher's comments on the level of difficulty of mathematical problems. In M. Hejný and J. Novotná (Eds) SEMT 132–134.

Novákova, H., and Novotná, J. (2011). A priori analysis in theory and teachers' practice. In J. Novotná and H. Moraová (Eds) SEMT 252–260.

Piñeiro, J. L. (2019). Conocimiento profesional de maestros en formación inicial sobre resolución de problemas en matemáticas (Doctoral dissertation). Universidad de Granada, Spain.

Piñeiro, J. L., Chapman, O., Castro-Rodríguez, E., and Castro, E. (2019). Exploring prospective primary school teachers' mathematical problem-solving knowledge. In J. Novotná and H. Moraová (Eds) SEMT 305–315.

Priolet, M. (2013). 'Reference Boxes': Teaching and learning with reference tools in elementary mathematics. In J. Novotná and H. Moraová (Eds) SEMT 242–249.

Priolet, M., and Novotná, J. (2007). Solving numerical problems in the 3th year of primary school: Teaching and learning with connections. In J. Novotná and H. Moraová (Eds) SEMT 217–225.

Ruppeldtová, J. (2009). An operator without a state is difficult. In J. Novotná and H. Moraová (Eds) SEMT 225–233.

Samková, L. (2017). Planning and conducting inquiry based mathematics course for future primary school teachers. In J. Novotná and H. Moraová (Eds) SEMT 354–364.

Sarrazy, B. (1996). La sensibilité au contrat didactique: Rôle des Arrière-plans dans la résolution de problèmes d'arithmétique au cycle trois (Unpublished doctoral dissertation). Université de Bordeaux 2, France.

Sarrazy, B., and Novotná, J. (2005). Didactical contract: Theoretical frame for the analysis of phenomena of teaching mathematics. In J. Novotná (Ed.) SEMT 33–45.

Schütz, T. (1993). On the teaching of problem solving in math. In M. Hejný and J. Novotná (Eds) SEMT 79–81.

Shabtay, G. (2019). The link between classroom discourse and students agency. In J. Novotná and H. Moraová (Eds) SEMT 401–408.

Shulman, L. S. (1986). Those who understand: Knowledge growth in teaching. *Educational Researcher*, 15(2): 4–14. doi.org/10.3102/0013189x015002004

Trch, M., and Zapotilová, E. (2001). Creating of positive climate in teaching mathematics. In J. Novotná and M. Hejný (Eds) SEMT 162–167.

Tržilová, D. (1995). Problem solving on spreadsheets. In M. Hejný and J. Novotná (Eds) SEMT 123–125.

Wallach, T., and Even, R. (2001). Teachers' interpretations of students' mathematics thinking and behaviour. In J. Novotná and M. Hejný (Eds) SEMT 167–171.

Weber, K., and Leikin, R. (2016). Recent advances in research on problem solving and problem posing. In Á. Gutiérrez, G. C. Leder, and P. Boero (Eds), *The Second Handbook of Research on the Psychology of Mathematics Education*, 353–382. Rotterdam, The Netherlands: Sense.

1.3
Cultural and linguistic heterogeneity in research presented at SEMT symposia

Jarmila Novotná
Charles University, Faculty of Education, Prague, Czech Republic;
Chercheur titulaire, CeDS, Université Bordeaux, France
Hana Moraová
Charles University, Prague, The Czech Republic

Introduction

The topic of cultural and linguistic heterogeneity is tightly connected to language and communication in a classroom. It focuses on situations where the language of instruction is not the mother tongue of one pupil, several pupils, or all pupils and the teacher. There are two possible situations, which seem to be very different but are analogous in the sense that, apart from the subject (in this case mathematics), the teacher must pay attention to language, communication, understanding of what is being presented, and what is happening in the classroom. Regardless of whether the language of instruction varies from the mother tongue, or whether there are foreign pupils with an insufficient command of the language of instruction, a lot of attention must be paid to possible communication obstacles and misunderstandings.

The importance of work with cultural and linguistic diversity in classrooms is associated with two very important changes. The first one is globalization—establishment of global networks of communication and globalization of all social, political, economical, and ecological processes (Novotná and Hofmannová 2011). Internationalization, establishment of international networks, and intensive international co-operation require communication in foreign languages. All this puts a great deal of pressure on schools that are expected to prepare educational programmes that will equip their students with excellent command of at least one foreign language. One of the ways of achieving this is building the so-called functional bilingualism—i.e. making a foreign language the language of instruction. This is why the CLIL method—Content and Language Integrated Learning—is becoming more and more common in schools across Europe, not only in mathematics lessons.

Another trend is a growing cultural and linguistic heterogeneity in classrooms. The multicultural nature of modern society constitutes one of the most significant changes that have influenced schools in many countries, especially at primary and middle school levels (Favilli 2015). Minorities are no longer silenced and migration is connected with an increasing number of foreigners in classrooms, which means

Jarmila Novotná and Hana Moraová, *Cultural and linguistic heterogeneity in research presented at SEMT symposia*.
In: *Elementary Mathematics Teaching*. Edited by: Brian Doig, Jarmila Novotná, Berinderjeet Kaur and David K. Pugalee,
Oxford University Press. © Oxford University Press (2024). DOI: 10.1093/oso/9780192869647.003.0003

that teachers need to respect the different cultural and linguistic needs of very heterogeneous groups of pupils. Considerable attention must be paid to not only which methods are used to present mathematics, but also how it is made comprehensible and accessible to learners with adverse life experiences and insufficient knowledge of the language of instruction.

This chapter focuses on the impact these trends have had on presentations at SEMT symposia, as well as on what research results were presented. While this chapter respects the duality of the issue of the deliberate use of a language of instruction different from the mother tongue of learners in the classroom, with the aim of developing both content and language, and teaching mathematics in culturally and linguistically heterogeneous classrooms with minority and migrant pupils. However, this division is artificial, and in many cases the results presented at the symposia are relevant for both of the defined areas above.

Presentations before 2015

Presentations focusing on issues of cultural and linguistic heterogeneity became more frequent in SEMT symposia after 2015. Before 2015, the number of relevant presentations was relatively small and can be divided into the following three categories: presentations focusing on the impact of language and culture on teaching mathematics in general, presentations focusing on the use of CLIL in mathematics classrooms, and teaching and learning mathematics in an environment of limited language proficiency.

Impact of language and culture on teaching mathematics

Gardella and Tong (1999) focused on the acquisition of a language and its relationship to written or symbolic expression. Their research gave rise to a model for a similar development of mathematical understanding by children. They explored the learning of mathematics as a process parallel to the way children acquire language skills and look for parallels with second-language acquisition. They claimed that 'second language acquisition begins in similar fashion but can be complemented or interfered with by the level of first language acquisition' (p. 129). They integrated mathematics education research, second-language acquisition, and linguistics. They presented a model demonstrating how children should develop mathematics and language skills in parallel. The model requires a period of non-symbolic understanding before the formal written conventions can be successful. This article is directly linked to both linguistic and cultural issues.

Pugalee (2001) focused on the relationship between language and mathematics learning. He claimed that most teacher preparation programmes do not include explicit objectives addressing this relationship. In order to increase learners' mathematical experiences, the teacher has to know how mathematics teaching and learning relate to the four major communication modes: spoken language (speaking and listening), written language (reading and writing), graphic representation (diagrams, tables, pictures, graphs), and active demonstration (performance and physical involvement). Given relevant tasks, language becomes an essential tool for developing students' mathematical literacy. Pugalee claimed that when teachers implement communication practices in their teaching, they have to be able to uncover and build students' common mathematical knowledge. Ball and Bass (2001) identified three responsibilities of the teacher, which Pugalee found crucial to providing appropriate language experiences facilitating mathematical understanding and reasoning: uncover students' current base of knowledge, establish and extend that base of knowledge, and model as well as guide the construction of acceptable mathematical arguments.

Poirier (2005), in her presentation on culture, language, and mathematics, focused on the social dimension of mathematics. In many countries, mathematics is taught to children from different countries, different cultures, who had started to learn mathematics in their home countries in their mother tongue. In her presentation, Poirier looked at how the environment, the culture, and the language influence the development of some mathematical concepts. She presented some teaching activities both in regular classrooms and in the classes of pupils of various cultural backgrounds. The ideas presented were illustrated, inter alia, by an example of from an Inuit community in northern Quebec.

The issue of the impact of language on mathematics teaching and mathematics skills was also studied by Villamor, Angós, and Correa (2009). They examined, first, the numerical skills of two-to-four-year-old children of two different linguistic profiles: Basque and Spanish. The distinguishing feature between the two communities is that Basque people have two official languages: Spanish, which is common for both, and Basque. The sample also included bilingual children whose native language was Spanish, but who were educated in Basque. Of the children considered, 25 per cent were Basque monolingual, 25 per cent Spanish monolingual, and the remainder bilingual. In the context of the research presented, the authors considered several important differences between the Basque and Spanish contexts. The data from the research suggest that there exists a stage of development in numerical thinking that is conditioned by the language profile in which children learn to count. The authors found that 70% of children with a Spanish-only language profile knew the number two or higher, but only 30 per cent of children learning Basque at home and at school reach the same level. This research clearly shows how important it is, when teaching mathematics, to take into account the cultural and linguistic background of learners.

Increasingly, the multicultural, and multilingual, faces of all European countries, one of the rapid changes in the twentieth and twenty-first centuries, are linked to new demands on mathematics education and mathematics teachers. As a reaction to this challenge, Ulovec et al. (2013) carried out an international research study about multiculturalism in mathematics education in six European countries (Austria, Czech Republic, France, Greece, Italy, Norway). The aim of the research was to collect information about the working conditions of teachers and the teaching materials available for teaching mathematics in multicultural contexts. The authors identified teachers' approaches, experiences, and beliefs about teaching in multicultural settings. Attention was also paid to the methodological tools that teachers found suitable for facilitating the appreciation of the positive aspects of different cultures and creating favourable conditions for inter-cultural dialogue in the classroom. The presentation analyses the answers of teachers from the six European countries, and compares the differences in the experience, opinions, and needs of teachers in these countries.

Another notable presentation, connected to the issue of the impact of language on teaching mathematics, was that of Savard and Polotskaia (2013), although its primary objective was not this impact. However, the presentation showed results that are closely connected to the issue and cast light on it. The authors focused on problem solving in the early years of elementary school, and observed teachers' perceptions of the possible negative impact that their instructions might have on pupils' understanding. They used an ethno-mathematical model to identify different contexts involved in the tasks, and their impact. They also studied how teachers could help their pupils differentiate between the everyday context and the mathematical structure of the task. The analysis of 12 experienced teachers' conversations showed that teachers paid attention to socio-cultural factors, mathematical contexts, and language when planning classroom activities. However, it was less clear exactly what mathematical concepts they were targeting, and what mathematical structure the problems would have. The citizenship context seemed to be ignored when selecting a problem.

Content and Language Integrated Learning

CLIL refers to the teaching of a non-linguistic subject, such as mathematics, through a foreign language. CLIL works with equilibrium between content and language learning. In CLIL, subject understanding and thinking manifested by the language of the subject are developed through the foreign language. At the same time, foreign language is developed through the non-language content, such as mathematics.

The context for teaching mathematics is strongly influenced by the teacher. That is why it is essential that research should focus on educating teachers to use the CLIL method in their lessons. The context for teaching mathematics is strongly influenced by the teacher. That is why it is essential that research should focus on educating teachers to use the CLIL method in their lessons. Novotná, Hadj-Moussová, and Hofmannová (2001) focus on the professional competencies of prospective CLIL teachers. The target group of their research are students involved in pre-service teacher education. If pupils are to succeed in a challenging way of learning, and use simultaneous construction of mathematical knowledge and language skills, it is the teachers who first have to change their attitudes towards this new educational approach (Novotná and Hofmannová 2011), which means they have to overcome inner barriers. The authors of SEMT presentations focused on two types of inner barriers (Rogers 1996) that a CLIL teacher has to overcome: cognitive and affective. Barriers of the *cognitive* type result from the fact that the newly acquired knowledge and skills are ranked into cognitive categories formed during the preceding learning experiences; if the new and old knowledge do not agree, then the pre-knowledge either acts directly as a block to new learning or can result in creating misconceptions. *Affective barriers* are mainly those that originate from the learner's personality and strongly influence the learning process in both positive and negative ways. Other types of learning barriers are the result of outside causes, e.g. personal worries, financial problems, tiredness, or unfavourable circumstances of study. The barriers occur in every type of education. Some of them are increased through CLIL, others decreased. The decrease of barriers can be expected mainly in the area of fear of failing. The CLIL teacher is led towards sensitivity to the learner's personality. Referring to Vygotsky (1962), the competencies of newly qualified CLIL teachers should comprise the following methodological support categories: *verbal*, *visual*, *meta-cognitive* (dealing with content), and *peer* and *affective* (dealing with learners).

Two years later, Novotná and Hofmannová (2003) presented their research focusing on the teaching of mathematics in English to Czech learners. Concentrating on the perception stage of information processing, they presented context-dependent learner comprehension strategies. In a seminar for prospective mathematics teachers, they prepared and carried out two simulations illustrating what school learners experience when they are taught new subject matter in a foreign language. Both simulations were then discussed with participants. The experiments confirmed that learners always join the perceived stimuli with the knowledge and skills they already have stored in their mind. Moreover, the perception process significantly depends on the context. Learning isolated vocabulary items is unacceptable. The cognitive strategies are universal, and not domain-dependent.

CLIL, as a motivating teaching strategy, was discussed in Tejkalová's presentation (2007). She designed a lesson focusing on interpreting a diagram that allowed multiple communicative situations linked with pupils' everyday life experiences. She carried out the lesson in English in four different Czech classes. During the lesson, pupils showed intrinsic interest in mathematical concepts linked with the diagram. In the post-lesson questionnaire, they highlighted the enrichment in both English and mathematics. The experiment proved to be highly motivating. In the continuation of her research on CLIL, Tejkalová (2009) focused on co-operative teaching in CLIL. Among other things, she observed the following benefits of co-operation between mathematics and language teachers in CLIL: co-operation helps to enrich the methods that teachers use, and offers intense reflection. Co-teaching offers the teachers the opportunity to see their pupils in a different context, and results in more consistent teaching in both content and language, as well as creating a setting where pupils implicitly learn to co-operate. Tejkalová concluded that co-operative teaching not only helps to fulfil the dual objectives of CLIL teaching, but also fosters a co-operative environment in the classroom.

Assessment and evaluation, when teaching in CLIL, is a less researched area. However, in 2013, it was the topic of a SEMT presentation by Šteflíčková. How to assess in CLIL is one of the key questions connected to the use of this method. It must be borne in mind that a CLIL lesson has two equally important objectives: language and content goals. Integrated teaching should use integrated assessment and evaluation, and Šteflíčková showed how tests could integrate assessment of both language and content.

Teaching and learning mathematics in an environment of limited language proficiency

The presentation by Clanché and Sarrazy (1999) is based on a mathematics teaching experiment carried out during the first year of primary school with children aged six and seven in the jungle of New Caledonia. They show that transfer of everyday life activities cannot lessen the difficulties encountered by the pupils when they have to appropriate knowledge imposed by their teacher. When analysing an episode, designed and observed by the authors of the presentation, they worked in the intersection of two frames: a didactic frame (theory of didactical situations—Brousseau 1997) and an anthropological frame (Bensa 1996). The switching of contexts between an everyday one and the strictly school one (Goffman 1974) shed more light on didactic effects: 'The relationship between the teaching of mathematics and the pupils' everyday life experiences should be analysed for the better more as a breaking point (or an obstacle) than as a continuity (or facilitation)' (Clanché and Sarrazy 1999, p. 41). The authors concluded that:

> The anthropo-didactic frame that we called upon in this research is part of the desire to understand by avoiding the double snag according to which anthropology would be reduced to didactics and reciprocally. We bet there is a way there, which is new, for the anthropology of education. (p. 45)

In 2005, Seifert et al. made a presentation focusing on the relationship between language and mathematical learning. They stated that teachers face challenges when they must consider content issues as well as issues related to the classroom setting. These challenges are even deeper when working with Limited English Proficiency (LEP) pupils. In the presentation, the authors describe an effective curricular and instructional intervention targeting LEP pupils in mathematics, which took place in a North Carolina district (USA). As it was found that the third-grade students show a lack of improvement, the intervention targeted third-grade teachers. The instructional model included several components related to both the language and mathematics development of LEP pupils. It concerned several components: meaningful activities integrating lesson concepts with language practice, use of different techniques making concepts clear (e.g. modelling, visuals, demonstrations, gestures, or body language), and opportunities for interaction where discourse promoted elaboration of responses and lesson concepts. The test designed to determine whether the pupils' memorization moved to the application of process skills in mathematics investigations was used for analysing the results of the intervention. It showed that the innovative curriculum, together with the emphasis on the relationship between language and mathematics learning, had a positive impact on LEP pupils' mathematical performance.

In the same year, Petrová (2005) presented a part of her longitudinal study focusing on qualitative aspects of learning mathematics in a foreign language. She focused on communication obstacles and specifics based on the limited language proficiency of 10-to-11-year-old pupils and their teacher. The research questions were: what types of misunderstandings based on the use of a foreign language as the language of instruction in mathematics can occur? Is it possible to identify their causes? Which communication specifics does bilingual teaching and learning of mathematics have? Petrová used action research, and analysed her own teaching. In her presentation, she referred to an experiment in the field of probability, carried out in Great Britain. She taught in English, which is her second foreign language. She worked with two groups of children—native English speakers and children whose mother tongue was not English. Pupils were of mixed abilities in both mathematics and their level of mastery of English. No major difficulties were encountered during the experiment, as all groups solved the problem successfully. Petrová detected difficulties of two types among the pupils—language problems (unknown mathematical terminology, language misunderstanding, mathematical misunderstanding) and problems in the communication specifics of teaching and learning in a foreign language (ineptitude in speech, limited language production of pupils).

Selected presentations from 2015 to 2019

The number of presentations related to cultural and linguistic heterogeneity has grown in the past three SEMT symposia. Therefore, this section is not an exhaustive list of presentations; its goal is to outline the major trends in the presentations in this area.

Let us start by thinking about the rapid increase in the number of presentations related to the issue of teaching in culturally and linguistically diverse environments. Without any doubt, it reflects the changes in the current world and contemporary societies. A growing trend in migration has put teachers into situations where their classes are no longer heterogeneous, either linguistically or culturally. Their pupils come from very distant regions with different traditions, languages, social roles, and educational backgrounds. In their former education, they were not only doing different mathematics, but using completely different methods. Their understanding of what it means to be doing mathematics is diverse. Naturally, schools, teachers, educators, policy makers, and other stakeholders in education must respond to this situation. At the same time, systems of education, schools, and teachers are not prepared for this challenge, as teachers do not have adequate training or experience, and they lack materials, methods, and examples of good practice. This creates a natural need to focus research attention on this area.

Another reason for the growth in the number of presentations at SEMT is the themes of individual SEMT symposia. An overview of SEMT themes from 2001, when the practice of defining a theme for each symposium was introduced, was published in the SEMT 19 proceedings. The following are the themes from the past four SEMT symposia:

SEMT 2013: *Tasks and tools in elementary mathematics*
SEMT 2015: *Developing mathematical language and reasoning*
SEMT 2017: *Equity and diversity in elementary mathematics education*
SEMT 2019: *Opportunities in learning and teaching elementary mathematics*

We can observe a clear shift in attention towards language and diversity. The themes of 2015 and 2017 invited presentations connected to cultural and linguistic heterogeneity in mathematics education (in contrast to 2013, for example).

In this section, the presentations selected are divided into influence of language, CLIL, and Socio-Culturally Inclusive Mathematics Education. Some of these develop research presented in the previous section above, and others enhance the area with new trends in more recent research.

Influence of language

The role of communication in the learning of mathematical language, and reasoning, is a broad domain, with several perspectives. Communication in mathematics lessons and about mathematics has been the centre of attention for a relatively long time. Undoubtedly it is essential that pupils are able to understand what is happening.

In the context of this chapter, presentations that focus on the role of language and communication in the context of cultural diversity have been selected.

One perspective on the use of language in a mathematics classroom is that of Martin et al. (2015). The presentation was on the results of a three-month-long pilot study in four fourth-grade classrooms in the south-eastern United States. The students were introduced to using writing while solving problems, with the aim of expressing their process and sharing their conceptual thinking about particular mathematical ideas. The writing was meant to be a learning tool for the students to scrutinize their own mathematical thinking and identify challenges. The research questions were: how do students use writing as a tool for learning? And what did students' writing reveal about their learning?

The findings of the study showed that students' depth of understanding, and gaps in understanding, are observable in their writing and their lack of writing. Moreover, the writing showed varied problem-solving and organizational strategies (Martin et al. 2015). This presentation is relevant to the issue of cultural and linguistic heterogeneity, since it shows the link between the ability to formulate thoughts in a language and the ability to solve a problem successfully. A teacher must understand that a lack of knowledge of the language of instruction will block the process of thinking about mathematics, reasoning, and successful solving.

In the same year, Stott and Graven (2015) also focused on the development of learners' verbal reasoning in a slightly different level to that of than Martin et al. The aim of the presentation was to describe the research processes designed to adapt and implement number talks as a key teacher development strategy leading to progression in their local project schools. Stott and Graven (2015) built on the widely reported prevalence of one-to-one counting methods as the dominant calculation strategy across the four operations even in the intermediate grades (4–7) in South Africa, and tried to introduce number talks with the aim of simultaneously developing learners' mathematical talk (verbal reasoning) and to push explicitly for the use of more efficient calculation strategies. In other words, they believed that development of the ability to speak about mathematics (Martin et al. 2015), and to write about mathematics, would promote mathematical learning and improve it qualitatively.

And in the same year, Symeonidis, Tatsis, and Kaldrymidou (2015) presented a study of communication in a mathematics classroom and its impact on emotions, attitudes, and performance in decision-making processes. They also focused on the established socio-mathematical norms in mathematics classrooms and the effects that they have on pupils and their attitudes, concluding that the impact is significant. They understood a classroom as a small community, consisting of students and teachers interacting, where there are some social and cultural factors that often remain invisible, and may significantly affect a decision, or even the outcome, of an interaction. Some students are born leaders, others are mild and submissive: these, and other characteristics, affect the way students work and communicate, and this affects their performance in mathematics lessons (Symeonidis, Tatsis, and Kaldrymidou 2015). This research is very important for the area of cultural heterogeneity. When the classroom becomes culturally, and linguistically, heterogeneous, communication norms are very varied, and may be a little incomprehensible to some of the learners, with the effect of their getting lost in the classroom discourse and processes.

The importance of communication in a mathematics classroom, and, also of communication about mathematics in the classroom, existed at the beginning of project *Lexicon*. This international project, whose results were repeatedly presented in SEMT symposia (in all forms of SEMT presentation), takes a broader perspective on what is being communicated in mathematics classrooms all around the world, and how this interaction is described in the terminology of any given country. The comparisons of lexicon terminology developed by the project partners clearly showed that classroom practices around the world are different, which is reflected in the lack of terminology, in the same area, developed in other countries.

In 2017, Clarke et al. made a presentation titled 'Speaking in, and about, the mathematics classroom'. In three sessions the following aspects of speaking in and about mathematics classrooms were addressed: students speaking inside the classroom, teachers speaking about the classroom, and participants speaking about video extracts from mathematics classrooms. The fact that the SEMT participants came from different countries meant that the participants shared their experiences, and the differences, in their descriptions, as well as the difficulties they came across when describing a mathematics lesson. Discussion raised the differences in the discourse in, and about, mathematics lessons in different cultures.

CLIL

One of the important questions concerning the implementation of CLIL in mathematics lessons is for which age, and for which cognitive level, it is suitable. The opinions of the public, as well as of some teachers, are that CLIL might be difficult, even impossible, to use with young learners at primary school. Several authors claim that combining a foreign language with a non-linguistic subject, when working with young learners, increases the demands on pupils too much; others, on the contrary, state that young learners are able to combine a foreign language as a means of communication, and the language of the non-linguistic subject, when solving problems (for English and mathematics, see Alvarenga 2012).

Moraová and Novotná (2015) asked the following questions: is it possible to implement CLIL into teaching even in the case of pupils with very limited target language competence—even at the A1 level of the Common European Framework of Reference for Languages? What problems will teachers have to solve in the course of the lesson? What will they have to take into account when planning the lesson (or the sequence of lessons)? How will the pupils perceive the CLIL lessons?

They conducted a six-month teaching experiment in a primary school in Prague in a third-grade class (eight-to-nine-year-old pupils). The experience from the teaching experiment suggests that CLIL lessons of mathematics at this level are not only possible but also appreciated both by the teacher and the pupils. Pupils' improved performances, observed during the experiment, were very likely the consequence of a general development of competence in mathematics and in English. Pre-service mathematics teachers observed the lessons, and their comments focused a lot on the good and relaxed atmosphere in the lessons, and the pupils' willingness to co-operate. They realized that the pupils were not producing much language, but most of them were able to follow the lesson without much difficulty. They pointed out that there were no open questions and most communications were of just numbers. The authors of this presentation claimed that with carefully selected areas of mathematics, and well-planned lessons, pupils are able to follow the lessons in a foreign language, carry out tasks, and learn mathematics, all at the same time. Of course, CLIL lessons in mathematics, at these early stages, mean very little spontaneous communication in English among the pupils. But their receptive skills are developed, and productive skills simply need more time, and will come eventually.

In her presentation, Procházková (2015) put forward a suitable activity for CLIL at primary level: language showers. A language shower is a short, typically game-like activity, which introduces the foreign language into the content of the classroom. Its main role is to challenge the pupils, and boost their linguistic confidence, by offering them an activity in which they succeed. Procházková showed several authentic mathematical language showers created by primary teachers. She pointed out the CLIL principles they employed, and addressed the possible problems they might cause when brought into a classroom. The presentation was complemented by a summary of CLIL practices reflected in language showers in a primary mathematics classroom: adequacy, multiple foci, authenticity, active learning, scaffolding, and co-operation both among the pupils and between the teachers.

Future mathematics teachers' competencies, needed for planning CLIL lessons, were investigated by Moraová and Novotná (2017). The authors presented the results of a detailed analysis of pre-service teachers' mathematics lesson plans. They focused on how higher-order thinking skills (remembering, understanding, applying, analysing, evaluating, and creating; Anderson and Krathwohl 2001) and their development are addressed in the lesson plans and whether the pre-service teachers are aware of the need to target more than just lower-order thinking skills. They also investigate the presence of the *4Cs* (content, communication, cognition, and culture; Coyle 1999). The analysis of lesson plans showed that a minority of future mathematics teachers described, explicitly, thinking skills; on the other hand, implicitly, a vast majority of the lesson plans analysed included activities supporting development of higher-order thinking skills. It was also found that participants were able to address Coyle's third C even in the case of young learners with limited language command.

In her presentation, Baschek (2019) identified the linguistic and technical knowledge pupils acquire in bilingual mathematics classes while using ICT. She explained which competencies bilingual classes should achieve, and how bilingual learning and ICT fit together. Baschek claimed that it has been proved that the foreign language learning of pupils improves in bilingual classes, because of the higher frequency of language contact. Her research focused on how the use of ICT can boost this learning.

Socio-culturally inclusive mathematics education

Inclusive mathematics education means mathematics education for all pupils regardless of their various 'disabilities', whether learning difficulties, health issues, language, or culture. With respect to the thread of research, that we follow in this chapter, means we should discuss here those presentations that dealt with inclusive approaches in classrooms with pupils from different cultural and linguistic backgrounds.

This topic is relevant for two different types of countries: those that have a history of marginalizing one of their groups (a lot of research attention to the topic can be observed in South Africa), and those that are now in the situation of having migrant pupils in their classrooms.

A country that falls into the first category is South Africa. In her presentation 'Blending elementary mathematics education research with development for equity: An ethical imperative enabling qualitatively richer work', Graven (2017) presented research conducted in a nation of extreme socio-economic and educational inequality, as a consequence of apartheid, and where performance and opportunity gaps still persist along racial lines. In her presentation, Graven revealed the design of the South African Numeracy Chair Project (SANCP) that was set up to enable a powerful dialectical relationship between research and development through merging the two in a network that would create multiple opportunities for dialogue across stakeholders. Graven argues that the creation of collaborative spaces for dialogue and active participation among all participants is critical in the joint enterprise of finding sustainable ways to meet the educational challenges, for bringing equity into education.

Stott (2017), in her presentation 'Interactive gesturing: Deepening understanding of mathematical activity with young learners in a South African context', presented an analysis of video data collected from paired-learner, task-based interviews, with learners aged between nine and ten years, participating in after-school mathematics clubs in South Africa. In this research, Stott explored the idea that if the non-verbal part of communication of learners is taken into account, our understating of what unfolds during a mathematical learning activity deepens, especially in the context of a mathematical learning activity where the language of learning and teaching are not the mother tongue. Stott shows that spontaneous, in-the-moment, interactive gestures could serve as an important mediation strategy for extra-curricular practitioners, as well as pre- and in-service teachers, in situations where we cannot rely on sufficient language command.

Yet another presentation from South Africa came in 2019, when Venkat addressed the topic 'Improving early number learning in contexts of disadvantage'. Venkat's focus is not on language and culture, but on the conditions of the South African context and how this socio-cultural disadvantage can be overcome. She presented an overview of an intervention focused on improving early number learning in a South African context of disadvantage involving large classes and limited resources.

In the same year, Višňovská, Cortina, and Vale (2019) in their presentation focused on designing specific learning trajectories in classrooms of disadvantage, building on examples from Mexico and South Africa. The authors' experience from working in under-resourced schools and classrooms seems to be favourable. Teachers with whom the authors co-operated were in many ways typical of the settings of their work, yet accomplished unparalleled learning for their students. The authors suggest that what are essential are the conditions for teacher support, which make use of the products of design research, locally viable and accessible to them.

A further presentation of research conducted with marginalized groups was Hunter and Miller's (2017) presentation 'Using cultural tasks to develop growing pattern generalisations for young culturally diverse students'. In their research they focused on two groups of students from different cultures: Pāsifika and Maori. The aim of the study was to explore how contextual Pāsifika and Maori patterning tasks could potentially support young children to develop their understanding of algebraic growing patterns. An experiment undertaken with 27 Year 2 students (six-year-olds) participating in ten lessons focusing on growing patterns was presented. Analysis of the data revealed that the contextual growing tasks assisted these young students to generalize growing patterns. It was apparent that building on both rich Pāsifika and Maori patterns, used within craft design and dance, is a good way for these students to begin exploring growing patterns. Similarly, this has also been found in studies with young Australian Indigenous students, exploring environmental growing patterns before traditional mathematical patterns. Acknowledging that a student brings their own cultural knowledge to the classroom provides an opportunity for culturally diverse students to make more meaningful connections to mathematics.

Another presentation, which focused on learners with low sociographic status, and what it means for them in mathematics lessons, was Goetze's 'Language- and mathematics-integrated intervention for understanding division and divisibility' (2017). The presentation was of a study with 45 third- and fourth-graders from primary schools whose catchment areas had a low sociographic status. This study focused on difficulties in division. As the author argued, linguistic sensibility is crucial for success here, as division, in particular, is distinguished by specific language structures. Children must decode verbal information, be able to differentiate between everyday words and technical terms, and recognize the connection between mathematical symbols and language. As Goetze states, this is a problem faced by children who have weaker language competence than that of mainstream schoolchildren. In the presentation, individual misconceptions in division serve as a basis for initiating a sustainable understanding of factors among the children in a language-sensitive manner.

Kasmer, Harrison, and Billings (2017), in their presentation titled 'Equitable learning opportunities: Textbook language accessibility in English medium classes', looked at the accessibility of mathematics textbooks written in English, trying to find how textbooks might have an impact on the mathematical learning experience of multilingual learners in Tanzania, where the medium of instruction in elementary schools is English. They were well aware of the fact that in mathematics classrooms across the globe, many students attempt to move between the language spoken at home and the language spoken in school. Learning mathematics in a culturally heterogeneous classroom increases the demands on students, as they simultaneously navigate mathematical content while negotiating the nuances of a language different from their mother tongue. The presenters argued that language demands in mathematics textbooks, used in this multilingual environment, are higher, as their accuracy, accessibility, and clarity are prerequisites to equity and understanding in linguistically heterogeneous classrooms. This presentation provided instances of improper grammar, mathematical inaccuracies, or unclear instructions in elementary mathematics textbooks and analysed the impact that this has on learning mathematics.

In her presentation, Haydar (2019) built on the fact that the student populations in US schools continue to become more diverse, with an emergent bilingual population representing the fastest-growing group in the country. This group has diverse linguistic experience, and thus also different learning needs. It is essential, in this situation, that teachers are encouraged to inquire into who their students are, and to understand what resources they bring into the classroom, in order to forge innovative ways of working with them. Haydar reported in her presentation on a study of prospective and beginning teachers learning to conduct clinical interviews to enhance emergent bilingual students' access to mathematics. The study was conducted within the context of a teacher education programme in New York City. The focus of this study was the method of clinical interview used to work on the skills of noticing and enhancing the mathematics of emergent bilingual students.

Internationally, several studies show that migrant pupils' learning outcomes are significantly lower than those of their classmates. Suitable materials for teachers in such heterogeneous classes are scarcely available. Production of them is an important research theme in the domain of mathematics education in multicultural and multilingual classes.

In 2015, Moraová, Novotná, and Favilli's presentation focused on the specifics of teaching mathematics in classes with differences in pupils' cultures and languages. In the three presentation sessions the participants not only saw examples of lessons, and environments supporting multicultural and multilingual approaches, in teaching elementary mathematics, but they had also the opportunity to discuss and develop their own teaching materials and units for such environments. Attention was also paid to the difficulties that non-native speakers encounter when reading and writing in a foreign language, including possible conflicts arising from different meanings of words that are common to both natural and mathematical languages.

Ulovec (2019) presented more examples of mathematics materials, using ICT, suitable for culturally heterogeneous classrooms developed by mathematics educators in co-operation with ICT specialists from Austria, Italy, and Slovakia. The materials were piloted and the analysis of these episodes, together with interviews with teachers, confirmed their usefulness. While Ulovec's presentation (2019) focused on evaluating the quality and usefulness of the materials developed, and their quality, Ulovec and Novotná (2019) focused on the conception of materials for heterogeneous classes. Building on results from a literature review, their own experience, and piloting some of the materials developed, they created a list of recommendations for teachers who planned to create quality materials tailored for their classes.

Conclusion

The topic of cultural and linguistic heterogeneity in mathematics classrooms has attracted the attention of researchers, who have presented their work at SEMT symposia. As suggested above, the topic became more central after 2015, both due to the focus of the last three SEMT symposia, but also as a consequence of the fast changes in the contemporary world.

Presentations on communication in mathematics classrooms in general have not been included in this chapter unless they explicitly referred to work with pupils from different social, cultural, and linguistic backgrounds. Attention was paid to presentations that focused on the influence of language, and socio-culturally inclusive mathematics settings.

The research interest in the area, and results from the SEMT presentations, confirm the need for paying attention to how pupils understand in a mathematics classroom. Lack of knowledge of the language of instruction must be taken into account, and teachers must look for ways to facilitate understanding of what is being communicated. This, of course, holds for all pupils in a mathematics classroom, not just pupils from different linguistic backgrounds.

Research presented at SEMT has also shown that the issue of cultural and linguistic heterogeneity must be included in pre-service teacher education. A novice teacher without this training is very likely to face immense obstacles when entering the real classroom of culturally and linguistically heterogeneous pupils. Their training must equip them with the tools needed to overcome possible obstacles. They must know what to expect, how to develop suitable teaching materials, and how to facilitate understanding for pupils whose mother language is not the language of instruction. Similarly, in-service teachers need support within in-service teacher training to equip them with the needed skills and tools.

Educators should also focus on development of materials suitable for heterogeneous classrooms. Attention must be paid to the language, wording, and cultural background of what is presented, so that minority pupils are not left out, so that they can recognize the mathematics that is presented. This is easier for communities of a monolithic background with a common culture, than for heterogeneous groups from different parts of the world. However, we have to try hard to look for ways of helping them if we want them to succeed in the new home, and new culture.

Cultural and linguistic heterogeneity brings not only new demands on the teacher, but also many positive aspects. If we learn how to handle diversity, we will be able to learn from each other. New ideas, practices, and motives will be introduced to mathematics classrooms for the benefit of everyone. Inclusive practices may mean more thinking about what is being taught, and how, and what methods are used. Teachers in heterogeneous classrooms, where the language of instruction is not the mother tongue of all pupils, pay more attention to how they explain: they tend to use more active methods, individualize, differentiate, be illustrative, and check understanding.

This topic, cultural and linguistic heterogeneity, is very likely to grow in importance in the years to come. Future SEMT symposia will very likely bring new questions, and propose new solutions and approaches in this area.

References

Alvarenga, D. M. de Figueiredo (2012). Developing young learners' logical/deductive thinking skills and second language skills through a CLIL approach. http://run.unl.pt/handle/10362/8651

Anderson, L. W., and Krathwohl, D. R. (Eds) (2001). *A Taxonomy for Learning, Teaching, and Assessing: A Revision of Bloom's Taxonomy of Educational Objectives*. New York: Addison Wesley Longman.

Ball, D. L., and Bass, H. (2001). *What Mathematical Knowledge Is Entailed in Teaching Children to Reason Mathematically? Knowing and Learning Mathematics for Teaching.* Washington, DC: National Academy Press.

Baschek, E. (2019). Using ICT in bilingual mathematics classes – an example. In J. Novotná and H. Moraová (Eds) SEMT 455–457.

Bensa, A. (1996). De la micro-histoire vers une anthropologie critique. In J. Revel (Ed.), *Jeux d'échelles*, 37–70. Paris: Gallimard, Le Seuil.

Brousseau, G. (1997). *Theory of Didactical Situations in Mathematics 1970–1990* [translated by M. Cooper et al.]. Amsterdam: Kluwer Academic Publishers.

Clanché, P., and Sarrazy, B. (1999). Contribution to the study of the links existing between real daily experience and daily scholar experience as regards the teaching of mathematics: Example of an 'additional structure' within a first year group in Kanak school. In M. Hejný and J. Novotná (Eds) SEMT 41–49.

Clarke, D., Mesiti, C., Novotná, J., Moraová, H., and Chan, M. (2017). Speaking in and about the mathematics classroom. In J. Novotná and H. Moraová (Eds) SEMT 477–479.

Coyle, D. (1999). Theory and planning for effective classrooms: Supporting students in content and language integrated learning contexts. In J. Masih (Ed.), *Learning Through a Foreign Language: Models, Methods and Outcomes*, 46–62. London: CILT.

Favilli, F. (Ed.) (2015). *Multiculturalism, Migration, Mathematics Education, and Language.* Pisa: M^{3}EaL Project. http://m3eal.dm.unipi.it

Gardella, F., and Tong, V. (1999). Implications of language development in the learning of mathematics. In M. Hejný and J. Novotná (Eds) SEMT 129–133.

Goetze, D. (2017). Language- and mathematics-integrated intervention for understanding division and divisibility. In J. Novotná and H. Moraová (Eds) SEMT 177–187.

Goffman, E. (1974). *Frame Analysis.* New York: Harper and Row.

Graven, M. (2017). Blending elementary mathematics education research with development for equity: An ethical imperative enabling qualitatively richer work. In J. Novotná and H. Moraová (Eds) SEMT 20–31.

Haydar, H. (2019). Elementary mathematics teachers celebrating student voice: The clinical interview again and again. In J. Novotná and H. Moraová (Eds) SEMT 192–201.

Hunter, J., and Miller, J. (2017). Using cultural tasks to develop growing pattern generalisations for young culturally diverse students. In J. Novotná and H. Moraová (Eds) SEMT 215–225.

Kasmer, L. Harrison, D., and Billings, E. (2017). Equitable learning opportunities: Textbook language accessibility in English medium classes. In J. Novotná and H. Moraová (Eds) SEMT 265–272.

Martin, C., Wang, C., Lambert, R. G., Polly, D., and Pugalee, D. K. (2015). Supporting learning mathematics through writing: formative assessment and instructional tool. In J. Novotná and H. Moraová (Eds) SEMT 240–249.

Moraová, H., and Novotná, J. (2015). Teaching maths in English at primary school level: Utopia, nightmare or reality. In J. Novotná and H. Moraová (Eds) SEMT 249–257.

Moraová, H., Novotná, J., and Favilli, F. (2015). Including multicultural elements in teaching and learning mathematics at elementary level. In J. Novotná and H. Moraová (Eds) SEMT 372–374.

Moraová, H., and Novotná, J. (2017). Higher order thinking skills in CLIL lesson plans of pre-service teachers. In J. Novotná and H. Moraová (Eds) SEMT 336–345.

Novotná, J., Hadj-Moussová, Z., and Hofmannová, M. (2001). Teacher training for CLIL – Competences of a CLIL teacher. In J. Novotná and M. Hejný (Eds) SEMT 122–126.

Novotná, J., and Hofmannová, M. (2003). Context-dependent learner comprehension strategies: Mathematics taught in English to Czech learners. In J. Novotná (Ed.) SEMT 117–120.

Novotná, J., and Hofmannová, M. (2011). *The Onset of CLIL in the Czech Republic; Selected Texts from 2000–2008.* Saarbrücken, Germany: LAP LAMBERT Academic Publishing.

Petrová, J. (2005). Communicational obstacles and specifics in bilingual teaching/learning of mathematics caused by the limited language proficiency of the children and of the teacher. In J. Novotná (Ed.) SEMT 268–276.

Poirier, L. (2005). Culture, language and mathematics. In J. Novotná (Ed.) SEMT 342–343).

Procházková, L. (2015). Language showers in primary mathematics. In J. Novotná and H. Moraová (Eds) SEMT 386–387.
Pugalee, D. (2001). Mathematics and language: Identifying requisite pedagogical knowledge. In J. Novotná and M. Hejný (Eds) SEMT 135–139.
Rogers, A. (1996). *Teaching Adults. Blocks to Learning.* Maidenhead, UK: Open University Press.
Savard, A., and Polotskaia, E. (2013). Word problem solving task management and students' access to mathematics: Cases in elementary education. In J. Novotná and H. Moraová (Eds) SEMT 290–297.
Seifert, A., Temple, B. A., Pugalee, D. K., Perez, T., and Douville, P. (2005). Mathematics, language, and learning: Focusing on curriculum and instruction to support limited English proficient students. In J. Novotná (Ed.) SEMT 285–291.
Šteflíčková, A. (2013). Assessment and evaluation in CLIL. In J. Novotná and H. Moraová (Eds) SEMT 367–368.
Stott, D. (2017) Interactive gesturing: Deepening understanding of mathematical activity with young learners in a South African context. In J. Novotná and H. Moraová (Eds) SEMT 393–403.
Stott, D., and Graven, M. (2015). Adapting number talks to foreground mathematical progression in South African classrooms. In J. Novotná and H. Moraová (Eds) SEMT 311–322.
Symeonidis, N., Tatsis, K., and Kaldrymidou, M. (2015). Norms in decision-making and in shaping attitudes, emotions and performance. In J. Novotná and H. Moraová (Eds) SEMT 322–331.
Tejkalová, L. (2007). On motivating features of CLIL. In J. Novotná and H. Moraová (Eds) SEMT 294.
Tejkalová, L. (2009). Co-operative teaching in CLIL. In J. Novotná and H. Moraová (Eds) SEMT 313.
Ulovec, A. (2019). Developing learning software for migrant students in a European collaboration project. In J. Novotná and H. Moraová (Eds) SEMT 487–489.
Ulovec, A., Moraová, H., Favilli, F., Grevholm, B., Novotná, J., and Piccione, M. (2013). Multiculturalism in theory and teachers' practice. In J. Novotná and H. Moraová (Eds) SEMT 332–330.
Ulovec, A., and Novotná, J. (2019). Concepts for teaching units in mathematics for migrant or minority students. In J. Novotná and H. Moraová (Eds) SEMT 489–491.
Venkat, H. (2019). Improving early number learning in contexts of disadvantage. In J. Novotná and H. Moraová (Eds) SEMT 40–52.
Villarroel Villamor, J. D., Nuño Angós, T., and Miñón Correa, M. (2009). The awareness of the first numbers during the infancy: Differences between children Basque- and Spanish-speaking children. In J. Novotná and H. Moraová (Eds) SEMT 242–248.
Višňovská, J., Cortina, J. L., and Vale, P. (2019). Building learning opportunities in classrooms of disadvantage: Rethinking the learning trajectories. In J. Novotná and H. Moraová (Eds) SEMT 452–454.
Vygotsky, L. (1962). *Thought and Language.* Cambridge, MA: MIT Press.

1.4
Professional development of teachers as an agent of change in primary mathematics teaching

Alena Hošpesová
University of South Bohemia, České Budějovice, Czech Republic
Berinderjeet Kaur
National Institute of Education, Nanyang Technological University, Singapore

Introduction

Primary school teachers, in most countries, are not specialized in teaching mathematics. Study programmes (in university or other institutions of higher education) include pro-paedeutics to all subjects taught at primary school level, and their didactics, pedagogy, psychology, special education, and school practicum. Mathematics teacher education (MTE) involves various processes and practices that lead to one's ability to teach mathematics at a given level of schooling. It is not limited to undergraduate education, as it also includes activities of teachers, teacher educators, and other persons in education. Professional development (PD), a critical component of MTE, is a process that attempts to increase teachers' knowledge and/or enhance their flexibility, autonomy, and adaptation, to the teaching context, or as a reflective process of better understanding teaching practice (Climent and Carrillo 2002).

Practices of MTE and PD are complex, as they are institutionally and culturally specific. Career structures for teachers affect their motivation to take part in PD programmes and these differ from place to place. In addition, at times, relationships between university and school pose challenges too. Furthermore, the scale of PD is another critical aspect. Small-scale PD work with schools necessarily creates studies of special cases, while large-scale work runs the risk of producing surface, or naïve, interpretations of 'reform' messages.

The theme of the SEMT 2001 symposium was 'What is meant by the competence and confidence of people involved in the teaching of elementary mathematics?' Since then, activities at SEMT symposia have focused on a wide range of issues related to varied activities of teachers, teacher educators, and researchers; their influence on participants' professional competency and/or teaching; and the impact of contextual variables connected with them. During the SEMT 2001 symposium, a presentation

Alena Hošpesová and Berinderjeet Kaur, *Professional development of teachers as an agent of change in primary mathematics teaching*. In: *Elementary Mathematics Teaching*. Edited by: Brian Doig, Jarmila Novotná, Berinderjeet Kaur and David K. Pugalee, Oxford University Press. © Oxford University Press (2024).
DOI: 10.1093/oso/9780192869647.003.0004

by Tall (2001), a context for the needs of teachers and how these could be met, were noted as follows:

> ... [teachers] need to have a real insight into how mathematics develops cognitively ... It means an awareness of how children do mathematics in different ways and why they encounter conflicts and difficulties ... starting to reflect on one's own experiences to see why certain things were difficult, or even impossible, at the time ... This requires time and sharing of ideas, not only during a training course but on a continuing basis throughout one's career, so that the new teacher can be inducted into the culture of helping children learn and the experienced teacher can grow reflectively through a life-time of improvement. (p. 35)

In the last 20 years, one way in which the needs of primary mathematics teachers have been addressed during SEMT symposia was through deliberations on PD for teachers. This chapter attempts to review presentations related to PD for teachers made at the symposia. To identify the main ideas discussed during the symposia, we adopted an evolutionary approach. We first scanned the list of abstracts of the presentations of past symposia, and identified more than 50 contributions specific to mathematics knowledge for teaching, and ways through which PD of teachers took place. We next looked for commonalities among the presentations, and identified three broad areas that could facilitate our review of presentations related to the PD of mathematics teachers.

The three broad areas were:

Development of teachers' mathematical knowledge for teaching
Context, content and structure of PD programmes
The impact of PD programmes on participants' instruction

Development of teachers' mathematical knowledge for teaching

Over the last 20 years we have seen, not only at SEMT, but also globally, increasing efforts by researchers and teacher educators to develop teachers' knowledge through PD, enabling teachers to use effective instructional practices that will improve students' learning outcomes. A number of presentations at SEMT symposia had links to the concept of mathematical knowledge for teaching (MKT) (Ball, Thames, and Phelps 2008). MKT at the primary school level is a multi-faceted paradigm that requires a solid understanding of the mathematical content and a firm grasp of pedagogical methodologies. Teachers also must possess the ability to understand students' thinking, and pose productive questions, while making important instructional decisions in the moment. Kasmer and Billings (2011), for example, showed how to incorporate prediction questions into mathematics lessons to support the quality of teaching and learning of mathematics. According to their findings, prediction questions can serve as a vehicle for PD to help transfer the knowledge from a PD experience into the classroom, and promote reform-teaching practices.

Several other presentations also documented teachers' MKT and analysed its structure. Ribeiro, Carrillo, and Monteiro (2009) advocated for a 'comprehensive' approach to each topic of the mathematics curriculum. According to them, teachers should not only have an in-depth knowledge of each topic, but also the rich range of techniques to deliver it to the students, in order to make it understandable to them, by anticipating their thinking, difficulties, and motivations. Later, in 2013, Ribeiro, Mellone, and Jakobsen (2013) presented particular tasks for teachers from the field of rational numbers, aimed at developing MKT, which led to interpretations and making sense of students' productions.

A similar view was adopted by Cabassut (2017), who presented exploratory research on the beliefs of the actors in the primary school (trainees, teachers, trainers, etc.) in relation to the teaching of modelling. The results showed that teachers' diversity in modelling can be represented by four teacher types, which explain the difficulties encountered in this teaching, in relation to time, assessment, lesson organization, context, student's involvement, and resources.

Attention was also paid to the evaluation of MKT. Chestnut (2011) identified pedagogical content scaffolds as a means for measuring MKT, by comparing the cognitive and teaching patterns of teachers of mathematics. Marcinek and Partová (2011) adapted measures of MKT developed at the University of Michigan: they validated these measures for use outside the United States (namely, for use in the Slovak Republic) and piloted them *via* a study (a sample of six practising primary teachers and one teacher educator).

Presenters at SEMT also often discussed the relationship between MKT and teachers' other forms of knowledge. Vinner (2011), in his presentation, recommended that in addition to MKT at primary school level, attention should also be given to wider aspects of teachers' work. He suggested discussion on the rationale of teaching mathematics as a compulsory subject, and the goals of education, as important for all. He also cautioned against topics that were beyond the zone of proximal development (ZPD) of the teachers.

Cheeseman's (2009) presentation reported on the close relationship between MKT and teacher behaviour in the classroom, and pupils' mathematical knowledge. The intensive communications that highly effective teachers have with pupils during their mathematics lessons often challenge young children to examine their mathematical thinking, and support their construction of new concepts. These teachers' behaviours reveal a genuine respect for children's thinking, a deep knowledge of the way children typically learn mathematics, and the use of highly skilled mathematical pedagogies.

In the Niess and Pugalee (2011) presentation the concept of MKT was expanded into an interdisciplinary construct, *Assessment of technological, pedagogical and content knowledge* (TPACK), which involved reasoning about cognition, observations, and inferences. They stated that current assessments often address simplistic components of this construct.

Bardelli's (2011) presentation examined the link between a teacher's interpretation of an error and its remediation, as a part of pedagogical content knowledge (PCK) of teachers that is fundamental if teachers want to draw on student thinking to direct their teaching. The teachers in his study showed great trust in the power of explanation of concrete models, but, according to Bardelli's interpretation, they were not able to use their mathematical knowledge in a more effective way.

Some presenters illustrated and connected their findings with difficult topics from primary mathematics. Fung (2011) illustrated the forms and depths of MKT *via* three examples that could possibly surface during rigorous preparation, discussion, and delivery of teaching. Macháčková and Tichá (2007) presented on one of the most difficult topics in mathematics education, the notion of fraction, and showed what meanings (part-whole, measure, operator, quotient, or ratio) are used by pupils, future teachers, and in-service teachers when working with fractions. As a diagnostic tool they used the problems posed by respondents.

Less prevalent were presentations related to the knowledge of kindergarten teachers. Kaslová and Marchini (2011), through their comparison of the official 'curricula' of the Czech Republic and Italy, showed that the aims of the teaching and learning in their kindergartens are similar. They stated that the time devoted to mathematics in teacher education is too little in both countries and proposed that both pre- and in-service teacher education include, in particular, teachers' awareness of children's preconceptions, and a range of tools supporting their teaching. The *Senso-Math Preschool* programme was presented by Hassidov and Ilany (2015). The goal of the programme was to enhance mathematical development in preschool children through the intervention of trained facilitators. The results indicated a positive change in the attitudes of the facilitators to their work, and a positive shift in the development of preschool teachers.

Competency of teachers

Some authors did not refer to MKT or derived concepts, but they did refer to teachers' professional qualities which are required in all aspects of their profession and which are linked to their development. At times the term *(professional) competency* was used. We interpret professional competency of a teacher as a concept that includes knowledge, skills, attitudes, experience, values, and personal characteristics. There is emphasis on a dynamic concept of competency that is expressed by orienting towards and coming nearer to the desirable parameters of the teaching profession (Spilková 2001).

Our understanding of the notion of 'teacher's competency' was strongly influenced by a Czech psychologist, Helus, who distinguishes four basic kinds of competency that form the basis of the teacher's self-confidence (Helus 2001: partly quoted, but our translation):

1. Pedagogical competency
2. Subject-didactic competency
3. Pedagogical-organizational competency
4. Competency in qualified pedagogical (self-) reflection

Almost all of these competencies were apparent during presentations at SEMT.

Pedagogical competency

This consists of creating conditions for the development of pupils' prerequisites by the effective organization of educational influences, by motivating students' own educational activities, and by exploiting their potential, removing mental blocks and barriers, mastering diagnostic operations, getting insight and empathy, and designing procedures for effective pedagogical intervention.

Wallach and Even (2001) stressed the creation of this competency while examining procedures of pedagogical intervention by fostering teachers' interpretations of students' talking, acting, and behaving when they are engaged in mathematical problem solving. The teachers in an in-service course solved several mathematical problems, after which each teacher video-recorded a pair of students from her, or his, own class solving one of the problems. Then the authors administered a questionnaire and interviewed the teachers. Results published here showed that 'hearing students' (understanding what students are saying, showing, feeling, and doing) is intertwined especially with teachers' own understanding of the subject matter, the solution to the specific problem, their conceptions of what is acceptable as a solution in mathematics, and their commitment to their students' success.

Doig and Groves (2003) drew on data from several sources and argued that, while teachers of young children are reasonably accurate in their predictions of when the majority of children will achieve mastery of specific objectives in mathematics, they are much less likely to be aware of the conceptions of high-achieving children, and that their classroom activities constrain these students' learning.

Part of pedagogical competency is also the sensitive use of knowledge of children's thinking. Lubinski et al. (2005) worked with six US primary-grade teachers in the framework of Cognitively Guided Instruction (CGI). These teachers, to various degrees, based their instructional decisions on their knowledge of their pupils' mathematical thinking. However, changes in their decision making in the environment of their mathematics classrooms reached a plateau after several years. The study showed how we can help teachers continue to grow in their knowledge of CGI as they conduct PD sessions for their colleagues.

Subject-didactic competency

This consists of a skilled orientation towards the educational meaning of teaching a specific subject, and its implementation in relation to a specific class, and specific pupils. It includes proper mastery of the scientific foundations of the subject (respectively the areas of general cultivation, techniques and technologies of work, artistic areas, etc.). In organic connection with this, however, it also includes didactic creativity—that is, the ability to update the curriculum in its cognitive, motivational, aspects and cultivating a seriousness for the student's independent classroom cognitive activities, for the development of their thinking, for their life orientations and plans, and for their opinions and beliefs.

Tsamir and Tirosh in their presentation (2003) described the initial explorations of primary school teachers' perspectives on the role of students' mathematical errors in mathematics instruction. They focused on the participants' views of the merit of intentional presentations of error-triggering tasks, and teachers' perspectives regarding the soundness of the deliberate presentations of incorrect responses in such classes. They found that, while most teachers voted for presenting error-triggering tasks, only a few advocated presenting incorrect strategies in primary classes. In the following SEMT, Tsamir and Tirosh (2005) continued with the elaboration of a more general topic—tasks in mathematics teaching. They presented a task that was developed for primary school teachers in a PD programme and offered a communicative analysis of mathematical and didactical occurrences.

Pedagogical-organizational competency

This competency consists of a skilled orientation of controlling the relations and activities in the classroom aimed at creating an effective educational environment, together with a supportive and stimulating climate.

This competency was less apparent in the presentations. Nevertheless, Pugalee's presentation (2001) on communicative competency is an example. The author based it on the belief that language is inherently related to mathematics learning; yet, most teacher preparation programmes do not include explicit objectives addressing this relationship. In order for teachers to maximize the mathematical experiences of their students, they must have a working knowledge of how mathematics teaching and learning relate to the four major communication modes: spoken language (speaking and listening), written language (reading and writing), graphic representation (diagrams, tables, pictures, graphs), and active demonstration (performance and physical involvement). Each of these four modes involves two perspectives: receptive (processing of received communication) and expressive (an individual's use of language). Pugalee explained how these four modes, through both perspectives, can be expanded in mathematics classrooms, providing students multiple opportunities to investigate mathematics in ways that facilitate a conceptual development of mathematical thinking.

Competency in qualified pedagogical (self-) reflection

This competency places emphasis on the analysis of the teacher's own thinking, and dealings with students, ensuring it supports the students' ability to plan their own lifelong education. This competency was apparent in the presentation by Perkkilä (2001) that investigated relationships between primary school teachers' beliefs and their mathematics teaching practices, especially the use of textbooks. Based on the analysis and comparison of the diverse data (belief questionnaire (N = 230), video-recorded classroom observations, written lesson plans, interviews, and discussions about the video-recorded lessons with six teachers), it was found that there were inconsistencies between beliefs and teaching practices. The teaching practices were influenced by teachers' recollections of their own mathematics studies at school and their dependence on mathematics textbooks.

Tichá, Macháčková, and Hošpesová (2005) in their presentation identified qualified pedagogical reflection as the basis for teacher professionalization. They presented an example of the co-operation between two researchers (teacher educators) and one practising teacher on the preparation, realization, and analysis of the teaching experiment (*via* individual and collective reflections) to illustrate how collective reflection helps teachers to understand the environment of the mathematics classroom, and thus to improve the teaching and learning process significantly. Similarly, Ribeiro and Martins (2009) emphasized the importance of reflection in the *Program for Continuous Training in Mathematics for Portuguese Primary School Teachers.*

Groves and Doig (2005) noted that the first priority for changing teaching practice is to make problematic for teachers', aspects of what they do currently. As part of a project on improving mathematics and science teaching in the middle years of schooling, the participants were asked to rate their practice against components of effective teaching and learning and to rate each of these according to their perceived importance. Findings suggested that primary teachers endorsed the components as representing effective practice, scoring most components higher than their actual practice. Gaps were particularly evident for items relating to challenging students conceptually and higher-order thinking, with these becoming the basis for some of the action planning for change.

Context of professional development programmes, their content and structure

In-service teacher education programmes are nationally specific. PD programmes often reflect existing legislation in the country and also the unwritten cultural conditions, not only in education, but also outside it. In this section we examine the contexts, contents, and structure of PD programmes.

Context of PD programmes

Presentations related to PD programmes during the SEMT symposia illuminated cultural specificity that often contributed to the efficacy of the programmes. Fujii (2015) stated that Lesson Study, as a method of PD, came to the attention of educators primarily through the publication of *The Teaching Gap* (Stigler and Hiebert 1999). Since then, many mathematics teachers and teacher educators outside of Japan have become involved in Lesson Study. Fujii stated that it seems that there are many assumptions, and roles, that are well understood by Japanese teachers, which simply do not transfer easily to other countries. For example, the creation of a lesson plan is accepted without difficulty, but the collaboration among teachers which goes into creating that lesson plan is largely underestimated by non-Japanese adopters of Lesson Study, possibly because the effort involved is invisible to outsiders. Fujii clarified the process of lesson planning by focusing on designing and adapting tasks for the lesson.

The idea of cultural specifics was confirmed by the presentation of Doig and Groves (2015), who discussed the difficulties in creating an effective structured problem-solving lesson. They stressed that it is important to pay attention to the details of all aspects that together contribute to the effective achievement of the lesson goal. In the case of structured problem-solving lessons within Japanese Lesson Study, less understanding of the complete process leads to less effective lessons. This has often been confirmed internationally, when it has been observed that adaptations of Lesson Study do not have the same impact on teachers' learning as they do in Japan.

In her presentation, Graven (2017) emphasized the cultural specifics in the context of the situation in South Africa. PD failed in supporting teachers in implementing three post-1994 cycles of curriculum revision which largely discouraged teachers. She summed up her experience in these words:

> establishing a network of development programs is not just an ethical 'nice to do give-back' to research participants. These partnerships enable access to data and stakeholder perspectives that would be inaccessible without the relationships developed in these spaces of collaboration, thereby strengthening the design of more appropriate research projects and leading to richer and more valid research findings. In these collaborative spaces dialogue and active participation among all participants is critical in the joint enterprise of finding sustainable ways forward to the educational challenges. (p. 20)

Savard and Polotskaia (2013) showed, in the context of a collaborative research project dealing with the creation of additive structure, that the perception that teachers have about the task might negatively affect the instructions they give to their students. They used an ethno-mathematical model to analyse problem-solving activities implemented by primary school teachers. They identified contexts involved in the tasks and how those contexts might have influenced students' understanding. They studied how teachers could help their students differentiate between the everyday context of the task and its mathematical structure. In order to change their perception of problem-solving tasks and modify how they implement them in their classrooms, a learning activity for teachers was developed.

Kaur (2019) showed how a PD programme that was coherent with the needs of teachers, in Singapore schools, led to much success at a national level. The project developed teachers in designing and infusing mathematical tasks for reasoning and communication in their lessons. The programme adopted a three-phase developmental approach that engaged teachers in acquiring new knowledge, infusing their knowledge into practice, and developing their peers through collaborative learning activities.

Stott and Graven (2015) adapted and implemented number talks as a key teacher development strategy. Their aim in introducing the talks was to simultaneously develop learners' mathematical talk (verbal reasoning) and to push explicitly for the use of more efficient calculation strategies. At the end of the project, teachers confirmed the effectiveness of the teaching method, and the fact that they included it in their repertoire.

Content of PD programmes

The most frequently discussed topic was the diverse *use of problem solving* in PD and other inquiry approaches. Problem solving was used as a method or content in PD courses with different goals. Littler and Koman (2001) emphasized the link between the problems addressed by the pupils and their teachers and showed how to introduce problem solving in teacher (in-service) education. The same issue was also dealt with in several other presentations. Nováková and Novotná (2011) explored the potential of 'a priori analysis' as a professional tool that helps the teacher in the process of selection and decision making in problem solving. The presentation analysed the differences between what is included in a priori analysis in the *Theory of Didactic Situations in Mathematics*, on the one hand, and the reality of everyday teaching on the other.

Albert (2001) asserted that if teachers were to use modern methods, they would need to experience such learning themselves. She successfully confronted teachers with challenging problems. The use of such problems has been shown to build teachers' mathematical knowledge and their mathematical confidence.

The complementary activity to solving tasks is the creation of tasks. Bureš and Clanché (2009) discussed the question 'In which aspects is problem posing conceived as a mathematical activity?' They addressed two points of view: that of researchers and that of novice teachers (with two to five years' teaching experience). They started from general characteristics of what is meant as *mathematical activity* and then they focused on the teachers' interests and the benefits that they could gain from problem-posing activities.

Ovadiya (2019) followed the process of seven teachers, in the role of researchers, who were asked to design problems and tasks in tessellations for their primary pupils, where the designed problems should promote the linking of knowledge on tessellations to geometrical or mathematical concepts (e.g. multiplication). Ovadiya found seven features common to the posed problems dealing with pupils' transfer of knowledge, and different levels of transfer. As a summary it was stated that the topic (tessellations) motivated both teachers and students to connect mathematical and geometric knowledge.

Hošpesová et al. (2015) reported on a survey whose aim was to contribute to the research in the area of inquiry-based mathematics education (IBME), primary school teacher training, and problem posing. They asked pre-service and in-service primary teachers to pose problems that they believed enabled IBME. Later the teachers used the problems with their pupils, and wrote self-reflections on this activity. They summarized that if teachers were to face this task, they should recognize the need for deeper conceptual understanding and a broader repertoire of problem interpretation.

Savard (2011) conducted a study on using inquiry-based mathematics learning in science and technology classes, with 18 primary school teachers, with the aim of seeing whether teachers used the opportunity to teach mathematics when it was relevant to do so in the science and technology classes. However, even though there are many opportunities to integrate mathematics into learning situations, the teachers made virtually no reference to mathematical intentions or goals. Savard summarized that the teachers' sensitivity about mathematics was weak, and attributed this effect to their weak mathematical knowledge.

In recent years the content of PD has included topics like formative assessment and inclusive education. Biton et al. (2017) examined the primary mathematics teachers' views on assessment through a questionnaire. Their findings suggest that many share a dissatisfaction with the traditional method of assessment which boils down to only a numerical grading. The answer could be to extend formative assessment. Lang, Hawthorne, Sakon, Reta, and Schoen investigated this and presented their work in 2011. The teachers participating in their study implemented one or more of the formative assessment tasks. The research findings suggested that teachers are in need of an answer to the practical questions that bother them when assessing students. The authors suggested introducing programmes in which teachers would take an active part in creating alternative assessment tools.

The question 'How can we face the increased diversity of learners and make sure that all children work and progress on their individual level but at the same time learn with and from each other?' was asked by Korten (2017). Based on this question, the aims of the research project were the development of, and the research on, a teaching–learning arrangement for the *inclusive mathematics classroom*. She used a Design Research approach, and interpretative data analyses showed that co-operative–interactive learning situations of primary school children, with, and without, cognitive learning disabilities, can be fruitful for all participants. Children had progressed according to their individual abilities triggered by 'impulses', which occurred in the interaction.

Haydar (2019) presented on a study of prospective and beginning teachers learning to conduct clinical interviews to enhance emergent bilingual students' access to mathematics. She showed how teachers develop and, or, modify interview protocols to support their teaching of mathematics to children whose first language is not English. The presentation zoomed in on a case study of a primary teacher who took these methods further and developed a structure for interviewing as a pre-lesson-planning assessment and for incorporating peer interviewing.

Scherer, Nührenbörger, and Ratte (2019) considered coping with heterogeneity in mathematics classrooms a big challenge for teachers. In particular, considering students with special needs in inclusive classrooms requires further development of classroom instruction. For teachers' PD, in-service courses for mathematics are necessary for both special education teachers and teachers in regular schools. As many special education teachers are out-of-field teachers for mathematics, a specific programme for laying a mathematical and didactical foundation is advisable. In their presentation a concrete in-service course was sketched out, and corresponding experiences and findings were discussed.

Structure of PD programmes

Not many presentations focused on the structure of PD programmes. In his presentation, Wollring (2003) called for the co-ordination of teacher education and school-related research. According to him, this approach supports in-service teachers at primary school level for those who ask for support in constructivist-orientated teaching. The framework he discussed was based on a special co-operation between university and primary schools, which was anchored as a special topic in the school programme and based on a specific concept of teacher training by the university. In Wollring's opinion, teachers in primary schools need particular support in teaching geometry.

While Wollring spoke of institutional co-operation, Višňovská and Cortina (2017) reported a case study that showed collaboration between a researcher and a teacher. The teacher, while collaborating with the researcher, enacted a research-based instructional sequence on fractions and measures in her under-performing fifth-grade classroom. During the process of the collaboration, the teacher accepted the view that it was important to support all students' understanding, and strongly endorsed equity in access to mathematical ideas. This influenced her planning and classroom decision making.

The impact of PD programmes on participants' classroom instruction

The goal of almost all PD programmes is to have an impact on teachers' classroom instruction. However, not many such programmes document this aspect in detail, as often researchers lack the luxury of being able to track the development of teachers in the programmes over a sustained period of time. Nevertheless, in several SEMT presentations there were glimpses of outcomes of PD programmes that had an impact on participants' classroom instruction.

Priolet (2013) summarized her research results based on the implementation of the didactic frame devoted to teaching and learning solving word problems with numerical data. When comparing the practices of teachers who participated in this research, involving eight French third-year primary school classes, she analysed the effects of 'reference boxes' on teaching and learning such problems.

Young-Loveridge and Mills (2011) reported on research on a lesson designed to teach the Equal Additions strategy for subtraction. Participants were nine teachers, each with a group of 5th or 6th graders (9- to 11-year-olds). Teachers stuck closely to the lesson description in the resource book, and, despite having a strong commitment to conceptual understanding, many teachers ended up teaching the lesson procedurally. Few of the students chose to use Equal Additions in a later assessment.

Errera and Patkin (2001) presented on a model of primary school mathematics curriculum that included a teacher guide, student workbooks, and a computer-based learning program that integrated diagnosis and remediation methods, and responded dynamically to the performance of each student. Basic to the curriculum was situated learning through discovery activities that require students to apply the mathematical concepts introduced in order to solve everyday problems. The authors documented how the instructional materials developed as part of the programme, increased the teachers' enjoyment of mathematics instruction, assisted them in diagnosis and remediation, and resulted in improved student achievement.

Durmaz (2019) examined the effect of a teacher-training project called Using Children's Literature in Teaching Mathematics and Sample Applications. The study group of the research was made up of 32 primary and lower secondary teachers from different cities in Turkey. The participation in the project had a significant and positive impact on teachers' creativity-fostering behaviours, curriculum enrichment skills, and differentiation skills.

Bobis (2007) explored the effectiveness of a school-based model for sustaining change in primary schools via a survey with 49 PD facilitators and 23 principals. Data (quantitative and qualitative) showed agreement among participants that the facilitator model provided a mechanism by which extensive PD programmes could be maintained, even if the support of external experts was removed.

Bambico (2001) presented on a study carried out in the Philippines to investigate if there was a relationship between teachers' personal characteristics and their achievement in mathematics, as well as the improvement of their instructional skills as a result of the training in which they participated. The study also described the heart of the in-service training programme that was devoted to reinforcing the use of Practical Work as a springboard to developing an understanding of mathematical concepts. This was in keeping with the time-tested principle of the 'hands-on, minds-on, and hearts-on' approach.

Wang et al. (2013) presented on the activities of a project called Assessment Practices to Support Mathematics Learning and Understanding for Students. The presentation focused on the comparison between pupils whose teachers participated in the PD, which was targeted on the use of online assessment tools, and those whose teachers did not participate in the PD. Statistical models revealed no statistically significant differences between the treatment group and control group students. However, both classroom average and school average performances were found to have significant impacts on student performance, and the relationship between student performance and the time elapsed from the start to the end of the assessments.

Novice teachers are a specific group. Özdemir et al. (2015) stated that the reform movement for school mathematics in Turkey expected teachers to be 'self regulated individuals, who can control their cognition, motivation, and behaviours in order to achieve effective instruction' (p. 349). These skills are particularly essential for novice teachers. The presentation described a case study of one novice teacher's teaching activities and the underlying regulatory processes (monitor and control behaviours) behind her instructional decisions and practices. Results showed that, even though the teacher mostly monitored issues related to her students' knowledge, skill, performance, and participation, she rarely attended to her own decisions and actions during teaching. With regard to the control over the content of the instruction, she aimed to cover the curriculum and to solve as many, and as great a variety of, problems as she could. In terms of control over the course, it was observed that she was not flexible in using different instructional methods or techniques when she encountered a problem.

Bakos and Sinclair (2019) in their presentation investigated pupils' and teachers' interactions with an iPad touchscreen application (TouchTimes), which provided a visual and kinaesthetic way for children to learn about multiplicative relationships. TouchTimes (TT) offered a functional and symbolic model of multiplication that did not rely on repeated addition. Based on the Theory of Semiotic Mediation, Bakos and Sinclair analysed the potential of TT for supporting primary pupils' learning of multiplication. They also studied the semiotic potential of TT for primary school teachers' learning-to-teach multiplication, by analysing their own interactions with this technology, as well as their responses to videos of pupils using it.

Kaur (2019) in her presentation showed how a Grade 1 teacher successfully enacted a reasoning and communication lesson following participation in a PD programme. The teacher created mathematical tasks, which were used for review of numbers less than 10, and for the four operations of addition, subtraction, multiplication, and division. The teacher's self-evaluation of her lesson showed that her students enjoyed the lesson and she managed a high level of student engagement, while maintaining the desired cognitive demand of the tasks she created for use.

Conclusion

Teaching is a cultural activity (Stigler and Hiebert 1999). Presentations during symposia like SEMT provide platforms for presenters, and opportunities for attendees, to glean knowledge about the why, what, and how, of mathematics teaching, and how these may differ from country to country, or even from one classroom to another in the elementary grades. It is valuable for participants to be better informed about the situation and opportunities in other countries, as this can lead to a better understanding of their own situation, their own actions. Such a reflection can then become the starting point for change.

This chapter drew on past presentations at SEMT and provided a one-stop resource for teachers of mathematics (both pre-service and in-service) and researchers in some aspects of PD that focused on teachers of preschool and primary school. In no way are we, the authors, making any claim that the aspects considered are exhaustive or complete.

Nevertheless, it is critical to note that PD is a lifelong endeavour and teachers must continue to participate earnestly so that research may continue to create knowledge that is worthy of emulation across all boundaries. Lastly, more can be done to empower teachers to codify, verify, disseminate, and institutionalize their expert knowledge through carefully crafted partnerships between university scholars and classroom teachers.

Acknowledgements

The chapter was elaborated with the support of the project of the Grant Agency of the University of South Bohemia České Budějovice named Edukační obsah pro rozvoj matematického a informatického myšlení (Educational Content for the Development of Mathematical and Computer Thinking).

References

Albert, J. (2001). Working with primary teachers on challenge problems in mathematics. In J. Novotná and M. Hejný (Eds) SEMT 41–43.

Bakos, S., and Sinclair, N. (2019). Exploring the semiotic potential of TouchTime with primary teachers. In J. Novotná and H. Moraová (Eds) SEMT 53–62.

Ball, D. L., Thames, M. H., and Phelps, G. (2008). Content knowledge for teaching: What makes it special? *Journal of Teacher Education*, 59(5): 389–407.

Bambico, T. A. (2001). Influence of teachers´ personal variables to teachers´ mathematics achievement and instructional skills: A study on the effectiveness of in-service training in the Philippines. In J. Novotná and M. Hejný (Eds) SEMT 43–47.

Bardelli, M. (2011). The pedagogical content knowledge of primary mathematics teachers facing a student error. In J. Novotná and H. Moraová (Eds) SEMT 63–71.

Biton, Y., Hershkovitz, S., Hoch, M., Ben-David, B., and Fellus, O. (2017). Assessment issues that trouble mathematics teachers. In J. Novotná and H. Moraová (Eds) SEMT 63–71.

Bobis, J. (2007). Facilitating change as part of a large-scale numeracy program. In J. Novotná and H. Moraová (Eds) SEMT 76–84.

Bureš, J. and Clanché, P. (2009). Problem posing as a mathematical activity: The young teachers' perspective. In J. Novotná and H. Moraová (Eds) SEMT 81–88.

Cabassut, R. (2017). Diversity of teachers' beliefs on modelling through a French-Spanish comparison. In J. Novotná and H. Moraová (Eds) SEMT 119–128.

Cheeseman, J. (2009). Challenging children to think: Teacher behaviours that stimulate children to examine their mathematical thinking. In J. Novotná and H. Moraová (Eds) SEMT 11–23.

Chestnut, A. (2011). Using PC scaffolds evidenced in transcripts to measure teacher knowledge. In J. Novotná and H. Moraová (Eds) SEMT 95–103.

Climent, N., and Carrillo, J. (2002). Developing and researching professional knowledge with Primary Teachers. *CERME2*, 269–280. Prague, Charles University, Faculty of Education.

Doig, B., and Groves, S. (2003). Teachers' pre-conceptions of young children's conceptions of number. In J. Novotná (Ed.) SEMT 56–61.

Doig, B., and Groves, S. (2015). Reaching for the heavenly bead of teaching through structured problem solving. In J. Novotná and H. Moraová (Eds) SEMT 117–127.

Durmaz, B. (2019). A teacher training experiment: Using children's literature in teaching mathematics and sample applications. In J. Novotná and H. Moraová (Eds) SEMT 157–163.

Errera, A., and Patkin, D. (2001). The primary school math teacher's role in the computer era. In J. Novotná and M. Hejný (Eds) SEMT 64–68.

Fujii, T. (2015). Designing and adapting tasks in the Japanese lesson study: Focusing on the role of the quasi variable. In J. Novotná and H. Moraová (Eds) SEMT 9–18.

Fung, C. (2011). A practitioner's encounter with the mathematical knowledge needed for teaching in primary school. In J. Novotná and H. Moraová (Eds) SEMT 134–141.

Graven, M. (2017). Blending primary mathematics education research with development for equity – and ethical imperative enabling qualitatively richer work. In J. Novotná and H. Moraová (Eds) SEMT 20–31.

Groves, S., and Doig, B. (2005). Effective mathematics learning environments: Primary teachers' beliefs and practices. In J. Novotná (Ed.) SEMT 115–123.

Hassidov, D., and Ilany, B.-S. (2015). Successful cooperation between mathematical facilitators and preschool teachers, the aim of 'senso-math' preschool program. In J. Novotná and H. Moraová (Eds) SEMT 137–146.

Haydar, H. (2019). Learning to enhance emergent bilinguals' access to mathematics: elementary teachers experimenting with the clinical interview. In J. Novotná and H. Moraová (Eds) SEMT 192–201.

Helus, Z. (2001). Čtyři teze k tématu 'změna školy'. (Four theses on school reform, in Czech). *Pedagogika*, 51(1): 25–41.

Hošpesová, A., Samková, L., Tichá, M., and Roubíček, F. (2015). How primary school teachers posed problems for inquiry based mathematics education. In J. Novotná and H. Moraová (Eds) SEMT 156–165.

Kaslová, M., and Marchini, C. (2011). The mathematical knowledge needed for teaching to five year old pupils. In J. Novotná and H. Moraová (Eds) SEMT 182–189.

Kasmer, L., and Billings, E. (2011). Prediction questions in the primary classroom. In J. Novotná and H. Moraová (Eds) SEMT 189–197.

Kaur, B. (2019). Opportunity to learn: reason and communicate mathematically in the primary class. In J. Novotná and H. Moraová (Eds) SEMT 9–19.

Korten, L. (2017). The investigation of co-operative interactive learning situations in an inclusive arithmetic classroom. In J. Novotná and H. Moraová (Eds) SEMT 282–292.

Lang, L., Hawthorne, H., Sakon, F., Reta, A., and Schoen, R. (2011). Examining the effects of A K-3 mathematics formative assessment system. In J. Novotná and H. Moraová (Eds) SEMT 214–221.

Littler, G. H., Koman, M. (2001). Challenging activities for students and teachers – cast. In J. Novotná and M. Hejný (Eds) SEMT 113–118.

Lubinski, C., Warfield, J., Beauchamp, B., and Webster Moore, M. (2005). Conducting staff development for colleagues: The effect on K-3 teachers´ instructional decision making in mathematics. In J. Novotná (Ed.) SEMT 210–218.

Macháčková, J., and Tichá, M. (2007). Assessment of knowledge and conceptions through problem posing: the case of fractions. In J. Novotná and H. Moraová (Eds) SEMT 178–185.

Marcinek, T., and Partová, E. (2011). Measures of mathematical knowledge for teaching: Issues of adaptation of a U.S. developed instrument for the use in the Slovak Republic. In J. Novotná and H. Moraová (Eds) SEMT 229–236.

Niess, M. L., and Pugalee, D. K. (2011). Assessing K-8 teachers' knowledge for teaching with technology: A complex problem needing a comprehensive assessment system. In J. Novotná and H. Moraová (Eds) SEMT 244–252.

Nováková, H., and Novotná, J. (2011). A priori analysis in theory and teachers' practice. In J. Novotná and H. Moraová (Eds) SEMT 252–260.

Ovadiya, T. (2019). Posing problems and designing tasks to promote transfer of learning in geometry by teacher researchers: The case of tessellations. In J. Novotná and H. Moraová (Eds) SEMT 280–287.

Özdemir, Y., Bozkurt, E., Yıldız, P., and Gürel, R. (2015). Teacher regulation of instructional practices: A case study with a novice teacher. In J. Novotná and H. Moraová (Eds) SEMT 349–359.

Perkkilä, P. (2001). Elementary school teachers´ view of mathematics and their teaching practices. In J. Novotná and M. Hejný (Eds) SEMT 131–135.

Priolet, M. (2013). 'Reference boxes': Teaching and learning with reference tools in primary mathematics. In J. Novotná and H. Moraová (Eds) SEMT 242–249.

Pugalee, D. K. (2001). Mathematics and language: Identifying requisite pedagogical knowledge. In J. Novotná and M. Hejný (Eds) SEMT 135–139.

Ribeiro, C. J., Mellone, M., and Jakobsen, A. (2013). Giving sense to students' productions: A particular task in teacher education. In J. Novotná and H. Moraová (Eds) SEMT 273–281.

Ribeiro, C. M., Carrillo, J., and Monteiro, R. (2009). We teach what we know, but do we know what we teach? The practical 'distinction' between squares and rectangles in a primary school class. In J. Novotná and H. Moraová (Eds) SEMT 204–212.

Ribeiro, C. M., and Martins, C. (2009). The impact of a professional development program in teacher's practice. In J. Novotná and H. Moraová (Eds) SEMT 212–219.

Savard, A. (2011). Elementary teachers do not teach through mathematics. In J. Novotná and H. Moraová (Eds) SEMT 297–305.

Savard, A., and Polotskaia, E. (2013). Word problem solving task management and students´ access to mathematics: Cases in primary education. In J. Novotná and H. Moraová (Eds) SEMT 290–298.

Scherer, P., Nührenbörger, M., and Ratte, L. (2019). Inclusive mathematics – in-service training for out-of-field teachers. In J. Novotná and H. Moraová (Eds) SEMT 382–391.

Spilková, V. (2001). Professional development of teachers and student teachers through reflection of practice. *The New Hampshire Journal of Education*, 4: 9–14.

Stigler, J. W., and Hiebert, J. (1999). *The Teaching Gap: Best Ideas from the World's Teachers for Improving Education in the Classroom*. New York: Free Press.

Stott, D., and Graven, M. (2015). Adapting number talks to foreground mathematical progression in South African classrooms. In J. Novotná and H. Moraová (Eds) SEMT 311–322.

Tall, D. (2001). What mathematics is needed by teachers of young children? In J. Novotná and M. Hejný (Eds) SEMT 26–31.

Tichá, M., Macháčková, J., and Hošpesová, A. (2005). Mathematics classroom and collective reflection. In J. Novotná (Ed.) SEMT 307–315.

Tsamir, P., and Tirosh, D. (2003). Errors in an in-service mathematics teacher classroom: what do we know about errors in the classroom? In J. Novotná (Ed.) SEMT 26–34.

Tsamir, P., and Tirosh, D. (2005). Common nominator or common denominator? Teachers´ views on comparing fractions. In J. Novotná (Ed.) SEMT 316–323.

Vinner, S. (2011). What should we expect from somebody who teaches mathematics in primary schools? In J. Novotná and H. Moraová (Eds) SEMT 31–43.

Višňovská, J., Cortina, J.L. (2017). Learning to support all students' fraction understanding. In J. Novotná and H. Moraová (Eds) SEMT 430–440.

Wallach, T., Even, R. (2001). Teachers' interpretations of students' mathematics thinking and behaviour. In J. Novotná and M. Hejný (Eds) SEMT 137–144.

Wang, C., Cifarelli, V., Stephan, M., Martin, M., Pugalee, D. K., and Lambert, R. (2013). Assessment practices to support mathematics learning and understanding for students. In J. Novotná and H. Moraová (Eds) SEMT 331–338.

Wollring, B. (2003). Linking pre-service and in-service teacher training: Co-operative design and dissemination of working environments for primary mathematics. In J. Novotná (Ed.) SEMT 35–41.

Young-Loveridge, J., and Mills, J. (2011). The challenge for teachers of helping students acquire a range of strategies to solve addition and subtraction problems. In J. Novotná and H. Moraová (Eds) SEMT 352–360s.

SECTION 2
ELEMENTARY MATHEMATICS LEARNING

Berinderjeet Kaur
National Institute of Education, Nanyang Technological University, Singapore

This section focuses on the contexts within which learners learn mathematics. These contexts match the learning needs of learners, for example their dispositions such as age and special needs, or the situation they are in relation to the subject of mathematics. The four chapters in this section are different in that they address four different contexts: early childhood, special needs, mathematical problem solving extending to problem posing and enhancing creativity, and learning to solve word problems.

The first chapter, by Barkai, Levenson, Tsamir, and Tirosh, addresses early childhood contexts for learning mathematics. Though early childhood may include children up to Grade 3, the chapter focuses on children who have yet to enter Grade 1. It may be said that the early childhood context has been apparent in the SEMT conferences from the beginning. During the inaugural SEMT symposium in 1991, Urbańska (1991) reported on the numerical competence of children entering their 'K' class. In the next symposium, Gifford (1993) explored an appropriate mathematics curriculum for three-to-six-year-olds. In this chapter, Barkai, Levenson, Tsamir, and Tirosh have noted early childhood contexts for learning number concepts, geometry, and spatial visualization, and patterns and generalizations have been prevalent during past SEMT symposia. Their review of past research presented at the symposia shows that many of these childhood contexts for learning the aforementioned topics were the senses of touch, sight, and hearing. This reinforces that mathematical learning should use concrete (tactile), pictorial (visual), and audio (audible) representations for preschoolers in 'K' and the years below. The authors also identified an emerging focus of the research, that is out-of-school settings in childhood contexts, for early mathematical learning. Adults who are not preschool teachers involved in such a setting play a significant role in the formative years of a child's approach and disposition towards the learning of mathematics. Therefore, there is a need to study how these adults have an impact on 'young minds' in their beliefs of what mathematics is, and how they make sense of it.

Scherer, in the second chapter, notes that though the term 'special needs' has not been featured in the themes of SEMT symposia, nevertheless themes such as 'Confidence and Competence' and 'Equity and Diversity' have led researchers to address 'special needs' from the perspective of mathematics education at these symposia. This is apparent in the symposia proceedings. The author notes that, over the past 30 years of SEMT proceedings, special needs has covered a wide range of abilities and aspects: students with learning difficulties or disabilities on the one hand, and more or less extremely gifted students on the other hand, multicultural classrooms, challenges for migrant students, questions of language or language problems, and specific impairments like dyscalculia. The international context of SEMT has provided for comparisons of school systems in the different countries and glimpses of comparable approaches in addressing common issues and contextually situated findings. In what the author claims to be a state-of-the-art review of presentations at past SEMT symposia, the chapter examines high-quality instruction for all, inclusive of those with special needs, and highlights the role of language, cultural diversity, interactions, and teachers with regard to the learning of mathematics by special needs students. It also reviews how mathematical tasks that students work with to make sense of the mathematical knowledge they are constructing are critical. The abundance of examples the author has drawn on from past SEMT symposia show that indeed mathematics education for students with special needs has been addressed at SEMT for the last 30 years, albeit in a myriad of ways.

The third chapter, by Hershkovitz and Dana-Picard, examines a context for learning mathematics that is often deemed the core of school mathematics—problem solving, which is often followed by problem posing at the next cognitive level, and an outcome that is befittingly creative. The authors noted that over the last 30 years, research on problem solving has evolved and has been presented from different perspectives. The presentations, during past SEMT symposia, reviewed and reported in the chapter cover mathematical word problems—what they are, why they should be solved, and their solution processes. The multiple strategies that support solution processes, obstacles, and goals, such as Polya's (1945) four stages, students' difficulties, creativity and problem solving, and the use of technology to solve problems, are also detailed in this chapter. The authors conclude that research on mathematical problem solving will continue to flourish in future SEMT symposia, as mathematical problems link the real world to the world of mathematics.

The last chapter in this section, by Vondrová, is on learning to solve word problems with understanding. Past SEMT presentations on word problems are categorized into four broad areas, namely characteristics of problem solvers (pupils) and how they affect their ability to solve these problems, variables which make word problems difficult, the effect of different instructional methods on pupils' problem-solving ability, and the effects of classroom-related conditions on pupils' ability to solve word problems. Among the presentations on word problems, there appears to be a significant focus on the effects of instructional methods on pupils' problem-solving ability. These methods include reading with understanding the stem of a word problem, often resolving the conflict between information given in the problem and pupils' life experiences, representing the information given in a word problem in the form of a graphic representation as a first step in the solution process, using *solving maps* as a tool for meta-cognition when solving word problems, engaging pupils in problem posing while completing an incomplete word problem, and conceptualizing word problems as modelling tasks. The chapter concludes with an affirmation that research on word problems presented during SEMT symposia paralleled global trends, and contributed towards literature on word problems in mathematics education. Needless to say, a challenge was also identified. It was to open new lines of thought related to word problems in school mathematics.

References

Gifford, S. (1993). What is an appropriate mathematics curriculum for three to six year olds. In M. Hejný and J. Novotná (Eds) SEMT 38–40.
Polya, G. (1945). *How to Solve It.* Princeton University Press.
Urbańska A. (1991). On the numerical competence on children beginning their 'K' class. In M. Hejný and J. Novotná (Eds) SEMT 38–40.

2.1
Early childhood contexts for learning mathematics

Ruthi Barkai
Tel Aviv University, Israel
Esther Levenson
Tel Aviv University, Israel
Pessia Tsamir
Tel Aviv University, Israel
Dina Tirosh
Tel Aviv University, Israel

Introduction

The Symposium on Elementary Mathematics Teaching (SEMT) focuses on elementary school mathematics. Thus, one might ask why this chapter is dedicated to early childhood contexts. One might also ask what is meant by early childhood. We begin with the second question. There is no consensus regarding what is considered early childhood. Some studies specializing in early childhood (e.g. Downton et al. 2020) refer to children from birth to eight years of age, although birth to age three might refer to infancy, rather than early childhood. According to the United Nations Children's Fund (UNICEF) (UNICEF Programme Division 2017), early childhood is from birth to age five. Going back to the first question, there are several reasons for including early childhood studies in an elementary mathematics symposium. First, some countries have suggested a mandatory mathematics curriculum for children aged three to six. For example, in Israel (Israel National Mathematics Preschool Curriculum 2010), there is a mandatory preschool mathematics curriculum that sets separate standards for children aged three to four, four to five, and five to six. In Australia, the Foundation Stage (prior to first grade) sets achievement standards in the areas of numbers, geometry, and data collection. A second reason for considering early childhood mathematics is that in several countries (e.g. the United States of America) kindergarten classes (for children aged five to six) are part of the elementary school system and children learn in the school building. Third, it is now well known (e.g. Sarama and Clements 2009) that children enter Grade 1 with mathematical skills and conceptions, as well as misconceptions. Specifically, Evans, Strnadová, and Wong (2007) showed that children enter school with varying degrees of number sense. Thus, early childhood mathematics research can inform elementary school mathematics educators.

Ruthi Barkai et al. *Early childhood contexts for learning mathematics*. In: *Elementary Mathematics Teaching*. Edited by: Brian Doig, Jarmila Novotná, Berinderjeet Kaur and David K. Pugalee, Oxford University Press. © Oxford University Press (2024). DOI: 10.1093/oso/9780192869647.003.0005

The notion of including early childhood studies in SEMT symposia goes back to the first conference in 1991, when Urbańska (1991) delivered a report regarding the numerical competence of children beginning their 'K' class (K referring to kindergarten). At the second symposium meeting, Gifford (1993) asked, 'What is an appropriate mathematics curriculum for three to six year olds?' (p. 38). In her review of what was then current research, she concluded that, prior to Grade 1, children should be encouraged to count verbally to large numbers, estimate amounts of objects, and develop strategies for addition and subtraction. Notably, there was no mention of mathematical content other than number competencies, such as geometry or spatial orientation. Gifford did refer to teaching methods, claiming that curricula should pay more attention to social contexts, with the adult acting as a role model rather than an instructor. Finally, and relevant to this chapter, Gifford noted that research had little effect on published guidelines for early childhood educators and that many early childhood educators avoided engaging children with mathematical activities. This is yet another reason why it is important to publish and disseminate early mathematics studies at symposia such as SEMT, where teacher educators and researchers gather and share information that can influence future practice.

While acknowledging, as mentioned above, that early childhood may include children in Grades 1 to 3, the intention of this chapter was to review studies presented at the SEMT symposia related to children who have not yet entered the first grade. The age of these children is usually up to, and including, six years old. Some studies refer to children in kindergarten (the year before first grade), and some refer more generally to preschool, which can include two or three years prior to first grade. Several studies (e.g. Funkhouser, Risley, and Gannon 2003) consider children of different ages: those in preschool, kindergarten, and upper grades. In this chapter, we include studies that have at least some participants who have not yet entered the first grade. Also note that, since the terms kindergarten and preschool are used differently in different studies, we used the term kindergarten for studies that took place in the United States of America, where the year before the first grade is considered the first year of formal schooling but is exclusively called kindergarten. For all other studies, we use the term preschool to refer to all ages prior to first grade.

The organization of this chapter is as follows. Based on three major mathematical content strands, the first three sections review contexts for learning number concepts, geometry and spatial reasoning, and patterns and generalization. The following section summarizes those studies, mentioning some additional content domains. The last section reports on our own research connected to early childhood contexts for learning mathematics, and highlights ways in which SEMT has been involved in this research.

Contexts for learning number concepts

When discussing counting activities, it is important to differentiate between verbal counting and object counting. Verbal counting refers to saying the number words in the proper order and knowing the principles and patterns in the number system as coded in one's natural language (Baroody 1987). Additional verbal competencies including counting backwards, skip counting (e.g. counting by 2s, 5s, and 10s), and counting forward from some number other than one. Groves (1999), in her plenary address, described how calculators can be used to encourage preschool children's verbal counting skills. She described how the 'built-in constant function on the calculator allowed children to begin with any number and count on by any number they chose' (p. 10). In other words, children were practising counting forward from some number other than one and reaching very large numbers (as suggested by Gifford, in her 1993 report). Some children moved on to counting by 5s, 10s, or 100s, with one child recognizing that, if counting by 10s, he could reach the number 1050. Another child used the calculator to count by 9s, and noticed that the units digit decreases by one each time, while the tens digit increases by one. These children were not only noticing patterns in the number sequences, as suggested by Baroody (1987), but were learning to appreciate mathematics as the science of patterns (Schoenfeld 1994). Gifford went on to describe children who used the calculator to help them count backwards, even reaching negative numbers. This led to further inquiries related to real-life situations that may be described with negative numbers, such as animals that live underground.

From calculator technologies, we move forward to digital technologies. Beginning with a workshop in 2017, Tirosh et al. (2017) invited participants to explore together and then discuss findings of a study that investigated the affordances and constraints of tablet applications (apps) for promoting young children's knowledge of counting. In their survey of 15 game-like apps, they found that almost all of the apps counted out loud in sync with the user's touch, offering repeated auditory opportunities for hearing the counting words in order. Counting from some number other than one was possible in some of the apps. For example, in one app, when counting eight fish, if the user placed three fingers on three fish, the app would say three, and then, when touching one fish at a time, you would hear four, five, six, seven, eight. However, if one first touched three fish, and then two, and then three, you would hear three, five, eight. Thus, counting by ones may not necessarily be promoted. None of the apps specifically promoted skip counting. However, in some apps, if one consistently touched two or three objects at a time, skip counting could be promoted. None of the apps promoted counting backwards. Barkai et al. (2019) reported on an analysis of the apps in terms of their affordances and constraints for promoting children's object counting, specifically focusing on cardinality (i.e. knowing that when counting objects in a set, the last number mentioned represents the number of objects in that set), and one-to-one correspondence (i.e. assigning one counting word to each object). While one-to-one correspondence was supported by nearly all of the apps, cardinality was supported by only a few.

One of the strong points of *apps* is the visual representation of number. Visual contexts have also been investigated in SEMT symposia as a way of promoting number concepts. Kaslová (2017) carried out three studies (in addition to a pilot study) during different time periods where preschool children were requested to modify the representation of the number six (⁝ ⁝). In her first study, during the years 1995–1998, six strategies for changing the configuration of the six dots were detected. During the years 2000–2010, besides those six strategies, an additional three strategies were detected, and in the final stage, between 2000 and 2014, new drawings were again found. Kaslová inferred that the different results, especially the increase in children's variations, was due to the changes in teaching styles, from direct transmissive styles to more constructivist teacher strategies, encouraging more variability among children's work.

Visualization of number can be especially important for cognitively delayed children. Funkhouser, Risley, and Gannon (2003) showed that the use of 5-frame cards can support children's subitizing. After a five-week intervention with the 5-frame cards, children who were in kindergarten to second grade were able to recognize and match the numbers zero to five. Furthermore, using the visual representations, the children were also able to add sums up to five with increased independence.

Subitizing was also mentioned in a SEMT workshop organized by Poirier and Bisaillon (2015). They suggested that developing preschool children's ability to subitize would support their view of numbers being made up of groups, which may in turn help children deal with large numbers.

During early childhood, along with using their sense of sight, children use movement and their sense of touch to learn new ideas. Guidoni (1999) suggested that preschool children may learn about number, number operations, and even properties of numbers by using their bodies. For example, he pointed out that children willingly play games that require them to take steps forward and backwards. He suggested that, within the context of such movement games, children are able to feel and see that if they take two steps forward and then three steps, they reach the same place as if they had originally taken five steps forward. The commutative property of addition may be played out as children see and feel that taking three steps forward followed by two steps forward is the same as taking two steps forward followed by three steps forward.

Sight and the tactile sense are also employed when using concrete manipulatives. Several SEMT presentations employed manipulatives as contexts for either assessing or promoting young children's number knowledge (Kaslová 2017; Pratt 1997; Urbańska 1991). Perhaps because manipulatives are so frequently employed, a workshop dedicated to their use was presented at SEMT in 2015. There, Back and Griffiths (2015) defined manipulatives as 'objects that can be handled and moved, and are used to develop understanding of a mathematical situation' (p. 362). During this workshop, participants engaged with various types of unstructured manipulatives, such as collections of toys, and others designed for teaching, such as Dienes blocks. With these manipulatives, participants could explore how young children may solve simple addition and subtraction problems, investigate patterns, and even justify solutions.

Context for learning geometry and spatial reasoning

Early childhood is a time for exploring both 2D and 3D shapes. Children draw pictures and play with blocks. They navigate their way from room to room. In fact, Kuřina, Tichá, and Hošpesová (1997, 1999) reported how children's drawings may be used as a basis for both evaluating and developing children's geometric reasoning and communication. In one of their experiments, preschool children were requested to draw a cup from different points of view, looking at the cup from the front, from above, and from below. While half of the children were able to satisfactorily complete the task, others produced the same picture for all three requests, even though they had actually looked at the cup from the prescribed view. In other words, they did not draw what their eyes were seeing, but instead drew the image of a cup stored in their mind.

Related to visual perspective is spatial orientation. For young children, space is explored while conducting motor activities such as moving along paths, locating objects in a particular space, and using related language to describe movement and location (e.g. forward, behind, and above) (Ancona, Montone, and Pertichino 2005). While these types of activities may take place in everyday undertakings, guided activities offer children increased opportunities for using spatial language, as well as their spatial sense. One type of movement activity that pre-kindergarten children may enjoy is dance. According to Blundell et al. (2017), 'Dance elements such as body awareness, pathways, and creating shapes with the body all build spatial awareness in young children ... During dance or movement exercises, a child's understanding of spatial language enables him/her to move one's body through space' (p. 74). Furthermore, dance often involves movements where children cross the midline of their bodies (e.g. crossing a hand or foot to the opposite side of the body), movements that help build increased brain functions, which enable them to perform more advanced tasks.

Movement and spatial awareness may also be promoted through games. Ancona, Montone, and Pertichino (2005) devised games so that children could explore space within a playful context. In one game, a simulated tour of Italy was carried out, with the children setting up the course, having to consider distances and angles. In another game, children had to toss a coin at a prescribed distance, encouraging their skill for measuring, as well as their understanding of topological notions such as closer, further, higher, and lower. A simple game of hopscotch encouraged children's awareness of inside–outside, while introducing them to congruent geometrical figures.

Recognizing and naming shapes is also a major part of early geometric experiences. At this age, young children begin to perceive attributes, but need guidance in order to assess which attributes are critical for identifying a figure and which are not (van Hiele 1959). Both Tsamir et al. (2011) and Kopáčová and Žilková (2015) studied preschool children's shape recognition and level of reasoning, although with different populations and different methodologies. Kopáčová and Žilková used a multiple case study method and investigated nine children aged between 4 years 8 months and 6 years, while Tsamir et al. investigated the effects of an intervention conducted with 130 preschool children, aged 5–6 years old, out of which 67 children were labelled as abused and/or neglected by the social welfare department of their municipality. Tsamir et al. focused on triangles, pentagons, and circles; Kopáčová and Žilková focused on triangles, squares, rectangles, and circles. Using case study methodology, Kopáčová and Žilková were able to analyse children's difficulties when identifying the shapes. They found that some children do not know the name of a shape, while some know names of shapes, but connect names to the wrong figures. The intervention carried out in Tsamir et al.'s study included card games, board games, and colouring activities. These activities encouraged children to distinguish between critical and non-critical attributes of triangles. The difference between 'pointy' vertices and 'rounded' vertices and the difference between straight and curved lines were discussed. Findings indicated that the intervention was beneficial for all children, but were specifically significant for the neglected and abused children. While before the intervention the knowledge of children who were abused and/or neglected significantly lagged behind that of their peers, after the intervention the knowledge gap was narrowed.

The block corner of a preschool class can be an excellent context for exploring three-dimensional geometry. As an example, Franke (1999) investigated preschool and first-grade children's ways of sorting 24 geometric solids made of wood. Working in pairs, the preschool children distinguished between quadratic prisms, round figures, acute figures, and poles. First-graders found additional ways of sorting the figures, although none of the pairs sorted the figures according to the number of corners. Franke concluded that van Hiele levels 0 and 1 could be seen in the work of all children, independent of age. Children first registered geometric figures as a whole and arranged them according to similarities and prototypes. They then continued to analyse other features as well (level 1). More recently, Nováková (2019) described preschool children exploring geometric constructions using Magformer shapes (magnetic building blocks). Some children used the magnetic shapes to build three-dimensional shapes, while others built 2D figures. Like Franke (1999), Nováková found that some children began by describing their construction as a whole (e.g. 'a house'), reminiscent of van Hiele level 0, and then focused more on the details. Constructing and describing their constructions allowed children to explore the relationships between two- and three-dimensional figures.

Related to geometry, spatial orientation, and patterning are tiling and tessellation, which were the focus of studies conducted by Swoboda, and reported at SEMT in 2005, 2007, and 2013. In her first reported study, Swoboda (2005) presented four tiles to children aged between five and seven. Each tile had different drawings of curves. Children were requested to construct a floor from the tiles. Referring to van Hiele (1986), Swoboda categorized children's creations into 'feeble' or 'rigid' structures. Feeble structures are difficult to describe using geometrical transformation language. They are characterized by locally similar concentrations of curved lines, local repetitions, concentrations of similar shapes, and local incomplete symmetries. These structures may be an essential step leading to more rigid structures. Rigid structures, on the other hand, can be extended, may be part of a finer structure, may be part of a more inclusive structure, and may be isomorphic with another structure. While this study considered structure and patterns, as Swoboda (2005) stated, 'Work during which a child starts to order the space is the geometrical activity indeed. Many ways of organizing the space can by described by the language of geometrical relations' (p. 306). In her follow-up study, Swoboda (2007) used the same task, this time focusing on geometric relations such as translation, axial symmetry, and rotation. Finally, in 2013, while still interested in geometrical figures, Swoboda's study focused more on patterning than on geometry, and will be discussed in the next section.

Contexts for exploring patterns and promoting generalization

The importance of engaging young children with pattern activities is supported by several mathematics educators (e.g. Greenes, Ginsburg, and Balfanz 2004). Educators have suggested that exploring and analysing patterns may promote children's deductive reasoning skills, providing opportunities to observe and verbalize generalizations, as well as laying the foundation for early algebraic thinking (Clements and Sarama 2007; Greenes, Ginsburg, and Balfanz 2004). At the kindergarten level, educators have specifically noted that exploring repeating patterns may serve as a context for promoting children's appreciation of underlying structures (Starkey, Klein, and Wakeley 2004). These patterns may be visual (seeing the pattern—e.g. seeing a triangle, circle, triangle, or circle), audio (hearing a pattern—e.g. beep, bop, beep,

beep, or bop), or movement (e.g. clap, clap, kick, clap, clap, or kick). Swoboda's (2013) experiment with nine preschool children included both visual and auditory patterns. Each child was given a repeating pattern made up of geometric figures to look at, while the teacher rhythmically called out the figures presented. The child was then requested to continue the pattern by drawing the next figure in the pattern. Her findings suggested that the teacher's verbal rhythmic utterances supported the children's ability to complete the task. Furthermore, she noted that one girl used her own voice as she worked to complete the task, while two others added movement, by pointing to each figure in its turn. More recently, Tirosh et al. (2015) investigated preschool children's abilities to recognize similar and different structures of repeating patterns, when the patterns were presented through different media (beads on a string and picture cards). When the patterns were presented in the same media, children had greater success in recognizing the structure than when the patterns were presented in two different media. As a result of their study, Tirosh et al. suggested that 'akin to presenting both examples and non-examples when introducing some concept, presenting children with both similar and different structures, and having them compare the structures, may help them discern between critical and non-critical attributes when it comes to comparing patterns' (p. 340).

Mulligan (2013) pointed out that activities involving repeating patterns are but one type of activity among many others that can promote structure awareness. In her plenary address, she outlined the Pattern and Structure Mathematics Awareness Program, a programme that aims to promote generalization in early mathematical thinking. Part of the programme involves having children copy, or reproduce, a model as the teacher uses probing questions to highlight important features of their drawings, comparing drawings to models, or to other children's drawings, and thus focusing children's attention on aspects of spatial and numerical structure.

As can be seen from both Swoboda's (2013) and Mulligan's (2013) presentations, exploring patterns may also be an opportunity to engage in geometrical concepts and spatial awareness. This was discussed explicitly by Hošpesová and Semerádová at SEMT in 2013. During free-play time, teachers observed children forming a pattern with cards picturing two-dimensional shapes and used this context as a way to introduce the children to new figures, such as the hexagon and ellipse. The children and teacher then went on to discuss the critical attributes of the hexagon, comparing the number of vertices of the hexagon to a triangle and rectangle.

Contexts for patterning may also be contexts for learning about numbers. According to Sarama and Clements (2009), for children of a young age, pattern exploration and recognition may encourage mathematical skills such as skip counting, and the use of arithmetic strategies, such as adding 'doubles'. Littler and Benson (2005) pointed out that '[i]n early number work children recognize the count of numbers more easily if the objects they are counting are in set patterns such as those on a die or domino' (p. 203). In one experiment, young children were asked to divide a cake with straight lines and find the largest number of cake pieces resulting from cutting the cake with 1, 2, 3, 4, 5 ... cuts. By drawing, or using plasticine 'cakes' and plastic knives, preschool children ended up with the sequence 2, 4, 6, 8 ... etc. Notably, although children tended to work on their own, when they 'discovered' a pattern, they shared it with other children, suggesting that patterns may also be a context for encouraging mathematical discourse among young children.

With the increased use of digital tablets and tablet applications (apps), it is not surprising that two recent presentations, a workshop and a research report, focused on the use of tablet apps for exploring patterns. In the same workshop for exploring counting apps described previously, Tirosh et al. (2017) invited participants to explore tablet applications (apps) for promoting young children's knowledge of repeating patterns. Specifically, participants discussed the affordances, and constraints, of engaging students with patterning tasks with and without apps, considering how the touchscreen technology can affect children's learning of pattern concepts. Also investigating applets were Benyak and Kopáčová (2019), who designed repeating pattern applets for children aged 3 to 7, and then observed the children's engagement with the applet, their choice of tasks, level of success, and interest. They found that children under the age of four experienced difficulties with drag-and-drop manipulation of the touchscreen and were unable to identify and complete patterns if the unit of repeat was anything but AB.

Conclusion

As seen in the preceding sections, over the years, SEMT participants have reported on and discussed various contexts for learning mathematics prior to beginning formal schooling. What many of these contexts have in common is that they often involve our senses, employing sight, hearing, and touch. For example, children's visual sense was used to encourage their knowledge of patterning (Mulligan 2013), as well as their knowledge of numbers (Funkhouser, Risley, and Gannon 2003). Children's auditory sense was employed as they listened to tablet apps counting numbers out loud (Tirosh et al. 2017). Children's sense of touch was used as a context for investigating three-dimensional figures (Franke 1999). Our senses are activated through activities such as making visual comparisons between two patterns (Tirosh et al. 2015), dancing and exploring space (Blundell et al. 2017), and drawing figures from different perspectives (Kuřina, Tichá, and Hošpesová 1997, 1999).

Perhaps what stands out most is that the same context may be used for promoting different mathematical concepts and skills. For example, within the context of calculator use, children were able to practise counting skills, as well as exploring patterns and generalizations (Groves 1999). A drawing activity may be used as a way to encourage pattern recognition, but also as a way to enhance geometrical knowledge and language (Hošpesová and Semerádová 2013). Dance and movement can lead to spatial awareness (Blundell et al. 2017), as well as encouraging an understanding of adding, and exploring number properties (Guidoni 1999). That the same context may be used to encourage different mathematical notions, such as counting, as well as shape recognition, was exemplified by Ginsburg and Uscianowski (2017) during their plenary address focusing on mathematical stories. They offered examples of stories, some specifically written to promote early mathematics and some passed down from generation to generation, possibly written to present a moral lesson, that could be used to encourage mathematical thinking.

While this chapter has focused on three central early childhood mathematics domains (number, geometry, and pattern), we note that some studies discussed other domains. For example, a context for learning size comparison (big, bigger, biggest) was described by Hošpesová and Semerádová (2013) and Ginsburg and Uscianowski's (2017). Children's descriptions of conservation of volume were reported by Pugalee, Williams, and Lamm (2003). Another presentation by Giordano, Gagliardi, and Recchi (2003) presented an interdisciplinary programme, using the context of light and shadow to study geometry and physics, for children from preschool to Grade 8.

Although we have described in this chapter various early childhood contexts for learning mathematics, it is often the adult who can help turn an everyday context, such as storytelling, into a mathematical experience (Ginsburg and Uscianowski 2017). Barkai et al. (2019) pointed out that, for tablet apps to make the most of their potential, a knowledgeable adult sitting with and guiding the child is important. While this chapter did not review SEMT presentations specifically related to preschool teachers' knowledge for teaching mathematics, or their professional development, we note here that several such studies were reported (e.g. Hassidov and Ilany 2015; Clarke 2001; Lubinski et al. 2005).

Extending contexts for early childhood mathematics

Most of the early childhood studies reported at past SEMT symposia took place in preschool or elementary school settings. In our own studies, we also investigated mathematical learning that took place in school settings (Tirosh et al. 2015). Recently, researchers have begun to recognize the importance of out-of-school settings as contexts for early mathematical learning (Anders et al. 2012). At SEMT, our own workshop (Tirosh et al. 2017) and presentation (Barkai et al. 2019) regarding the affordances and constraints for learning numerical and patterning concepts with tablet applications grew out of our recognition that parents and other caregivers may be using apps to promote young children's mathematical skills at home. We argue that parents have a central role in promoting young children's early mathematical competence, a role that can affect children's transition to formal school mathematics. Thus, we are concerned with helping adults guide their young children's mathematical growth.

Recently, we have begun to investigate the knowledge and beliefs of adults, who are not preschool teachers, regarding out-of-school contexts for promoting mathematics in early childhood (e.g. Levenson et al. 2020). Building on our framework for developing preschool teachers' knowledge for teaching mathematics (e.g. Tsamir et al. 2015), we devised an adapted framework which considered the mathematical knowledge adults need in order to promote children's knowledge, as well as the knowledge needed for engaging children with playful mathematics. This framework also took into consideration beliefs regarding what mathematics children and the adults who interact with them should know, and how children can engage with mathematics. Findings suggested that, in general, participants had positive beliefs regarding supporting children's engagement with various numerical activities, although some differences between participants with different backgrounds (e.g. connections with young children, professions) were found. Looking further into adults' beliefs, we found that most adults believe that it is important for an adult to be involved in developing preschool children's quantitative reasoning (Tsamir et al. 2020). The reasons for this positive claim included mostly the cognitive aspects of learning, such as the necessity to correct children's mistakes, but also some affective aspects, such as raising children's motivation to learn number concepts. A majority of adults also claimed that it was important for an adult to receive guidance in order to foster children's quantitative reasoning. Reasons for needing guidance included the recognition that there must be special methods for fostering mathematical thinking at a young age. Reasons for not considering that guidance is needed included a belief that helping children comes naturally to most adults, and that only professional early childhood educators need guidance.

Currently, we are investigating adults' awareness and suggestions for supporting young children's early numerical competencies. Based on our findings, we will then plan workshops for adults interested in receiving guidance for supporting young children's mathematical knowledge. We hope to share our findings at future SEMT symposia.

Acknowledgements

This research was supported by the Israel Science Foundation (Grant No. 1631/18).

References

Ancona, R. L., Montone, A., and Pertichino, M. (2005). Street games for the acquisition of spatial orientation in primary school. In J. Novotná (Ed.) SEMT 202.

Anders, Y., Rossbach, H. G., Weinert, S., Ebert, S., Kuger, S., Lehrl, S., and von Maurice, J. (2012). Home and preschool learning environments and their relations to the development of early numeracy skills. *Early Childhood Research Quarterly*, 27(2): 231–244.

Back, J., and Griffiths, R. (2015). Making numbers: Using manipulatives to teach early arithmetic. In J. Novotná and H. Moraová (Eds) SEMT 361–363.

Barkai, R., Levenson, E., Tsamir, P., and Tirosh, D. (2019). Counting 'game' for young children: Affordances and limitations of tablet applications. In J. Novotná and H. Moraová (Eds) SEMT 63–72.

Baroody, A. J. (1987). *Children's Mathematical Thinking: A Developmental Framework for Preschool, Primary, and Special Education Teachers.* New York: Teacher's College Press.

Benyak, J., and Kopáčová, J. (2019). Pattern applets: Working with applets promoting pattern recognition. In J. Novotná and H. Moraová (Eds) SEMT 72–80.

Blundell, N., Bentley, K., Temple, A., and Pugalee, D. K. (2017). Kinesthetic and creative approaches to pre-k spatial geometry learning: A qualitative case study. In J. Novotná and H. Moraová (Eds) SEMT 72–81.

Clarke, D. (2001). Understanding, assessing and developing young children's mathematical thinking: The experiences of the early numeracy research project. In J. Novotná and M. Hejný (Eds) SEMT 7–14.

Clements, D., and Sarama, J. (2007). Effects of a preschool mathematics curriculum: Summative research on the Building Blocks project. *Journal for Research in Mathematics Education*, 38(2): 136–163.

Downton, A., MacDonald, A., Cheeseman, J., Russo, J., and McChesney, J. (2020). Mathematics learning and education from birth to eight years. In J. Way, C. Attard, J. Anderson, J. Bobis, H. McMaster, and K. Cartwright (Eds) *Research in Mathematics Education in Australasia 2016–2019*, 209–244. Singapore: Springer.

Evans, D., Strnadová, I., and Wong, M. (2007). Examining number sense: An exploratory study of students with additional learning needs. In J. Novotná and H. Moraová (Eds) SEMT 101–109.

Franke, M. (1999). What children know about geometric figures. In M. Hejný and J. Novotná (Eds) SEMT 89–92.

Funkhouser, P. C., Risley, R. M., and Gannon, G. (2003). Developing mental representations of number and computation in students with special needs. In J. Novotná (Ed.) SEMT 66–69.

Gifford, S. (1993). What is an appropriate mathematics curriculum for three to six year olds. In M. Hejný and J. Novotná (Eds) SEMT 38–40.

Ginsburg, H. P., and Uscianowski, C. (2017). Stories, stories, and more math stories. In J. Novotná and H. Moraová (Eds) SEMT 9–19.

Giordano, E., Gagliardi, M., and Recchi. M. (2003). Between mathematics and physics: The shadows. In J. Novotná (Ed.) SEMT 181.

Greenes, C., Ginsburg, H. P., and Balfanz, R. (2004). Big math for little kids. *Early Childhood Research Quarterly*, 19: 159–166.

Groves, S. (1999). Calculators, budgerigars and milk cartons: Linking school mathematics to young children's reality. In M. Hejný and J. Novotná (Eds) SEMT 7–11.

Guidoni, P. (1999). Steps towards a 'Resonance' approach in the dynamics of explaining/understanding elementary mathematics. In M. Hejný and J. Novotná (Eds) SEMT 139–143.

Hassidov, D., and Ilany, B.-S. (2015). Successful cooperation between mathematical facilitators and preschool teachers, the aim of 'senso- math' preschool program. In J. Novotná and H. Moraová (Eds) SEMT 137–146.

Hošpesová, A., and Semerádová, S. (2013). Didactical situations in the preschool period. In J. Novotná and H. Moraová (Eds) SEMT 130–138.

Israel National Mathematics Preschool Curriculum (INMPC) (2010). Retrieved October 2020 from: https://edu.gov.il/minhalpedagogy/preschool/subject/math/Pages/math-curriculum.aspx

Kaslová, M. (2017). Diversity of results in research in the domain of pre-school mathematics at kindergarten. In J. Novotná and H. Moraová (Eds) SEMT 255–264.

Kopáčová, J., and Žilková, K. (2015). Developing children's language and reasoning about geometrical shape: A case study. In J. Novotná and H. Moraová (Eds) SEMT 184–192.

Kuřina, F., Tichá, M., and Hošpesová, A. (1997). How to evaluate and utilise our children's geometrical experience. In M. Hejný and J. Novotná (Eds) SEMT 19–32.

Kuřina, F., Tichá, M., and Hošpesová, A. (1999). On children's everyday experience and geometrical imagination. In M. Hejný and J. Novotná (Eds) SEMT 73–77.

Levenson, E., Barkai, R., Tirosh, D., and Tsamir, P. (2020). Studying adults' beliefs related to engaging young children with number conceptions. In M. Inprasitha, N. Changsri, and N. Boonsena (Eds) *Proceedings of the 42nd Conference of the International Group for the Psychology of Mathematics Education*, 323–332. Khon Kaen, Thailand.

Littler, G. H., and Benson, D. A. (2005). Patterns leading to generalizations. In J. Novotná (Ed.) SEMT 202–210.

Lubinski, J., Warfield, B., Beauchamp, M., and Webster Moore, M. (2005). Conducting staff development for colleagues: The effect on K-3 teachers' instructional decision making in mathematics. In J. Novotná (Ed.) SEMT 210–218.

Mulligan, J. (2013). Inspiring young children's mathematical thinking through pattern and structure. In J. Novotná and H. Moraová (Eds) SEMT 45–55.

Nováková, E. (2019). Reflection of video recordings as a part of the creation process of prospective kindergarten teachers' professional portfolio. In J. Novotná and H. Moraová (Eds) SEMT 271–280.

Poirier, L., and Bisaillon, N. (2015). The role of language and communication in the learning of math in primary school: Some examples. In J. Novotná and H. Moraová (Eds) SEMT 41–51.

Pratt, K. (1997). Raising standards in numeracy. In M. Hejný and J. Novotná (Eds) SEMT 135–136.

Pugalee, K. D., Williams, K., and Lamm, M. (2003). Children's descriptions of volume conservation: An investigation of language and children's mathematical understanding. In J. Novotná (Ed.) SEMT 126–129.

Sarama, J., and Clements, D. (2009). *Early Childhood Mathematics Education Research: Learning Trajectories for Young Children*. New York: Routledge.

Schoenfeld, A. (1994). What do we know about mathematics curricula? *The Journal of Mathematical Behavior*, 13(1): 55–88.

Swoboda, E. (2005). Structures of van Hiele's visual level in work of 5–7 years old children. In J. Novotná (Ed.) SEMT 299–307.

Swoboda, E. (2007). Intuition of geometrical relations in plane in works of 4–7 years old children. In J. Novotná and H. Moraová (Eds) SEMT 249–257.

Swoboda, E. (2013). How to introduce young children to the repeating geometric patterns. In J. Novotná and H. Moraová (Eds) SEMT 314–322.

Starkey, P., Klein, A., and Wakeley, A. (2004). Enhancing young children's mathematical knowledge through a pre-kindergarten mathematics intervention. *Early Childhood Research Quarterly*, 19: 99–120.

Tirosh, D., Tsamir, P., Levenson, E., Tabach, M., and Barkai, R. (2015). Kindergarten children's recognition of pattern structure. In J. Novotná and H. Moraová (Eds) SEMT 331–341.

Tirosh, D., Tsamir, P., Levenson, E., and Barkai, R. (2017). Using tablets in preschool: Patterns and number concept. In J. Novotná and H. Moraová (Eds) SEMT 486–488.

Tsamir, P., Tirosh, D., Barkai, R., Tabach, M., and Levenson, E. (2011). Investigating the effects of an intervention on the geometric knowledge of abused and neglected children. In J. Novotná and H. Moraová (Eds) SEMT 398–407.

Tsamir, P., Tirosh, D., Levenson, E., Tabach, M., and Barkai, R. (2015). Preschool teachers' knowledge and self-efficacy needed for teaching geometry: Are they related? In B. Pepin and B. Rösken-Winter (Eds), *From Beliefs to Dynamic Affect Systems: (Exploring) a Mosaic of Relationships and Interactions*, 319–337. Switzerland: Springer International Publishing.

Tsamir, P., Tirosh, D., Levenson, E., and Barkai, R. (2020, July). Adults' beliefs regarding adult intervention in fostering number concepts among young children. Paper presented at EARLI, The European Association for Research on Learning and Instruction (SIG 05). Bar Ilan University, Israel.

UNICEF Programme Division (2017). Unicef's Programme guidance for early childhood development. Retrieved 5 October 2020 from: https://www.unicef.org/sites/default/files/2018-12/UNICEF%20Programme%20Guidance%20for%20Early%20Childhood%20Development%202017.pdf

Urbańska A. (1991). On the numerical competence on children beginning their 'K' class. In J. Novotná (Ed.) SEMT 38–40.

van Hiele, P. M. (1959). The child's thought and geometry. In D. Fuys, D. Geddes, and R. Tischler (Eds), *English Translation of Selected Writings of Dina van Hiele-Geldof and Pierre M. van Hiele*, 243–252. Brooklyn, NY: Brooklyn College, School of Education.

van Hiele, P. M. (1986). *Structure and Insight. A Theory of Mathematics Education*. London: Academic Press Inc.

2.2
Students with special needs: Learning mathematics successfully

Petra Scherer
University of Duisburg-Essen

Introduction

When looking at learning processes in mathematics education, diverse learners can be found, and there exists a wide spectrum of different interests, capabilities, and difficulties. Within this spectrum, there are students with special needs, and those needs might cover specific talents as well as difficulties or disabilities, handicaps, and impairments. The question arises as to how to address this group or these groups of students when learning mathematics (Scherer et al. 2016). The answer might be given from the perspective of mathematics education on the one hand or special education on the other hand. This chapter will focus on mathematics education and follow a constructivist view on teaching and learning (von Glasersfeld 1995a, 1995b). With respect to students with special needs, research results concerning concepts and models for instruction are multifaceted, based on the specific contents and mathematical topics as well as the underlying view of mathematics.

The following chapter will illuminate to what extent students with special needs were addressed in research reports, working groups, or workshops during the 30 years of SEMT. Firstly, a short retrospective view of the early SEMT symposia will be given, followed by a focus on the state-of-the-art research represented in the 2017 SEMT symposium. The actual situation shows that, in the meantime, models of integration and inclusive education have developed and most of the students with special needs learn in regular classrooms. The chapter will also illustrate suitable learning offers for students with special needs.

Looking back on the beginnings of SEMT

At first glance, the symposium topics do not explicitly show the term 'special needs', but the symposium themes *Confidence and competence* (2001) or *Equity and diversity* (2017), for example, lead one to assume that this topic is addressed, and indeed the proceedings show the importance of students with special needs, and represent the clear connection to these students.

Having a closer look at the symposia structure and individual contributions, one can notice changes over time: in the early years, thematic grouping of research

Petra Scherer, *Students with special needs: Learning mathematics successfully*. In: *Elementary Mathematics Teaching*. Edited by: Brian Doig, Jarmila Novotná, Berinderjeet Kaur and David K. Pugalee, Oxford University Press.
 DOI: 10.1093/oso/9780192869647.003.0006

reports, topic groups, or working groups was common. Later on, you find different formats like research reports, short reports, posters, or workshops. The thematic linking to 'students with special needs' can be found with every format, as a rough overview on the beginnings of SEMT with selected examples will show: although the terms 'special education' or 'students with special needs' do not appear explicitly in the beginnings of SEMT, the consideration of individual learning processes was visible over the years. In the preface of the proceedings of SEMT 1991, Koman and Novotná emphasize the focus on problem solving:

> Putting an increased emphasis on problem-solving is not only a mathematical matter, but also a very important pedagogical and psychological matter. The success of our students, both before and after graduation, is closely connected with matters like motivation, accepting mathematical problems, desire to explore, feeling pleasure in solving problems, using one's own strategy in solving problems and many other things. (Koman and Novotná 1991, p. 7)

Here, mathematical talents are in the focus, and this represents one facet of 'special needs'. Moreover, from the very beginnings of SEMT, the constructivist view is obvious and with this the individuality of learners. Explicitly in 1993 the focus was put on psychological aspects and one paper addressing gifted students can be found: Teacher and mathematically gifted pupils (Tomásková 1993).

In 1995, in the working group *Different Topics*, several contributions dealt with students' learning difficulties, such as the specific group of 'dyscalculic' students, or typical mistakes with selected mathematical contents, or in general with mathematics, as shown by the contribution titles thus:

- William Ebeid (1995): Kuwaiti experience in dealing with mathematically talented children in elementary schools
- Grassmann (1995): Mistakes in Arithmetic—What Can We Do?
- Pokorna (1995): The Development of Mathematical Imagination of Dyscalculic Children
- Siwek and Kusioll (1995): The Idea of Functional Mathematics Teaching. Some Proposals on Solving the Students' Difficulties

In 1997 the relevance of diagnosis and diagnostic tools became obvious and the number of related contributions increased. Two topic groups were organized: *Evaluation of learning and teaching* and *Methods and types of evaluation*. The former also differentiated between the focus on learning on the one hand and teaching on the other hand. Examples of contributions were:

- Gray (1997): Misguided notions of mathematical development
- Novotná (1997): Using geometrical models and interviews as diagnostic tools to determine students' misunderstandings in mathematics
- Kubinová (1997): Students' projects from the point of view of diagnostics
- Doig (1997): Innovative techniques for evaluating children's mathematical understandings in large scale testing programs: Snapshots of Australian practice

Over the years, the spectrum of terms and research foci became wider. Contributions at SEMT symposia show that diversity and special needs cover a wide range of abilities and aspects: students with learning difficulties or disabilities on the one hand and more or less extreme gifted students on the other hand, multicultural classrooms, challenges for migrant students, questions of language or language problems, specific impairments like dyscalculia. The international context of SEMT illustrates in some respect differences concerning the different school systems in the different countries, but at the same time there are comparable processes and common findings.

With the symposium topic in 2017 of *Equity and diversity*, diverse learners and, with this, students with special needs were addressed explicitly. Olive Chapman, SEMT Chairman, stressed the importance in her presentation: 'This year, the symposium is framed by the theme, equity and diversity in elementary mathematics education, which is timely and important given the growing diversity of elementary classrooms in today's world and the need for teachers to engage all students meaningfully in the mathematics classroom and the learning of mathematics' (Chapman 2017, p. 7).

As expected, many contributions were concerned with the topic of students with special needs, and addressed this theme in a variety of ways. Moreover, the focus on 'inclusion' could be found, putting the international Convention of the Rights of Persons with Disabilities into action (United Nations 2006). A variety of research projects, research questions, and methods can be found, and the next section will integrate selected presentations from SEMT 2017, aiming at highlighting the state of the art in research.

Students with special needs: Special instruction or high-quality instruction for all?

In her presentation, Levenson (2017) focused on mathematical creativity and on factors that may affect the promotion of mathematical creativity in heterogeneous classes. She highlighted the teacher's as well as the students' role and asked questions such as 'What does it mean for students to be mathematically creative?' 'Can we observe mathematical creativity in mixed-ability classrooms?' 'How can we support mathematical creativity at the primary school level?' She stressed the point of ensuring that all students have access to developing mathematical creativity, and not only those who are excellent in mathematics. Her findings showed that providing access to mathematical creativity for all students might support excellent mathematical learning for all.

This specific focus on creativity is in accordance with a modern view of teaching and learning mathematics. There is a wide consensus for the constructivist approach as a way of viewing 'knowing' (von Glasersfeld 1995a) and learning in a social environment (Wittmann 2001). Actually, for mathematics education, inquiry-based learning and productive practising are seen as main elements of this paradigm. Productive work is to be understood in contrast to bare reproduction of knowledge. It should enable pupils to think, to construct, and to extend their knowledge (von

Glasersfeld 1995b). With respect to heterogeneous learning groups, several studies and experiences in general have confirmed that investigative learning combined with productive practising is appropriate for all learners—especially for low achievers and children with special needs (Ahmed 1987; Peltenburg 2012; Scherer 1997; Scherer 1999; Scherer et al. 2016; Trickett and Sulke 1988; van den Heuvel-Panhuizen 1991). But, there is still a need for a change of attitude towards students with special needs. This means, for instance, that the teacher has to trust the children to solve complex problems (Cobb, Wood, and Yackel 1991, p. 174; see also Peltenburg and van den Heuvel-Panhuizen 2012). According to this view, all learners should be confronted with complex learning environments. This means, for example, that so-called holistic approaches should be offered, which require all learners to see relationships between numbers, shapes, and so forth, in order to understand mathematical structures (Trickett and Sulke 1988). As a consequence, for instance, from the very beginning of the first year of schooling the numbers up to 20 should be offered and dealt with instead of introducing the numbers one after the other (Scherer 2013).

Agreeing that all students should have access to high-quality mathematics instruction, with regard to students with special needs, a number of questions arise. For instance, how much similarity is possible and how much specialization is necessary? With this question, further dimensions come into play: to what extent do students with special needs need a special mathematical content? Sometimes it is argued that context-related mathematics, the connection to real-life problems, should be stressed. Research shows that context-related mathematics as well as pure mathematics can be suitable for students with special needs (Scherer 2007, 2009). To what extent do they need a special learning arrangement? To what extent do they need special teachers?

There is no doubt that students with special needs and specific conditions have to be considered. But with the given questions, mathematics should remain in the centre: not only methodical questions have to be discussed and, for example, special education teachers, or out-of-field teachers, have to be qualified for mathematics education. The proceedings of SEMT 2017 provide substantial and encouraging research, and some selected examples with specific foci will be given here.

Focus on language: In ensuring that all students—especially students with special needs—have access to mathematical contents, language is a crucial factor. A contribution that focused on language was given by Götze in 2017. She presented an intervention project involving third- and fourth-graders from primary schools with low socio-graphic status. Individual misconceptions of division served as a basis for initiating a sustainable understanding of factors among the children in a language-sensitive manner. The data showed that the deliberate support of the connection between multiplication and division appears to be a fruitful approach for developing an understanding of division and divisibility. The technical vocabulary supported in mathematical discourses appeared to be a central requirement for success, and it was simultaneously possible to foster an understanding of division and divisors together because a potential language barrier could be deliberately avoided right from the start.

Focus on cultural diversity: In their study Hunter and Miller (2017) worked with Pāsifika and Maori students whose cultures have a rich background of mathematics, including a strong emphasis on patterns used within craft design. As there had been limited studies which investigated Pāsifika and Maori students' understanding of growing patterns, they explored how Pāsifika and Maori contextual patterning tasks could, potentially, support young children to develop their understanding of algebraic growing patterns. A teaching experiment was undertaken with Grade 2 students, and analysis of the data revealed that the contextual growing tasks assisted these young students to generalize growing patterns.

Other contributions focused on the general design of suitable learning environments, for example offering different levels of representation. The concrete design could be followed by analyses of learning processes on a micro level, for example gaining insight on interactions of students with special needs, as two contributions illustrate.

Focus on interactions: Arguing that participation in mathematical interaction is widely accepted to be important for students' meaningful learning of mathematics, in their presentation Schöttler and Häsel-Weide (2017) asked how interaction could be initiated when students show different levels of understanding in inclusive classrooms. The aim of the study was to design learning environments that allow students with different levels of mathematical understanding to extend their decimal system understanding by collaborating and discussing. Further analyses will investigate the extent to which the joint negotiation of meaning leads to an extension of the understanding when the students actively participate as equal partners on a content level of the negotiation.

Also, Hähn (2017) analysed students' interactions. Examples illustrated that the analysis of the learners' participation and mathematical processes could describe and categorize collective learning situations. The presented data showed that an intended but failed co-operative learning situation could be both an obstacle and an opportunity for learners with learning disabilities. In further analyses, co-existing learning situations should be differentiated: under which conditions is a learner's attempt to enhance another one's idea an obstacle or an opportunity for a deeper content-related understanding of the common subject matter, and how could a teacher's support help to initiate a co-operative learning situation?

Other contributions put their focus on the teachers and teaching situations for students with special needs.

Focus on teachers: In their presentation, Jirotková and Slezáková (2017) stated that in mathematics lessons a teacher often considers a pupil's answer wrong if it deviates from the expected answer. In the best case, teachers admit they do not understand their pupil's thinking, but do not discard the answer as wrong. Even in this case, though, they refuse to communicate about it any further. The fear of misunderstanding pupils often becomes an obstacle to initiation of class discussions. The study focused on the phenomenon of misunderstanding in the teacher–experimenter pupil interaction. The analysis of the data clearly showed that the misunderstanding is not caused by the pupil but by the teacher in the role of the experimenter. This finding was a strong motivation to study the phenomenon of misunderstanding in classroom interaction in further research.

Also, in their presentation, Thanheiser and Philipp (2017) focused on teachers. They stated that effective teachers must simultaneously attend to many facets of the classroom, including the mathematics topic to be learned, the way students think about the topic, and the cultural contexts in which the topic may be situated. In their study they focused on that to which prospective teachers attend in a teaching scenario involving a culturally based non-standard algorithm. In this scenario, a teacher dismissed a multi-digit subtraction algorithm because it was different from the standard algorithm. The results showed that, while most of the prospective teachers disagreed with the teacher's dismissal of the algorithm, their disagreement focused on student thinking, or mathematical arguments, but not on culturally based arguments.

With the presentation given by Graven (2017), an additional important research field was addressed, namely the wider systemic field including the different stakeholders. In this contribution, the situation for mathematics education in South Africa was reported. Among other things, it focused on the education research context, and particularly mathematics education research, which mostly tells deficit stories of both learner performance and teacher practice. The South African Numeracy Chair Project (SANCP) was presented that was set up to enable a powerful dialectical relationship between research and development. This successful network offered multiple opportunities for dialogue across stakeholders, dialogue that has focused on mutual learning towards addressing the challenges of elementary mathematics learning.

This short insight into SEMT 2017 has revealed the broad field of research concerning students with special needs in mathematics education. In the following section a concrete example for a learning arrangement will be illustrated, showing the potential as well as specific requirements when working with students with special needs.

Suitable learning arrangements: Tasks between openness and mathematical structures

As mentioned above, one crucial requirement for coping with heterogeneity and meeting the special needs of students in mathematics education is offering the individual construction of knowledge. Offering learning tasks on different levels of difficulty, predetermined by the teacher, bear the risk that some learners will be overtaxed or misjudged or fixed at a specific level as viewed by the teacher. In this way, the individual construction of knowledge might be prevented. Research shows that learning environments—with a sufficient level of complexity—following the concept of a *natural differentiation* (ND) can reduce this risk (Wittmann 2001; Krauthausen and Scherer 2013, 2014). The example that will be presented in this section shows this ND in action, which means that learners can choose their level of working by *themselves*, and that it is not predetermined. For the problem 'Which number does not belong?' the opportunities for students with special needs and possible learning arrangements will be discussed.

The concrete task is taken from Klavir and Hershkovitz (2008): 'Which of the following numbers 15, 20, 23, 25, does not belong? Explain why.' Analysing the mathematical potential of this task, number properties and number relations, as well as patterns and structures, are visible (see Krauthausen and Scherer 2014, p. 100). As students should also argue for and explain their decisions, central objectives for primary mathematics as formulated in standards and curriculum are required (National Council of Teachers of Mathematics 2000). Those objectives—the connection of content-related and process-related competencies—might also be relevant for students with special needs (see also Barkatsas and Seah 2015).

One can start with this task in Grade 2, but there is great potential for higher grades, up to teacher education (Klavir and Hershkovitz 2008; Krauthausen and Scherer 2014, p. 100). Klavir and Hershkovitz stress the task's open character, with the opportunity to work on an individual level and following individual strategies to deepen knowledge and develop mathematical thinking (Klavir and Hershkovitz 2008, p. 1), which is also true for students with special needs (see also Grossman 1975; Sullivan, Clarke and Clarke 2013). The task fulfils the criteria for an ND (Wittmann 1990; Krauthausen and Scherer 2013, 2014), as it contains a wide spectrum of mathematical activities and can be used without further variations in different grades and on different levels. Students with special needs, as well as heterogeneous learning groups, can also work on and solve this task.

The openness is combined with problem solving and requires problem-solving strategies. Moreover, the students have to choose and construct the mathematical structure by themselves. The concrete mathematical content will vary, dependent on the grade and on the individual existing knowledge and competencies. In this way, individual learning opportunities for diverse learners are possible.

The decision as to which of the numbers do not belong is based on characteristics one of the numbers shows but the others do not. Or it is based on characteristics three of the numbers show but the fourth does not. Decisions and arguments can focus on different mathematical topics and structures (Krauthausen and Scherer 2014):

- *Place value*: 20 does not belong, because the one's digit is not filled; 15 does not belong, because it's the only number with the ten's digit 1; 15 has the only odd tenth digit.
- *Divisibility/multiples*: 15 is the only number divisible by 3; 23 is the only number that is not divisible by 5; 25 is the only square number.
- *Figured numbers*: 15 is the only triangular number; 20 is the only rectangular number; 25 is the only square number.
- *Further number properties*: 20 is the only even number; 23 is the only prime number; 23 is the only number with successive digits.

Those different decisions, based on different mathematical relations (partly overlapping), can be expressed in individual terms and language, and these individual opportunities meet students' special needs. The abovementioned mathematical relation, that 15 is the only number divisible by 3, can be expressed differently:

- 15 does not belong because it cannot be divided by 3 without remainder.
- 15 does not belong because it does not appear in the table of 3.
- 15 does not belong because it is not a multiple of 3.
- 15 does not belong because it does not appear when counting in steps of 3.

How to work with students in a concrete lesson? If the plan is for all students to work with one and the same problem, one has to check whether the prerequisites are given for all students. The second question is if all students are familiar with open problems, including a variety of solution strategies and solutions as well as a flexibility of thinking. Several didactical and methodical decisions are proposed to meet students' special needs.

Becoming familiar with open problems: Open problems contain opportunities for students with special needs (Grossman 1975; Sullivan, Clarke, and Clarke 2013) but, at the same time, represent challenges.

Following a constructivist view for teaching and learning, these types of tasks and problems should be used regularly and with different levels of complexity. Less complex types of open problems, like 'Find you favourite tasks!' or 'Find tasks with the result 10 (or 100, 1000, ...)!' can give confidence to students and offer first steps in exploration. Students find different numbers of tasks, and sometimes they find tasks unsystematically with doubles, or make mistakes. But they also might work systematically or develop a systematic strategy while working on the problem, or be stimulated to sort their tasks. The problem, tasks with the result 10, covers different number relations and structures. The students discover the commutative property (4 + 6 = 10 and 6 + 4 = 10) as well as the operative principle (6 + 4 = 10, 7 + 3 = 10, 8 + 2 = 10 ...). These discoveries can be illustrated by different representational levels: enactive, iconic, symbolic (see Bruner 1969). As illustrated in Figure 2.2.1, students can place circular counters, of, say, white and grey, and can find with the systematic arrangement all possible solutions.

Pre-teaching: When planning a common starting point for all students, including students with special needs, the teacher has to be sure about the prerequisites, and it might be that some activities should be done beforehand. Those preparatory activities can be understood as a kind of pre-teaching (Brunsch 2020; Thornton et al. 2015): investigating, describing, and explaining number properties and characteristics could be done within the task 'Wanted!' Students should make comments on a selected number. For the number 24, the following characteristics and relations can be given: 'It's an even number', 'It belongs to the table of 4 (or 6, or 8)', 'It has two tens (four ones)' or 'If you add 6, you will get 30.' The different aspects can be illustrated with material to support the understanding.

Adaptations on the pre-numeric level: To clarify the specifics for this kind of problem, one can use a 'number-free' problem and present Figure 2.2.2 with four objects: a black cup, a black plate, a knife, and a black shirt.

Decisions with accompanying explanations can be the following:

- The shirt does not belong, because all other things are used for breakfast.
- The cup does not belong, because all other words have five letters.
- The knife does not belong to the others, because it isn't black.

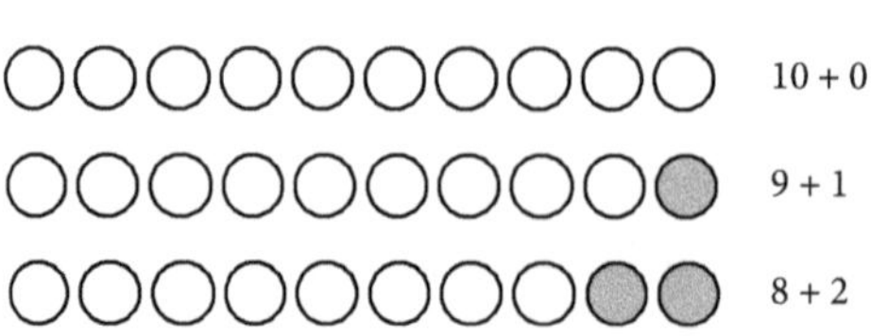

Figure 2.2.1 Tasks with some results for 10

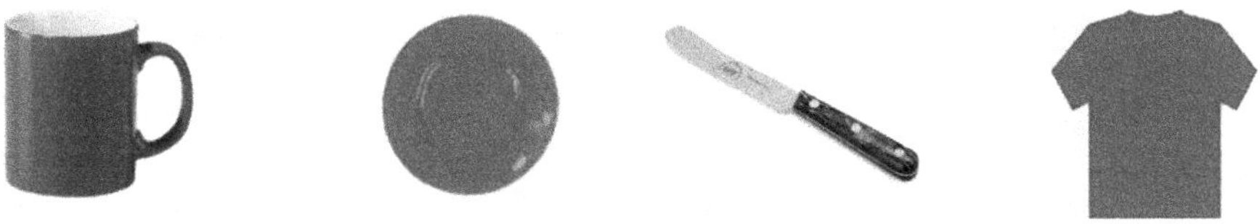

Figure 2.2.2 Variation of the problem without numbers

In this way, learners are made sensitive to the openness, the fact that there exists more than one correct answer and that the context is flexible by an individual framing (colour, function, context, letters, etc.; see Krauthausen and Scherer 2014).

Starting with this adapted problem would offer all learners the chance to participate, even those who are working on the pre-numerical level. The teacher has to consider in advance what opportunities a specific learning environment contains and if it fits with the students' competencies.

Targeted individual support: It might be that a specific mathematical content has to be worked on separately. Some learners might be familiar with the given numbers, but the problem's complexity or constructing the mathematical relations by themselves could be too challenging. Even and odd numbers could be worked on, on the iconic as well as on the symbolic level with suitable material (see Figure 2.2.3): even numbers are represented as double rows and odd numbers as double rows with an extra dot. With those insights, the given numbers (15, 20, 23, 25) could then be thought over again.

Further activities can be illustrated on the number line, which may lead to a better orientation with numbers up to 20, or up to 100, especially with even and odd numbers. According to a specific rule, numbers can be marked, for example starting from 1 always two steps further: 1, 3, 5, 7, 9, 11, 13, 15, 17, 19, or starting from 2 always two steps further: 2, 4, 6, 8, 10, 12, 14, 16, 18, 20, or starting from 1 always three steps further: 1, 4, 7, 10, 13, 16, 19.

Therefore in accordance, for understanding place value, working with material could be helpful: The given numbers can be represented with Dienes materials, and with the clear structure of tens and ones the differences and similarities of the four numbers can be worked out.

Transfer: Having successfully worked on the problem, another example can be offered: 'Which of the following numbers 7, 12, 16, 20, does not belong? Explain why.' With this, the teacher might see if the students are able to apply former strategies and ideas.

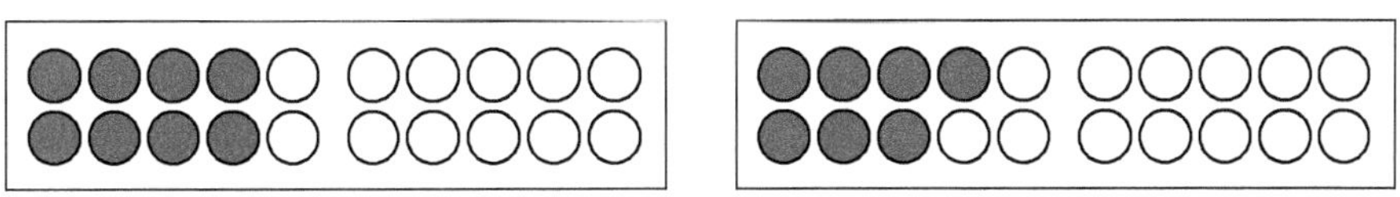

Figure 2.2.3 Even and odd numbers in the field of 20

The second example contains comparable mathematical structures but also new aspects; the focus on place value is again possible:

- 20 does not belong, because the position of the ones is not filled.
- 20 does not belong, because it's the only number with a 2 at the tens position.
- 7 does not belong, because it's the only number with an odd number in the ones.

Also, the topic of divisibility and multiples is obvious:

- 7 is the only number divisible by 7.
- 7 is the only number that is not divisible by 4.
- 12 is the only number divisible by 3 (or by 6).
- 16 is the only number divisible by 8.
- 16 is the only square number.
- 20 is the only number divisible by 5 (or by 10).

Again, arguing about even and odd numbers or about prime numbers is possible:

- 7 is the only odd number.
- 7 is the only prime number.

As a new focus, sums of successive numbers can be looked at:

- 7 is the only number that can be written as a sum of two successive numbers (3+4).
- 12 is the only number that can be written as a sum of three successive numbers (3+4+5).

It becomes clear that repetitions in the sense of transfer to other numbers are required as well as further investigations and explorations of patterns and structures.

Conclusion

Students with special needs learning mathematics were the topic of this chapter, and the traditional tension between special education and mathematics education has been indicated. Taking into account inclusive education, a closer orientation to mathematical education can be identified and the potential of selected teaching and learning concepts illustrated. Beyond this, the role of the teacher, their attitudes and beliefs, and the corresponding teacher education programmes was discussed (see also Hattie 2009; Mitchell 2014).

Thirty years of SEMT has shown a variety of attempts to foster successful mathematical thinking and learning for all students, especially students with special needs. Open problems, or tasks, such as the ones presented above are not only appropriate for the 'good' learners: children who have difficulties in calculating will be able to discover and use adequate strategies and relations between numbers as well. Such learning environments can contribute to the development of both the content-related competencies and the process-related ones (NCTM 2000) and should lead to successful learning processes in mathematics. Moreover, the illustrated structures and patterns form the necessary basis for understanding the nature of mathematics.

With inclusive settings, more and more learning environments are developed and related research is done. More research is needed to answer further open questions, such as the question of what role digital tools might play in the teaching and learning of mathematics (Schacht and Bebernik 2019).

Giving pupils the opportunity to test their limits can support motivation and self-confidence without overtaxing. Low achievers are often underestimated or misjudged, and we often bar the ways in which they—even if surprisingly—might succeed through a more open approach (see also Peltenburg and van den Heuvel-Panhuizen 2012). Supporting the children means making certain demands and aiming at long-term learning processes instead of just thinking of short-term success in learning.

The fact that some learners have difficulties in learning mathematics may lead to 'light' teaching and an insufficient mathematical education. The quality of teaching is of great importance, especially for low achievers: 'Good teaching that emphasizes the structure of a subject is probably even more valuable for the less able student than for the gifted one, for it is the former rather than the latter who is most easily thrown off the track by poor teaching' (Bruner 1969, p. 9). The concept of ND in accordance with a constructivist perspective represents a sustainable concept for teaching and learning mathematics in the primary grades, as illustrated above, as well as for education in general (Wittmann 2001).

References

Ahmed, A. (1987). *Better Mathematics: A Curriculum Development Study*. London: HMSO.

Barkatsas, A. T., and Seah, W. T. (2015). Learners' preferred mathematical task types: The values perspective. In A. Bishop, H. Tan, and T. N. Barkatsas (Eds) *Diversity in Mathematics Education: Towards Inclusive Practices*, 63–79. New York: Springer.

Bruner, J. S. (1969). *The Process of Education*. Cambridge, MA: Harvard University Press.

Brunsch, D. (2020). Preteaching als didaktisches Prinzip bei Schülerinnen und Schülern mit Lernschwierigkeiten in der inklusiven schulischen Bildung. *Zeitschrift für Heilpädagogik*, 71(9): 455–464.

Cobb, P., Wood, T., and Yackel, E. (1991). A constructivist approach to second grade mathematics. In E. von Glasersfeld (Ed.) *Radical Constructivism in Mathematics Education*, 157–176. Dordrecht, The Netherlands: Kluwer.

Doig, B. (1997). Innovative techniques for evaluating children's mathematical understandings in large scale testing programs: Snapshots of Australian practice. In M. Hejny and J. Novotná (Eds) SEMT 100–102.

Ebeid, W. (1995). Kuwaiti experience in dealing with mathematically talented children in elementary schools. In M. Hejny and J. Novotná (Eds) SEMT 126–129.

Götze, D. (2017). Language- and mathematics-integrated intervention for understanding division and divisibility. In J. Novotná and H. Moraová (Eds) SEMT 177–187.

Grassmann, M. (1995). Mistakes in arithmetic – What can we do? In M. Hejny and J. Novotná (Eds) SEMT 130.

Graven, M. (2017). Blending elementary mathematics education research with development for equity: An ethical imperative enabling qualitatively richer work. In J. Novotná and H. Moraová (Eds) SEMT 20–31.

Gray, E. (1997). Misguided notions of mathematical development. In M. Hejny and J. Novotná (Eds) SEMT 51–53.

Grossman, R. (1975). Open-ended lessons bring unexpected surprises. *Mathematics Teaching*, 71: 14–15.

Hähn, K. (2017). Analyses of learning situations in inclusive settings: A coexisting learning situation in a geometrical learning environment interactions. In J. Novotná and H. Moraová (Eds) SEMT 187–196.

Hattie, J. (2009). *Visible Learning: A Synthesis of Over 800 Meta-Analyses Relating to Achievement.* London: Routledge.

Hunter, J., and Miller, J. (2017). Using cultural tasks to develop growing pattern generalisations for young culturally diverse students. In J. Novotná and H. Moraová (Eds) SEMT 215–225.

Jirotková, D., and Slezáková, J. (2017). Do teachers understand their pupils. In J. Novotná and H. Moraová (Eds) SEMT 235–245.

Klavir, R., & Hershkovitz, S. (2008). Teaching and evaluating 'open-ended' problems. *International Journal for Mathematics Teaching and Learning.* URL http://www.cimt.plymouth.ac.uk/journal/default.htm

Koman, M., and Novotná, J. (1991). Preface. In Novotná, J. (Ed.) SEMT 7–8.

Krauthausen, G., and Scherer, P. (2013). Manifoldness of tasks within a substantial learning environment: Designing arithmetical activities for all. In J. Novotná and H. Moraová (Eds) SEMT 171–179.

Krauthausen, G., and Scherer, P. (2014). *Natürliche Differenzierung im Mathematikunterricht: Konzepte und Praxisbeispiele aus der Grundschule.* Seelze, Germany: Kallmeyer.

Kubinová, M. (1997). Students' projects from the point of view of diagnostics. In M. Hejny and J. Novotná (Eds) SEMT 88–91.

Levenson, E. (2017). Promoting mathematical creativity in heterogeneous classes. In J. Novotná and H. Moraová (Eds) SEMT 42–52.

Mitchell, D. (2014). *What Really Works in Special and Inclusive Education. Using Evidence-Based Teaching Strategies.* New York: Routledge.

National Council of Teachers of Mathematics (2000). *Principles and Standards for School Mathematics.* Reston, VA,: NCTM.

Novotná, J. (1997). Using geometrical models and interviews as diagnostic tools to determine students' misunderstandings in mathematics. In M. Hejny and J. Novotná (Eds) SEMT 61–67.

Peltenburg, M. (2012). *Mathematical Potential of Special Education Students.* Utrecht: Freudenthal Institute.

Peltenburg, M., and van den Heuvel-Panhuizen, M. (2012). Teacher perception of the mathematical potential of students in special education in the Netherlands. *European Journal for Special Needs Education*, 27(3): 391–407.

Pokorna, V. (1995). The development of mathematical imagination of dyscalculic children. In M. Hejny and J. Novotná (Eds) SEMT 144–146.

Schacht, F., and Bebernik, R. (2019). Teaching and learning geometry with digital tools. In J. Novotná and H. Moraová (Eds) SEMT 448–450.

Scherer, P. (1997). Productive or reproductive exercises: What is appropriate for low attainers? In C. van den Boer and M. Dolk (Eds) naar een balans in de reken-wiskundeles – interactie, oefenen, uitleggen en zelfstandig werken, 35–49. Utrecht: Freudenthal Institute.

Scherer, P. (1999). *Entdeckendes Lernen im Mathematikunterricht der Schule für Lernbehinderte: Theoretische Grundlegung und evaluierte unterrichtspraktische Erprobung*, 2nd edition. Heidelberg: Edition Schindele.

Scherer, P. (2007). Investigating children learning mathematics: Investigating mathematics. In J. Novotná and H. Moraová (Eds) SEMT 22–32.

Scherer, P. (2009). Low achievers solving context problems: Opportunities and challenges. *Mediterranean Journal for Research in Mathematics Education*, 8(1): 25–40.

Scherer, P. (2013). Natural differentiation in the teaching of mathematics for school beginners. *South African Journal for Childhood Education*, 3(1): 100–116.
Scherer, P., Beswick, K., DeBlois, L., Healy, L., and Moser Opitz, E. (2016). Assistance of students with mathematical learning difficulties: How can research support practice? *ZDM—Mathematics Education*, 48(5): 633–649.
Schöttler, C., and Häsel-Weide, U. (2017). Students constructing meaning for the decimal system in dyadic discussions: epistemological and interactionist analyses of negotiation processes in an inclusive setting. In J. Novotná and H. Moraová (Eds) SEMT 373–383.
Siwek, H., and Kusioll, L. (1995). The idea of functional mathematics teaching. Some proposals on solving the students difficulties. In M. Hejny and J. Novotná (Eds) SEMT 146.
Sullivan, P., Clarke, D. M., and Clarke, B. A. (2013). *Teaching with Tasks for Effective Mathematics Learning*. New York: Springer.
Thanheiser, E., and Philipp, R. (2017). Do prospective elementary teachers (in the United States) notice cultural aspects of mathematics in a teaching scenario. In J. Novotná and H. Moraová (Eds) SEMT 403–411.
Thornton, A., McKissick, B. R., Spooner, F., Lo, Y., and Anderson, A. (2015). Effects of collaborative preteaching on science performance of high school students with specific learning disabilities. *Education and Treatment of Children*, 38(3): 277–304.
Tomásková, Z. (1993). Teachers and mathematically gifted pupils. In M. Hejny and J. Novotná (Eds) SEMT 109–111.
Trickett, L., and Sulke, F. (1988). Low attainers can do mathematics. In D. Pimm (Ed.), *Mathematics, Teachers and Children*, 109–117. London: Hodder and Stoughton.
United Nations. (2006). *Convention on the Rights of Persons with Disabilities*. New York: United Nations.
van den Heuvel-Panhuizen, M. (1991). Ratio in special education. A pilot study on the possibilities of shifting the boundaries. In L. Streefland (Ed.), *Realistic Mathematics Education in Primary School: On the Occasion of the Opening of the Freudenthal Institute*, 157–181. Utrecht: Utrecht University, Freudenthal Institute.
von Glasersfeld, E. (1995a). A constructivist approach to teaching. In L. P. Steffe and J. Gale (Eds) *Constructivism in Education*, 3–15. Hillsdale, NJ: Erlbaum.
von Glasersfeld, E. (1995b). Sensory experience, abstraction, and teaching. In L. P. Steffe and J. Gale (Eds) *Constructivism in Education*, 369–383. Hillsdale, NJ: Erlbaum.
Wittmann, E. C. (1990). Wider die Flut der 'bunten Hunde' und der 'grauen Päckchen': Die Konzeption des aktiv-entdeckenden Lernens und des produktiven Übens. In E. C. Wittmann and G. N. Müller (Eds) Handbuch produktiver Rechenübungen, Band 1, 152–166. Stuttgart, Germany: Klett.
Wittmann, E. C. (2001). Developing mathematics education in a systemic process. *Educational Studies in Mathematics*, 48(1): 1–20.

2.3
Problem solving, problem posing, and enhancing creativity

Sara Hershkovitz
Levinsky College of Education, Tel Aviv, Israel
Thierry Dana-Picard
Jerusalem College of Technology, Israel

Introduction

Mathematical problems and word problems link the real world to the world of mathematics. Consequently, the array of problems is as vast as the array of real-life situations. Presenting problems in words requires facing the link between natural language and formal mathematical language. While we can use natural language to describe a situation in the many ways that linguistic wealth allows us, the language of mathematics is formal, short, and unequivocal. Part of the complexity of word problems is the form of verbal presentation, so that mathematics educators must recognize this complexity. Then, they must provide their students with ways and tools appropriate to them to solve various problems, taking into account the students' cognitive development, as well as their mathematical knowledge, their verbal knowledge, and their world knowledge. In certain cases, the students' literacy with mathematical software can be important.

Underlying the present work is the development of word problems as a research domain in mathematics education from the early 1970s until today, and reflected in the series of SEMT conferences since their inception, with about 120 presentations (research papers and workshops) to date.

We address first the general idea of solving word problems and present their analysis. Then we discuss *one-step problems*, which require computation involving only one of the four basic arithmetical operations (addition, subtraction, multiplication, or division). These problems are the most basic word problems and are the building blocks for many mathematical word problems. Understanding the basic building blocks and their complexity enables students to develop strategies for more complex ones. Finally, we consider solutions that are based not on the four basic arithmetical operations but on other strategies, which also are within the students' range of abilities. If one-step problems are the most basic situations that can be proposed, we also illustrate the topic with more complex examples.

In this chapter, problem solving and problem posing will relate to two main kinds of problems:

Sara Hershkovitz and Thierry Dana-Picard, *Problem solving, problem posing, and enhancing creativity*. In: *Elementary Mathematics Teaching*. Edited by: Brian Doig, Jarmila Novotná, Berinderjeet Kaur and David K. Pugalee, Oxford University Press. © Oxford University Press (2024). DOI: 10.1093/oso/9780192869647.003.0007

- *Real-world problems*: These can be presented in words, in pictures, or in asking the students to describe mathematically, or to ask about, a phenomenon in the real world.
- *Intra-mathematical problems*: These connect different mathematical domains, request new procedures, investigate mathematical ideas, and find regularities and generalizations, in a way that enables students to construct, and explore, the mathematical world.

Word problems in mathematics

What is a mathematical word problem?

Problems can be either within mathematics, thus creating theoretical mathematics, or real-world (applied). Solving an applied problem requires translating into a mathematical representation, then developing a solution. Descartes (1637) claimed that any problem must be translated into mathematical expressions (in his view an equation or a system of equations), and then this has to be solved. We should add that the solution has to be interpreted. In any case, mathematicians solve problems and the main usage of mathematics in the real world is in solving problems (mathematical literacy).

Pólya (1945), and later Schoenfeld (1992), and many other philosophers, researchers, and leaders of mathematics education, have emphasized that problem solving is the most central issue of mathematics instruction. Indeed, worldwide, problem solving is one of the main subjects in mathematics instruction at every level.

In recent decades, curricula have shifted from mastery of skills to emphasizing problem solving. In the United States, the National Council for Teachers of Mathematics (NCTM, 1989) emphasized the transition from memorizing procedures to searching for ways to solve problems. The NCTM also refers to the move from memorizing formulae to investigating constructs, and from solving exercises to formulating hypotheses. In many countries, the education authorities emphasize problem solving as a central issue (Gruszczyk-Kolczynska 2001, Israel Ministry of Education 2006, United States Common Core: Carmichael et al. 2010) in various contexts, multiple-step problems, and open-ended problems. All these require recalling previous knowledge and integrating different subjects.

This shift has led to so-called STEAM (Science, Technology, Engineering, Arts, and Mathematics) Education, where many subjects and skills are developed, involving at the same time new ideas, new concepts, analysing the mutual influence of processes and concepts, with new computing and technological skills being an integral part of the newly constructed mathematical knowledge (Artigue 2002). We elaborate on this issue later.

We focus at first on problems for primary school students, and then on problems for middle school and beyond. They are presented as descriptive texts with quantitative data, and the students are to discover other quantitative data. To answer the question presented in the problem, students must gather the data provided in the problem, and at times add more data from their own knowledge, or through searching appropriate resources. Then they can carry out the mathematical operations. There are three components that have to be taken into account: their language abilities, their world knowledge, and their mathematical knowledge. For younger students, the texts must be simple enough to avoid difficulties that could be associated with their reading and language abilities. Other types of problems are non-routine, or open-ended, problems, where students have to add to the given data using their own world knowledge or by gathering information relevant to the given situation. The students are required to set forth hypotheses to guide them towards a solution, and to be aware that changing a hypothesis can lead to a different solution. Such problems are reality based and take the constraints of reality into account.

On the other hand, word problems are written either around given numbers, or in order to enable students to learn the meaning of a given mathematical operation. The students either explore an idea proposed by the teacher or 'invent' their own. Kaslová (1999) addresses the tradition to introduce addition and subtraction via 'story problems', creating a model that is then 'described in mathematical language and completed by a statement using mathematical symbols'. Koman and Tichá (1999) and Bartoncová (1999) follow the same pathway and observe how students use their everyday experience in creating mathematical problems and 'link reality to mathematics'. Cifarelli, Stephan, and Pugalee (2015) observe that problem posing involves the process of reflection to consider imagined mathematical activity (Schoenfeld 1992; Silver 2013), a type of self-generated activity that may stimulate the students' overall abilities to mathematize, and develop understanding, within new situations. For Schoenfeld and Silver, problem posing helps students to develop reflection on their own mathematics, nurture their mathematical thinking, and become better problem solvers.

While Cifarelli et al. (2017) focus on addition–subtraction, Bicknell and Young-Loveridge (2015) illustrate this with the need to implement new ways of thinking to learn multiplication and division with young children, which requires skills different from those needed for addition and subtraction, and thus enhances new part–whole thinking.

In analysing simple word problems as they are presented to beginning students of mathematics, the problems are constructed from a text that describes an event from the 'real' world, and the way the event is described is based on commonly held assumptions, although, at times, they may be hidden.

Nesher and Katriel (1977) defined these hidden assumptions as follows:

- A mathematical word problem is an independent, self-contained freestanding text.
- For simple problems (beginners' level), the situation reality is irrelevant. A mathematical word problem meets all standards and can be solved even if it is not realistic, as long as the ties between the objects are appropriate.
- A word problem includes a question that has an answer. (The answer 'There is no answer' is a possible answer. They have to explain why there is no answer.)
- A mathematical word problem can be answered after using a mathematical operation or mathematical reasoning.

Students must understand the relationships among all the sentences in the problem before they can understand the entire text of a word problem. In particular, the order of the sentence is influential: the question may appear either before, after, or in between the data. The identity of the objects in the problem define the relationships among the sets to which the objects belong.

A word problem deals with quantitative attributes of objects, and the question relates to other objects. The objects' other attributes determine the problem's logico-mathematical structure. Some of these assumptions (related to the difficulties children have in solving word problems) are mentioned by Semadeni (1995) as well, and will be elaborated in a later section. With the solvers' maturation and their development of familiarity with the world, linguistic knowledge, and mathematical knowledge, the situations become more and more complex.

Nesher (1980) distinguished between the processes of solving 'simple' mathematical problems, of the type presented in mathematics lessons in school, and real-life problems. In a (simple) typical school mathematical problem the data are processed and presented concisely, in a manner aimed at conducting mathematical operations, whereas real-life problems require that a situation be analysed and the quantitative elements to support the problem be identified. Only after constructing the objects' space, and the quantitative relationships among the objects, can the solver address the problem derived from these data and plan a solution.

Example: How much does it cost to put up a fence around a swimming pool?

Even if the size of the pool is given, there are other considerations, such as whether the fence is attached to the pool or there is a space between them. If there is a space—what size is it? How high is the fence? How much does a metre of fence cost?

Kintsch (1986, 1994) studied comprehension of mathematical word problems. He distinguished between various forms of comprehension, beginning with the *textbase* that requires the student to reconstruct the story told in the given text. The student next moves to the *situation model*, which requires an understanding of the situation described in the problem. Having gone through these two stages, the student can construct a *mathematical model* to use for solving the problem. Kintsch found that there is a difference between students' ability to repeat the text of a story and their ability to use the data in this story as a foundation for operations or calculations. Furthermore, he found that both types of comprehension—that of the narrative and that of the situation—are affected by the complexity of the text. Thus, there were students who encountered difficulties in understanding the text of single-step problems or finding a mathematical solution, and therefore they changed the story in the text and simplified it. Even after solving the seemingly 'simpler' problem, these students did not feel that they had changed the situation. Hershkovitz and Nesher (2001, 2002) reached similar findings when studying the solution of two-step problems: from the onset, they were able to identify all those students who did not solve the problems correctly by hearing them tell the story in the problem. These students simplified the text and turned it into one they could solve.

Non-routine problems or *open-ended problems* are harder to define, but require a specific effort for understanding. In non-routine and open-ended problems, the emphasis is on the different possible solutions, or the possible different paths to a solution, and on the various solving strategies available. Kratochílová (1995) studied several examples.

Hershkovitz, Peled, and Littler (2009) tried to describe the attributes of open-ended problems, where not all problems contain all attributes:

- There are a variety of ways to solve the problems so that different students can solve them differently, whether finding a simple or a complex solution (emphasis is put here on 'solution' as a process, not as a final answer).
- The open-ended problems can be expanded to additional questions, which is easy to do using the 'what if' strategy (Brown and Walter 1990).
- The problems enable an examination of various possibilities, among them different conditions of some type of consistency.
- The problems raise discussion.
- The problems allow students to use various available mathematical subjects, and even expand their knowledge (p. 259).

There is a vast solution space, and the solutions can often be generalized. A nice example is given by Kopka (1995, p. 102): 'Problem 1 (Generator problem): Find the total number of squares in a 4×4 grid'. In his description of a two-phase solution process, he follows Frobisher's model from 1994.

Dana-Picard (2004) studies connections between different mathematical domains that can be used as bridges to broaden the scope of a given problem and to solve it. He describes them in a topological way, speaking about the 'cognitive neighbourhood' of a mathematical notion. The examples there are taken from advanced secondary school or undergraduate courses, but the underlying educational principle is valid at every level. This is important in the curriculum for pre-service teachers, and still more for the professional development of their educators, as it creates communication opportunities between the students (Urbańska and Rożek 1993; Gaulin 1993; Horalková and Kirilková 1993). Moreover, investigating and solving problems encourages teachers not to teach exercises, but to learn new mathematics and to conduct their teaching in a new way, which requires from students more verbalization and personal reflection (Albert 2001). An interesting illustration is given by Priolet and Novotná (2007): with a large group of third-grade classes (eight in France and two in the Czech Republic) they investigated whether explicit identification of mathematical, situational, and identifying connections among problems and their representations could contribute to improving pupils' performance.

Why solve mathematical word problems?

According to Silver (1997), problem solving is a focus of mathematics curricula worldwide. He maintained that problem solving is the main attribute of mathematics, and that it is the mathematician's work and the heart of *doing mathematics.*

Following Stanic and Kilpatrick (1988), Schoenfeld (1992) suggested five reasons to solve mathematics problems with school students of all ages: first, to demonstrate the importance of mathematics to everyday life. Second, to motivate students to seek new mathematical issues and ideas. Experiencing a problem exposes students to the ideas that should be learnt in relation to the given issue and the ways these ideas are manifested. Third, creativity and fun—showing that there is beauty and interest in mathematics, and thus encouraging students to pursue it (Levenson 2017; Bevan, Williams and Capraro 2019). Fourth, mathematics as a means to develop new skills—presenting a planned, graduated series of problems that allow students to develop knowledge and understanding of new subjects. Finally, strengthening mathematical abilities to achieve mastery. Trch and Zapotilová (1995) describe the steps involved as starting from the student's experience, use of means of illustration or drawing, verbal representation, symbolic and conceptual representation, development of problems that allow orientation in space, development of thinking, symbolic processing, algorithmic thinking. Vygotsky called this process exploration of the child's zone of proximal development (ZPD)—that is, 'the distance between the actual developmental level as determined by independent problem solving and the level of potential development as determined through problem-solving under adult guidance, or in collaboration with more capable peers' (Vygotsky 1978, p. 86). Evered (1995) describes teachers' professional development (PD) in collaborative groups exploring the contribution of the ZPD; in her study, she explored ability, self-confidence, and task execution in problem solving in groups. Her goal was to maximize the achievements in problem solving. Szetela (1993) notes that problem solving contributes strongly to the students' development of critical thinking, engages students in mathematical discourse, discussion, and reasoning, and develops mathematical communication.

The importance of problem solving has been emphasized by Bureš, Novotná, and Tichá (2009), and in the theoretical framework of Brousseau's theory of didactical situations (1997). As Freudenthal (1983) states:

> We assume that the students' recognition of the mathematical structures of a problem, their perception of different problem types and their development of diverse mathematical thinking are important conditions when using a given problem to generate new problems. In order to develop student abilities to perceive and recognize mathematical structure of a word problem, we focus our attention on a specific problem-posing activity that consists of generation of problems with the same mathematical model as a given problem. Considering the mathematical model of a given problem as a model as pre-image—an abstract representation of a set of concrete objects—we can find a set of objects (models as after-image) which are concrete representations of this abstract structure. (p. 51)

What is solving a mathematical word problem?

The theme *solution of a mathematical word problem* points to two different notions: (i) It may be the final answer (or answers) to the given question. For an intra-mathematics problem, it is a mathematical object (a number, a geometrical object, etc.). (ii) In other domains, the answer is a number together with a name: a category of objects (e.g. three apples), or a unit of measurement (for example 3 km), or something else. On the other hand, the process of obtaining that answer may be what is meant.

The study of the duality between the concept (point (i)) and the process initiated by Gray and Tall (1991) led to the notion of *procept*. In the present paper, and in the numerous papers presented at SEMT conferences, the main focus is, of course, on the process (see Example 2 later in this section).

Pólya (1945) defined solving a mathematical problem as a process having four stages:

1. *Understanding the problem*: At this stage, solvers ask themselves: what are the data in this problem? What must I find? What are the conditions and constraints in the problem? Have I solved a similar problem in the past?
2. *Planning and constructing the stages of the solution*: This may necessitate breaking down the problem into sub-stages that will enable the solver to solve the problem as a whole.
3. *Carrying out the plan.*
4. *Examining the solution*: This is where students examine their solutions against the original problem, and check whether the solutions they reached according to their hypotheses meet the requirements. If needed, they go back and correct where needed.

Pólya's framework has been implemented and followed by Gelfman, Demidova, and Matushkina (1995) for world problems, with 10-to-11-year-old students. They show that using this method increases the students' ability from problem to problem. They provide a series of examples (motion problems, either on foot or with bicycles), where students have to check the concrete relevance of their answers. We illustrate now the different steps with two examples—the first one is a world problem, and the second one is intra-mathematical.

Example 1

Forty-five children went for a hike. In order to get to the starting point of the trail, they ordered taxis, each taxi having room for a maximum of six children. How many taxis at most had to be ordered to transport all the children?

Many children identified the need for a division operation and recognized that, to find the minimum number of taxis, they needed to fill them up. So, their answer was 7.5. This answer may look accurate from a mathematical point of view, but does not return to reality—that is, to eight taxis.

Example 2

This example is from (Koman 1995):

Word puzzle Number Twins

Check the following relations:

12 x 42 = 21 x 24, 24 x 63 = 42 x 36, 46 x 96 = 64 x 69

The product does not change when reversing the digit order in both numbers of a pair; these pairs are called Number Twins.

Find out three other number twins.

Find a rule that makes it possible to find all 2-digit number twins. Justify the rule.

Describe in detail how you have searched this problem (including your failing attempts, your mistakes, *etc.*)

Suggest possible use in school practice. (p. 98)

The different attitudes and strategies were analysed by Koman (1995). He found that middle school students solve intra-mathematics problems, where the relevance of solutions can be critical. The standard process follows a series of logical inferences, without saying *if, then*. *A posteriori* verification is then needed to have a logical equivalence between the problem and the answer. We illustrate this with the following example that shows an intra-mathematics problem. It illustrates the need for a *post mortem* verification.

Example 3

A system of two linear equations in two real variables is given:

$$\begin{aligned} 2x + y &= 1 \\ 3x - 4y &= 7 \end{aligned}$$

We describe the process to solve the system as follows:

$$\begin{array}{lclclcl} 2x + y = 1 & & 6x + 3y = 3 & & & & \\ & \mapsto & & \mapsto & 11y = -11 & \mapsto & y = -1 \\ 3x - 8y = 7 & & 6x - 8y = 14 & & & & \\ 2x + y = 1 & & 8x + 4y = 4 & & & & \\ & \mapsto & & \mapsto & 11x = 11 & \mapsto & x = 1 \\ 3x - 4y = 7 & & 3x - 4y = 7 & & & & \end{array}$$

A standard student can say: the system has one solution, namely $(1, -1)$. Actually, what has been proved is: if the system has a solution, this solution is $(1, -1)$, but not more. In order to have a definitive answer, a reverse process has to be performed. Pólya (1945) calls it 'examining the solution', but we prefer to call it *checking whether the solution candidate is valid*. For this we substitute the numerical values into the original system of equations, and a check that it holds for the system. This verification consists in actually reversing the arrow's direction.

Pólya added heuristic strategies that help solve problems, such as making a drawing of the problem, working backwards, using analogies, breaking down the problem into its elements and reconstructing it differently, using induction, and using a systematic search that would enable generalizing. A way to work backwards has been developed by Meyer and Dana-Picard (1997). It consists in writing a flowchart, beginning from the end and proceeding towards the given data, and then the solution process works according to the direction of the flowchart. Mathematics educators viewed Pólya's work as being of great value, trailblazing the path by offering solvers coping methods when they reach an impasse. Cognitive psychologists (Funke 2010; Glaser, Linn, and Bohrnstedt 1997; Mayer 1992) also described these five stages as problem-solving stages and did not limit them to mathematical problems. However, English and Sriraman (2010), who reported a variety of ways to teach these strategies, claimed that, despite repeated attempts, there was no evidence of any meaningful success in doing so.

Mathematics education researchers have used different terms to define these solving processes.

Fuson, Hudson, and Pilar (1997) listed five stages:

1. Understanding the text and its relationship to reality by imagining the reality, drawing the reality, or having the solver describe the reality in the text in his or her words. (Canizzaro (1995) points out the influence of verbal structures on this understanding and the role of verbal activity in the solving process.)
2. Using mathematical tools (tangible models) and mathematical (graphical) schemas to represent the mathematical description.
3. Focusing and finding the missing elements—that is, presenting the process of reaching the mathematical solution (constructing the appropriate mathematical expressions).
4. Solving (calculating) using mathematical tools.
5. Examining the relationship between the mathematical solution and the description of the reality that the problem addresses.

On the same issue, see Novotná's 'atomic' analysis of the text comprehension (1995). With four versions of the same problem, Bruna and Havlíčková (2017) observed how changing mathematical, linguistic, and psychological parameters influenced the pupils' solutions of the problem. Soresi and Zucco (1995) studied details with children's comprehension: errors during reading, in comprehension, and in counting (see p. 162).

Davis and Maher (1990) claimed that problem solving involves several cycles, which are structured around five stages:

(1) *Data representation*: Constructing a mathematical representation of the data. The influence of mental representations on the activation of semiotic representations is analysed by Lemut and Dettori (1995). Their analysis is based on Duval's theory of semiotic representation (Duval 1993).
(2) *Knowledge representation*: Reviewing the solver's memory to reconstruct or rebuild the knowledge that will assist reaching a solution.
(3) Constructing a match between data representation and knowledge representation.
(4) Examining the match to see whether the solution is possible.
(5) Using technical means or algorithms that will enable a solution. Davis and Maher (1990) called this process mental representation of the way of solving the problem, and it, too, involved work with manipulatives or writing out the calculations.

Schoenfeld (1987) studied the processes that students undergo during problem solving. Based on this study, and in a manner similar to that of Pólya (1945), Schoenfeld

(1992) divided the process of problem solving into six stages: reading, analysis, investigation, planning, execution, and control. In examining the time that students spent on each of these stages and their sequence, Schoenfeld (1992) found that reading time is relatively short when compared with the other stages, and that most of the time is devoted to analysing the problem, with the following stages being short. He also found cases where, following planning, execution, and control, students had to go back to the analysis. When this happened, the second analysis took longer than the initial one. This procedure reveals processes of analysing alternative paths towards solving the problem and searching for them, requiring students to return to additional careful work from the analysis stage on.

Actually it seems that reading and analysis are intertwined. Meyer and Dana-Picard (1997) proposed a working approach from reading and analysis as follows:

- A first reading of the entire text without writing anything. This gives a general overview of the problem.
- A second reading, during which the solver begins the problem analysis and writes: 'what is given = . . .' and 'what is requested = . . .' (recall that it happens that the data appear in the text after the question). This can provide a basis for the planning (maybe a flowchart).
- The third reading and first step of execution are simultaneous, according to the planning in the previous step.

Students' difficulties

Mathematics is perceived as a difficult subject by many students, a perception which is documented in various works such as Gruszczyk-Kolczynska (2001), as well as for teaching trainees (Dofkova 2019). Most difficulties appear when dealing with problems and especially non-routine problems. Many articles point out these difficulties, and we will examine some of them. Hošpesová (1991) discusses where the difficulties are in problem solving: going from the text to the problem, within the learning process, or within the solution process?

We have seen in various works that dealing with word problems entails three components that the student must integrate: mastering the mathematical world, mastering the language, and familiarity with the realistic world that the problem describes, or that the student must know in order to connect it to the mathematical world. Each of these components is not trivial and each of them raises difficulties.

There are difficulties in reading, difficulties in making a connection between the story described and mathematical aspects. For example, the presentation of the verbal story is possible in different ways and in different orders. The data presentation of the given situation affects the complexity and the understanding of the text (Vondrová and Novotná 2017). Even when the child is familiar with the situation in reality, the transition from formalization to abstract thinking may raise difficulties: the quantities are given in numbers, but the connection between the quantities must be found in the words—and the integration between the two is required.

Semadeni (1995) treats these three components together as difficult, and emphasizes the dilemma:

- Postpone dealing with word problems to a later stage, when the student masters each of them individually?
- Or, at least, to make sure that the required mathematical knowledge has been already learnt?
- Or learn the required mathematics gradually along with solving the problems, and gradually develop the knowledge of the contexts—the knowledge of the world?

Irwin (1995) also mentions the difficulty of not mastering the mathematical field. She gives an example related to decimal numbers, or numbers with different units based on the decimal structure, and the relationship between the various names. For example:

> Our family was travelling on the motorway. My brother saw a sign that said our exit was 3 kilometres away. Next, he saw a sign that said it was 500 metres away. He said that we had gone too far and missed the exit. Was he right? What was he thinking? (p. 88)

Another difficulty, described by Gray (1995), was that even if the student has already overcome the above obstacles and obtained the mathematical expression that he must solve, there may still be a difficulty with the solution process. Gray points out the importance of paying attention to students' answers, in particular understanding why they answered what they answered, and to their solution processes. Young children often understand the given problem; however, the process of solving the exercise is based on basic strategies (like counting) and not necessarily on using mathematical operations.

Another typical problem can be in answering the question correctly but for incorrect reasons. For example, a student may answer a different question than the one they have been asked. For example, to answer the question 'find one seventh of fourteen', a student can give the correct answer 2. On closer inspection it is possible to find that he is thinking of the triple (2, 7, 14) from a reliance on multiplication and familiarity with (2, 7, 14) and not understanding what a seventh means.

Semadeni (1995) also points out that a problem may be more difficult if it has either superfluous data (the student has to relate to relevant data only) or missing data (the student has to complete from their personal knowledge).

Other researchers, such as Polotskaia et al. (2019), deal with a lack of knowledge during the transition from one stage of learning to another, such as the transition from elementary school to secondary school. They address the arithmetic–algebra transition and the difficulties students have because of different strategies used in different stages of their schooling. While 'in elementary school they rely on operational approaches and mainly they solve arithmetic problems, using concrete numbers and operations on them, in middle and secondary schools, they move to quantitative relationship and are expected to use algebra, i.e., equations, inequalities, functions, knowledge of structures and relationships' (Polotskaia et al. 2019, p. 316). Moreover, they observe students 'who have not developed more sophisticated arithmetic strategies as they continue to rely on concrete numerical data and on sequential operations, experience substantial difficulties' (Polotskaia et al. 2019, p. 316).

Researchers suggest various ways to deal with these difficulties. Gruszczyk-Kolczynska (2001) proposed gradually developing the form of presentation of word problems: the teacher illustrates the context with a drawing, and then the teacher tells a story using simple language. Only then do the students read and use their own means of illustration, drawings, and diagrams.

Word problems should also be graduated. The data order fits the order of operations required for the solution, and the question appears at the end. Generally, a problem is considered very difficult when a question is asked before the data are given. This happens because of reading difficulties in problems with conditional clauses that are more difficult to deal with. Experiencing this kind of problem greatly expands ways of teaching mathematics, and in general makes students' further mathematical learning easier. Irwin (1995) suggested bringing the students into conflict regarding the solution process, in order for them to deal with the solution process and make a decision.

Over the years, numerous articles have reflected teachers' attention to the types of difficulties and ways of working with students. Some of them appear in teachers' pre-service training, where they receive problems adapted to their level of knowledge and skills, but with difficulties similar to those that their students will encounter. Many such examples were presented at the 1995 SEMT symposium. See, for example, Hošpesová (1995). The same is true when using challenging problems: here the goal is to develop teachers' self-confidence in dealing with problems seen by their students, but at a higher difficulty level (Albert 2001).

Polotskaia et al. (2019) suggested using meta-cognitive reasoning while using efficient analysisof conceptual tools to solve problems. The acronym COMPLET stands for:

> Literal **Co**mprehension—Reading the text of the problem several times until the life situation is understood.
>
> **M**odelling—Representing quantitative relationships graphically until all of the relationships are visually represented.
>
> **Pl**anning—Based on the representation, developing a solution strategy and identifying arithmetic operations to calculate.
>
> **E**valuating and **Te**sting—Calculating and making sense of the numerical results that are compatible with the representation and the text of the problem. (pp. 319–320)

Creativity and problem solving

The well-known discussion on 21st Century Skills (Ananiadou and Claro 2009) emphasizes the following: problem solving, critical thinking, communication, collaboration, and creativity. However, these ideas have been in the problem-solving discourse at SEMT symposia since 1993 (see, e.g., the work of Szetela 1993; Trch and Zapotilová 1997 and others).

In fact, all those involved in problem solving, having experienced the so-called *routine problems*, or *school problems*, and having dealt with mathematics, linguistic knowledge, and suitable world knowledge, move on to challenging problems—either 'intra-mathematics' or world problems. In these situations, students are expected to find different solutions or different ways of solving, pose new problems, and/or expand a given problem exposing the students to different stimuli.

When we examine what is meant to be creative, we refer to the definitions by Guilford (1973) and Torrance (1969), who present four components:

1. *Fluency*: A person's ability to extract a large number of solutions that meet the limitations of the assignment.
2. *Flexibility*: A person's ability to shift from one way of thinking to another and extract solutions that refer to different categories.
3. *Elaboration*: A person's ability to elaborate the given idea, add details to it, develop it by means of a combination of additional ideas, and refine it.
4. *Originality*: A person's ability to approach the given problem in a new and unique way and extract unexpected and unconventional solutions.

All of these components exist in solving and posing problems of the types we have presented so far, and it is very important to expose students to various endeavours according to these properties. There is no doubt that teachers are required not to be satisfied with a final answer only, and certainly not with a single solution. Therefore, Dana-Picard and Hershkovitz (2019), working with technology for problems in plane geometry, required their students to use not only an *applet*, but also a text file, explaining their choices along the solution process, mentioning also what has been tried, or abandoned, and why. There must be space for discussion and extension of the field, as presented by Hershkovitz, Peled, and Littler (2009). We found that just dealing with such opportunities, and speaking with the students explicitly about the creativity characteristics, also contributes to creativity development. In further research, we have used these characteristics to evaluate students' solutions (Klavir and Hershkovitz 2008).

Problem solving with technology

Mathematics education has been interested in introducing technology into the teaching–learning process, and of course into problem investigation and solving. Naturally, this trend has been expressed in SEMT symposia.

In the early 1990s, the spreadsheet was widely used (Trižlová 1995). Bell (2001) displayed a detailed example as a basis for a comparison of spreadsheet usages, such as evaluating and recommending, designing and planning, finding and justifying generalizations, and of course problem solving.

Gravemeijer (1997) justified this direction in problem solving, citing the technological developments that make actual calculations easier, and therefore, he claimed, curricula, too, should emphasize problem solving. Lavoie (1994) said that the introduction of the handheld calculator was as important as the change from writing with a feather to writing with the iron quill: it enables transfer from the mind to the hand, whence freeing the brain for new thinking and tasks. Guin, Ruthven, and Trouche (1999) gathered a book with seminal papers on the introduction of calculators into the classroom. Trouche (2004) followed this and described the work organization required for teaching and learning with technology.

Dynamic Geometry Systems (DGS) appeared at first as a tool for verification of geometric properties, which would be proved formally afterwards. Yerushalmy and Schwartz began in the early 1980s to work in this field and developed The Geometric Supposer. This software saw continuous development and implementations (Schwartz and Yerushalmy 1987). Papadopoulos and Dagdilelis (2011)s distinguish several strategies for verification: visual, idiosyncrasy-based, copy–paste, etc. They also emphasize the dominant usage of the *dragging mode*. In reality, dragging points using the mouse together with the implementation of slider bar would quickly become central features of a DGS, and a comparison of their affordances, for an undergraduate problem, is given by Dana-Picard (2020). Other analyses by Hadas, Hershkowitz, and Schwarz (2000) and Healy and Hoyles (2001) describe the usage of technology in geometric problem solving. Later on, DGS became a very important instrument that is available for exploring, interactively, the classical theorems of plane geometry, as well as solving new problems.

During recent decades, when mathematics has been augmented by DGS, descriptions have been provided by, for example, Botana and Valcarce (2002) and Botana and Abánades (2014). We wish to mention that this new domain is relevant mostly for secondary school and beyond, but some of the tools, especially for plane geometry, can be used at an earlier stage. This has led to some networking between this kind of software and Computer Algebra Systems (CAS) to solve problems, not only numerically, but also using symbolic representations and computations. For example, Roanes-Lozano, Roanes-Macías, and Vilar-Mena (2003) and Dana-Picard (2020) provide instances of this networking. Both kinds of technology are efficient tools for visualization and, actually, the graphical capabilities have been developed for that purpose. Note that other kinds of visualization are important, and are a central part of problem posing and problem solving, namely visualization of numerical and algebraic situations, as described by Piñeiro et al. (2019).

The two kinds of technology mentioned above, namely CAS and DGS, enable exploration and experimentation of mathematical ideas. A CAS is mostly devoted to symbolic computations (calculus in one and several variables, algebraic manipulations with matrices and polynomials, statistics, etc.), but generally, nowadays, has well-developed graphical capabilities. Nevertheless, the dynamics of a DGS have a great influence on the possibilities to explore geometry. Dragging points enables one to check classical theorems in the geometry of the triangle, and of course, also more advanced constructions (Euler line, Steiner line, etc.). Students may explore, analyse, conjecture, and finally prove. Working with technology broadens the scope of tasks, which are well described by Horalková and Jirilková (1993). These authors applied their ideas to problems originating from art (Dana-Picard and Hershkovitz 2019; Dana-Picard et al. 2021).

Bureš, Novotná, and Tichá (2009) noted that the traditional usage of problem solving consists in using a pre-existing algorithm, whence its usage is only for assessment of the learnt material. Their work is based on a new process initiated by Novotná and Brousseau to 'change these student attitudes and to offer them the opportunity to look at problems as a "piece of work"'(p. 88). Technology is a wonderful instrument for this purpose.

Conclusion

The world of mathematics is a world of problem solving. Some problems are intra-mathematics, as in most textbooks in the past. Other problems are the so-called word problems, relating to concrete situations in the students' world. In this chapter, we have presented the full-extent of this world, from the basic problems that are also not easy in the early years when children start to study mathematics, to the complex problems older students meet in the school world, and beyond.

As noted, the range of possibilities for mathematical problems is very wide, and each type brings with it specific difficulties involving a variety of mathematical, conceptual, linguistic, and cognitive fields. It is clear that problem posing and solving is a central domain of work and research, and is constantly evolving along with the development of the understanding of the space of possibilities. The variety of tools for solving is large, and every new tool brings with it new opportunities. For example, the integration of technology, the integration of various domains of knowledge into STEAM Education, is continuously developing teaching and learning environments. They provide new points of view on intra-mathematics situations and on the concrete world. It seems that the future will invite us to further new opportunities that, at this stage, we cannot even imagine.

We believe that new research for each stage of development will be reflected in international collaborations at conferences, and that we will continue to study together.

References

Albert, J. (2001). Working with elementary teachers on challenge problems in mathematics. In J. Novotná and M. Hejný (Eds) SEMT 41–43.

Ananiadou, K., and Claro, M. (2009). 21st century skills and competences for new millennium learners in OECD countries. OECD Education Working Papers: 41. Paris: OECD Publishing. dx.doi.org/10.1787/218525261154

Artigue, M. (2002). Learning mathematics in a CAS environment: The genesis of a reflection about instrumentation and the dialectics between technical and conceptual work. *International Journal of Computers for Mathematical Learning*, 7(3): 245–274.

Bartoncová, L. (1999). How pupils complete a problem and create new ones from the point of view of their life experience. In M. Hejný and J. Novotná (Eds) SEMT 34–36.

Bell, A. (2001). What is problem-solving, for 9 year-olds? In J. Novotná and M. Hejný (Eds) SEMT 53–56.

Bevan, D., Williams, A., and Capraro, M. (2019). Strike a pose: The impact of problem-posing on elementary students' mathematical attitudes and achievement. In J. Novotná and M. Moraová (Eds) SEMT 80–88.

Bicknell, B., and Young-Loveridge, J. (2015). Using problem solving and representations to build number knowledge. In J. Novotná and H. Moraová (Eds) SEMT 63–72.

Botana, F., and Abánades, M. A. (2014). Automatic deduction in (dynamic) geometry: Loci computation. *Computational Geometry*, 47(1): 75–89.

Botana, F., and Valcarce, J. (2002) A dynamic-symbolic interface for geometric theorem discovery. *Computers and Education*, 38: 21–35.

Brousseau, G. (1997). *Theory of Didactical Situations in Mathematics* [translated by N. Balacheff]. Dordrecht, The Netherlands: Kluwer Academic Publishers.

Brown, S. I., and Walter, M. I. (1990). Problem posing in mathematics education. In S. I. Brown and M. I. Walter (Eds) *Problem Posing: Reflection and Application*. Hillsdale, NJ: Lawrence Erlbaum Associates.

Bruna, J., and Havlíčková, R. (2017). Parameters influencing difficulty of word problems: A case of four word problems in Grade 4. In J. Novotná and M. Moraová (Eds) SEMT 100–109.

Bureš, J., Novotná, J., and Tichá, H. (2009). Fostering students' mathematical understanding by developing the culture of word problems. In J. Novotná and M. Moraová (Eds) SEMT 88–96.

Carmichael, S., Martino, G., Porter-Magee, K., and Wilson, W. S. (2010). The State Standards – and the Common Core – in 2010. Thomas Fordham Institute, Advancing Educational Excellence. Retrieved October 2020 from https://files.eric.ed.gov/fulltext/ED516607.pdf

Cannizaro, L. (1995). On insight in problem solving activity. In M. Hejný and J. Novotná (Eds) SEMT 7–8.

Cifarelli, V., Stephan, M., and Pugalee, D. K. (2015). Examining the problem posing of third graders. In J. Novotná and M. Moraová (Eds) SEMT 92–100.

Cifarelli, V., Stephan, M., Pugalee, D. K., and Wang, C. (2017). Implementing an instructional sequence for solving net worth problems. In J. Novotná and M. Moraová (Eds) SEMT 148–158.

Dana-Picard, T., and Hershkovitz, S. (2019). Geometrical features of a Jewish monument: Study with a DGS. *Journal of Mathematics and the Arts*, 13(1–2): 60–71.

Dana-Picard, T. (2004). Three-fold activities for discovering conceptual connections within the cognitive neighbourhood of a mathematical topic. Proceedings of TIME-2004 (ACDCA Symposium in Montreal (Canada), J. Böhm (ed.). Linz, Austria: CD, bk teachware Schriftenreihe 41.

Dana-Picard, T. (2020). Envelopes and safety zone in an entertainment park: A Maltese Cross related to an offset of an asteroid. Proceedings of the 25th Asian Conference on Technology in Mathematics.

Dana-Picard, T., Hershkovitz, S., Lavicza, Z., and Fenyvesi, K. (2021). Introducing Golden Section in the Mathematics Class to Develop Critical Thinking from the STEAM Perspective. Preprint, submitted.

Davis, R. B., and Maher, C. A. (1990). The nature of mathematics: What do we do when 'do mathematics'? In R. B. Davis, C. A. Maher, and N. Noddings (Eds) *Constructivist Views on the Teaching and Learning of Mathematics*, 65–78. Reston, VA: National Council of Teachers of Mathematics.

Descartes, R. (1637). *La Géometrie*. Paris: Hermann, 1896; retrieved from Project Gutenberg. http://www.gutenberg.org/files/26400/26400-pdf.pdf

Dofkova, R. (2019). Perception of mathematics anxiety among prospective teachers while solving word and non-standard problems in mathematics. In J. Novotná and H. Moraová (Eds) SEMT 148–156.

Duval, R. (1993). Registres de représentation sémiotique et fonctionnement cognitif de la pensée. *Annales de didactique et de sciences cognitives*, 5: 37–65.

English, L., and Sriraman, B. (2010). Problem solving for the 21th century. In B. Sriraman and L. English (Eds) *Theories of Mathematics Education*, 263–290. New York, NY: Springer.

Evered, L. J. (1995). Problem solving and collaborative learning: A Vygotskian approach. In M. Hejný and J. Novotná (Eds) SEMT 73–76.

Freudenthal, H. (1983). *Didactical Phenomenology of Mathematical Structures*. Dordrecht, The Netherlands: D. Reidel Publishing Company.

Frobisher, L. (1994). Problems, investigation and investigative approach. In A. Orton and G. Wain (Eds) *Issues in Teaching Mathematics*, 150–173. London: Cassell.
Funke, J. (2010). Complex problem solving: A case for complex cognition? *Cognitive Processes*, 11: 133–142.
Fuson, K. C., Hudson, K., and Pilar, R. (1997). Phases of classroom mathematical problem-solving activity: The PCMPA framework for supporting algebraic thinking in primary school classrooms. In J. Kaput (Ed.) *Employing Children's Natural Powers to Build Algebraic Reasoning in the Context of Elementary Mathematics*. Mahwah, NJ: Lawrence Erlbaum Associates.
Gaulin, C. (1993). In-service courses for elementary teachers about problem solving in mathematics. In M. Hejný and J. Novotná (Eds) SEMT 73–74.
Gelfman, E., Demidova, L., and Matushkina, Z. (1995). Problems developing skills to solve word problems. In M. Hejný and J. Novotná (Eds) SEMT 82–85.
Glaser, R., Linn, R., and Bohrnstedt, G. (1997). *Assessment in Transition: Monitoring the Nation's Educational Progress*. Washington, DC: National Academy of Education.
Gruszczyk-Kolczynska, E. (2001). Why the solving of mathematical problems is so difficult for children? Intellectual and emotional conditions of mathematics education. In J. Novotná and M. Hejný (Eds) SEMT 14–25.
Gravemeijer, K. (1997). Commentary solving word problems: A case of modelling? *Learning and Instruction*, 7(4): 389–397.
Gray, E., and Tall, D. (1991). Duality, ambiguity and flexibility in successful mathematical thinking. *PME 15*, Assisi, 2: 72–79. http://homepages.warwick.ac.uk/staff/David.Tall/pdfs/dot1991h-gray-procept-pme.pdf
Gray, E. (1995). Tackling the problems: An explanation for success and failure. In M. Hejný and J. Novotná (Eds) SEMT 8–13.
Guilford, J. P. (1973). *Characteristics of Creativity*. Springfield, IL: Illinois State Office of the Superintendent of Public Instruction, Gifted Children Section.
Guin, D., Ruthven, K., and Trouche, L. (Eds) (1999). *The Didactical Challenge of Symbolic Calculator: Turning a Computational Device into a Mathematical Instrument*, Mathematics Education Library. New York, NY: Springer.
Hadas, N., Hershkowitz, R., and Schwarz, B. (2000). The role of contradiction and uncertainty in promoting the need to prove in dynamic geometry environments. *Educational Studies in Mathematics*, 44: 127–150.
Healy, L., and Hoyles, C. (2001). Software tools for geometrical problem solving: Potentials and pitfalls. *International Journal of Computers for Mathematical Learning*, 6: 235–256.
Hershkovitz, S., and Nesher, P. (2001). Pathway between text and solution of word problems. Proceedings of the Twenty-Sixth Conference of the International Group for the Psychology of Mathematics Education. Utrecht, The Netherlands.
Hershkovitz, S., and Nesher, P. (2002). Tools for fostering analysis skills: Strategies for solving arithmetical word problems. *ALEH*, 28: 15–25 (Hebrew).
Hershkovitz, S., Peled, I., and Littler, G. (2009). Mathematical creativity and giftedness in elementary school: Task and teacher promoting creativity for all. In R. Leikin, A. Berman, and B. Koichu (Eds) *Creativity in Mathematics and the Education of Gifted Students*, 16: 255–269. Rotterdam, The Netherlands: Sense Publishers.
Horalková, J., and Jirilková, D. (1993). Experience with preliminary geometry for future teachers of lower grades. In M. Hejný and J. Novotná (Eds) SEMT 74–76.
Hošpesová, A. (1991). Solving word problem by children aged six to ten. In M. Hejný and J. Novotná (Eds) SEMT 21–24.
Hošpesová, A. (1991). The analysis of children solutions of word problems. In M. ejný and J. Novotná (Eds) SEMT 85–87.
Hošpesová (1995). The analysis of children solutions of word problems. In M. Hejný and J. Novotná (Eds) SEMT 85–86.
Irwin, K. (1995). Using conflict in word problems to understand decimal fractions. In M. Hejný and J. Novotná (Eds) SEMT 87–89.
Irwin, K. (1999). Why everyday context hinders or aids understanding of decimals. In M. Hejný and J. Novotná (Eds) SEMT 60–63.

Kaslová, M. (1999). Pupils' creation of verbal problems: Method of penetrating into the world of mathematics. In M. Hejný and J. Novotná (Eds) SEMT 145–150.
Kintsch, W. (1986). Learning from text. *Cognition and Instruction*, 3(2): 87–108.
Kintsch, W. (1994). Text comprehension, memory, and learning. *American Psychologist*, 49: 294–303.
Klavir, R., and Hershkovitz, S. (2008). Teaching and evaluating 'open-ended' problems. *International Journal for Mathematics Teaching and Learning*. http://www.cimt.plymouth.ac.uk/journal/default.htm
Koman, M. (1995). Word puzzles – future teachers – pupils relay in teacher education. In M. Hejný and J. Novotná (Eds) SEMT 96–100.
Koman, M., and Tichá, M. (1999). How the children form phenomenon of dependence from their everyday experience. In M. Hejný and J. Novotná (Eds) SEMT 63–67.
Kopka, J. (1995). Problems and investigation in school mathematics. In M. Hejný and J. Novotná (Eds) SEMT 100–103.
Kratochílová, J. (1995). Pupils' strategies in the abracadabra problem. In M. Hejný and J. Novotná (Eds) SEMT 103–106.
Lavoie, P. (1994). Contribution à une histoire des mathématiques scolaires au Québec: l'arithmétique dans les écoles primaires (1800–1920) (Thèse de doctorat Faculté des sciences de l'éducation). Université de Laval, Québec.
Levenson, E. (2017). Promoting mathematical creativity in heterogeneous classes. In J. Novotná and M. Moraová (Eds) SEMT 42–52.
Mayer, R. E. (Ed.) (1992). *Thinking, Problem Solving, Cognition*. 2nd edition. New York: Freeman.
Meyer, J., and Dana-Picard, T. (1997). Problem solving in mathematics and physics. Proceedings of the 2nd International Conference on Teacher Education (Mofet Inst., Tel-Aviv), 343–352.
Ministry of Education Israel (2006). Division of Curricula, Mathematics Curriculum for Grade 1–6, all sectors (Hebrew).
National Council of Teachers of Mathematics (1989). *Curriculum and Evaluation Standards for School Mathematics*. Reston, VA: NCTM.
Nesher, P. (1980). The stereotyped nature of school word problems. *For the Learning of Mathematics*, 1: 41–48.
Nesher, P., and Katriel, T. (1977). A semantic analysis of addition and subtraction word problems in arithmetic. *Educational Studies in Mathematics*, 8: 251–269.
Novotná, J. (1995). Analysis of text comprehension in pupils' solutions to word problems via atomic analysis. In M. Hejný and J. Novotná (Eds) SEMT 114–117.
Papadopoulos, I., and Dagdilelis, V. (2005). Computer as a tool of verification in a geometry problem-solving context. In J. Novotná and M. Moraová (Eds) SEMT 260–268.
Piñeiro, J. L., Chapman, O., Castro-Rodríguez, E., and Castro, E. (2019). Students' visual representation of fractions and exponentiation. In J. Novotná and M. Moraová (Eds) SEMT 80–88.
Polotskaia, E., Savard, A., Calvacante, A., and Fellus, O. (2019). Middle school students' difficulties in problem solving and their roots in elementary education: what went wrong for them? In J. Novotná and M. Moraová (Eds) SEMT 315–324.
Pólya, G. (1945) (2nd Edition, 1973). *How to Solve It*. Princeton, NJ: Princeton University Press.
Priolet, M., and Novotná, J. (2007). Solving numerical problems in the 3rd year of primary school: Teaching and learning with connections. In J. Novotná and M. Moraová (Eds) SEMT 217–224.
Roanes-Lozano, E., Roanes-Macías, E., and Vilar-Mena, M. (2003). A bridge between dynamic geometry and computer algebra. *Mathematical and Computer Modelling*, 37: 1005–1028.
Schoenfeld, A. H. (1987). What's all the fuss about meta-cognition? In A. Schoenfeld (Ed.) *Cognitive Science and Mathematics Education*, 189–215. Hillsdale, NJ: Lawrence Erlbaum.
Schoenfeld, A. H. (1992). Learning to think mathematically: Problem solving, meta-cognition, and sense making in mathematics. In D. A. Grouws (Ed.) *Handbook of Research on Mathematics Teaching and Learning: A Project of the National Council of Teachers of Mathematics*, 334–370. New York: Macmillan.
Schwartz, J. L., and Yerushalmy, M. (1987). The geometric supposer: An intellectual prosthesis for making conjectures. *The College Mathematics Journal*, 18(1): 58–65. doi.org/10.1080/07468342.1987.11973012

Semadeni, Z. (1995). Developing children's understanding of verbal arithmetical problems. In M. Hejný and J. Novotná (Eds) SEMT 27–32.

Silver, E. (1997). Fostering creativity through instruction rich in mathematical problem solving and problem posing. *ZDM: International Reviews on Mathematics Education*, 29(3): 75–80.

Silver, E. (2013). Problem-posing research in mathematics education: Looking back, looking around, and looking ahead. *Educational Studies in Mathematics*, 83: 157–162.

Soresi, S., and Zucco, G. (1995). Effects of text characteristics on mathematical problem solving in Italian children. In M. Hejný and J. Novotná (Eds) SEMT 162–165.

Stanic, G., and Kilpatric, J. (1988). Historical perspective on problem solving in the mathematics curriculum. In R. Charles and E. Silver (Eds) *The Teaching and Assessing of Mathematical Problem Solving*, 1–22. Reston, VA: National Council of Teachers of Mathematics.

Szetela, W. (1993). Communication of critical thinking in problem solving in Grade 4. In M. Hejný and J. Novotná (Eds) SEMT 99–100.

Torrance, E. P. (1969). *Creativity: What Research Says to the Teachers*. Series No. 28. Washington, DC: National Education Association.

Trch, M., and Zapotilová, E. (1995). Nonstandard problems: The means of development of thinking and geometric imagination in the lowest school age. In M. Hejný and J. Novotná (Eds) SEMT 62–65.

Trch, M., and Zapotilová, E. (1997). Non-traditional mathematical tasks as a means of developing mathematical thinking of younger children and problems with their evaluation. In M. Hejný and J. Novotná (Eds) SEMT 74–78.

Trižlová, D. (1995). Problem solving on spreadsheets. In M. Hejný and J. Novotná (Eds) SEMT 123–125.

Trouche, L. (2004). Managing the complexity of human/machine interactions in computerized learning environments: Guiding students' command process through instrumental orchestrations. *International Journal of Computers for Mathematical Learning*, 9: 281–307. https://doi.org/10.1007/s10758-004-3468-5

Urbańska, A., Urbańska, E., and Rożek, B. (1993). Video records of student-teachers solving a problem as both a research and didactic means: Communication in the educational process. In M. Hejný and J. Novotná (Eds) SEMT 65.

Vondrová, N., and Novotná, J. (2017). The influence of context and order of numerical data on the difficulty of word problems for Grade 6 pupils. In J. Novotná and M. Moraová (Eds) SEMT 440–448.

Vygotsky, L. S. (1978). *Mind in Society: The Development of Higher Psychological Processes*. Cambridge, MA: Harvard University Press.

2.4
Learning to solve word problems with understanding

Naďa Vondrová
Charles University, Czech Republic

Introduction

There is no doubt that word problems (WPs) play an essential role in school mathematics curricula worldwide. Word problems represent a bridge between the world of mathematics and the everyday world. On the one hand, pupils are required to apply their mathematics knowledge when solving WPs, but on the other, they can acquire new knowledge (concepts and skills) through solving WPs. WPs may motivate pupils to learn mathematics and provide meaning for what pupils learn. The ability to solve WPs is essential for learners, not only in mathematics and outside it in other subjects, but mainly in their everyday life. There is no wonder, then, that the research on word problems is impressive. This chapter cannot even start accounting for the history of research on WPs, and thus the reader is advised to look at, for example, meta-analyses of studies (Hembree 1992; Daroczy et al. 2015; Verschaffel et al. 2020).

WPs have been defined in different ways. For Verschaffel, Greer, and De Corte (2000), they are:

> verbal descriptions of problem situations wherein one or more questions are raised the answer to which can be obtained by the application of mathematical operations to numerical data available in the problem statement. (p. ix)

In this chapter, we will only consider WPs to be problems that include some context (real, real-like, or imaginary)—that is, not all problems formulated in words will be considered. Before we continue with the synthesis of studies on WPs published in the SEMT proceedings, we will present some necessary concepts pertinent to WP solving.

The WP statement includes information of several types. Voyer (2011) distinguishes *solving information* (essential data), *situational information* ('playing a role in the development of a context that anchors the mathematical questions in a real life situation', p. 1075), *explanation information* (making the relationships among data explicit), and *superfluous information* (not helping pupils to understand the problem).

Taking into account that WPs include much more information than only the data, mathematical comprehension of the problem does not suffice for its successful

Naďa Vondrová, *Learning to solve word problems with understanding*. In: *Elementary Mathematics Teaching*. Edited by: Brian Doig, Jarmila Novotná, Berinderjeet Kaur and David K. Pugalee, Oxford University Press.
 DOI: 10.1093/oso/9780192869647.003.0008

solution. Reusser's (1990) model that distinguishes several levels of comprehension seems to be widely accepted in international research on WPs:

- *text comprehension* (resulting in a propositional structure of case frames 'by using a set of lexically based propositional meaning postulates', p. 7)
- *situation comprehension* (going beyond the text and resulting in the creation of a situation model)
- *mathematization* (resulting in the mathematical problem model made by 'mapping the semantic relations between the states and actions in the situation model onto an abstract structure of set relations', p. 9)
- *calculation* (ending with a numerical answer)
- *answer interpretation* (ending with a situational answer).

WPs can be seen as exercises of complex problem solving which call for the use of cognitive strategies (heuristics) as well as meta-cognitive (or self-regulatory) ones (Verschaffel et al. 2020). Thus, when solving WPs, pupils can fail both at the cognitive and meta-cognitive levels.

Researchers have investigated pupils' difficulties at all the comprehension levels. For example, much attention has been devoted to the strategies pupils employ when solving a WP. While it has been shown that pupils who give greater importance to situational information in a WP tend to be more successful in solving it (Voyer 2011), it has been amply documented that some pupils tend to skip the situation model level, and go straight from the text level to the mathematization, using the *keyword strategy* or a *direct translation strategy* (Nesher and Teubal 1975). This may have various causes, such as the numbers and keywords used in the WP text, a pupil's attempt to avoid cognitive overload (Sweller 2010; Páchová and Vondrová 2020), or the inability to make a situation model at all (e.g. due to the lack of comprehension of some words, or of the whole situation described by the WP). There may be didactic causes as well—researchers note that elementary teachers often promote the keyword strategy and textbooks often present WPs solvable with these strategies (e.g. Depaepe, De Corte, and Verschaffel 2010).

Given that research on WPs is abundant, the question arises as to what aspects of WP solving have been paid attention to by SEMT presenters.

The methodology of the analysis and structure of the chapter

We first scrutinized all SEMT proceedings for presentations dealing with WPs in the above sense. Moreover, we limited ourselves to only those presentations that included at least some elements of research *methodology*. Especially in the early days of SEMT, the presentations resemble extended abstracts, rather than full symposium presentations, and the description of the methodology is absent.

During the first reading of SEMT presentations on WPs, their division according to the specific types of WP targeted in them became apparent. For example, WPs including the division of a whole into parts (Novotná 1997), proportional WPs (Silvestre and da Ponte 2015), one-step and two-step additive and multiplicative WPs (Hershkovitz 2009), inconsistent language WPs (Soresi and Zucco 1995), and additive problems with operators only (Ruppleldtová 2007). However, for most of the presentations, such a classification could not be made.

The second reading led to the division of SEMT presentations on WPs into two categories: those studies focusing on factors making WPs more difficult for pupils, and studies aiming at ways to help pupils become better WP solvers. This enabled us to structure the chapter in the way used by Hembree (1992) in his attempt to provide a concise map of research on problem solving. Thus, the next section focuses on the characteristics of problem solvers in relation to WPs. The following section includes studies on conditions for harder and easier problems. The most attention by SEMT authors has been put on the effects of different instructional methods on problem-solving performance. The next section of this chapter, the longest, is sub-divided into five sections. By way of contrast, the impact of classroom-related conditions on problem-solving performance was studied by only a few SEMT presenters, and, thus, this section is the shortest.

Table 2.4.1 presents an overview of the SEMT presentations used as a basis for this chapter and their inclusion in the above categories. The last row includes presentations which are outside the categories, but which are clear studies on WPs. They all report on elementary pupils' use of visual representations, or the lack of it, when solving WPs. These studies are used to support some claims made in the later sections that deal with teaching experiments involving the development of pupils' ability to use graphical representations.

Except for Hembree (1992), two other meta-analyses will be used to compare SEMT attention to WPs with broader mathematics education research, namely Daroczy et al. (2015) and Verschaffel et al. (2020). It is necessary to stress, though, that, unlike these studies, we are only concerned with research with *elementary pupils* in this chapter.

Naturally, it is beyond the scope of this chapter to present the research methodology for the studies separately. Generally, it can be said that studies focusing on pupils and how they solve WPs (what strategies they use and mistakes they make) mostly utilize a qualitative research paradigm, with researchers analysing pupils' written solutions, or carrying out think-aloud interviews with pupils while solving WPs. Studies aimed at determining whether some variable influences WP difficulty are mostly quantitative and combine pupils' success rates in the WP's two variants, differing in the studied variable. In studies aimed at the effect of a teaching approach on pupils' ability to solve WPs, mostly simple teaching experiments were utilized (in some presentations, more developed design research methodology is reported) and only in some of them were there an experimental group and a control group.

To sum up, many of the SEMT presentations do not present sufficient details of their methodology for us to judge the validity of results. The conclusions presented below should be treated with caution. Thus, when applicable, some other research studies with elementary pupils will be referred to, to augment or support the claims made in the SEMT presentations.

Table 2.4.1 SEMT research studies on WPs

Category	Author and year of SEMT presentation
Characteristics of problem solvers	Navarra (1995) Navarra (1997) Gruszczyk-Kolczynska (2001) Peled (2015) Polotskaia, Savard, Calvacante, and Fellus (2019)
Conditions for harder and easier problems	Semadeni (1995) Soresi and Zucco (1995) Hershkovitz (2009) Bruna and Havlíčková (2017) Vondrová and Novotná (2017)
Effects of instructional methods on problem solving performance	Szetela (1991) Novotná (1997) Bartoncová (1999) Bednářová (1999) Kaslová (1999) Leppäaho and Ahtee (2005) Bureš, Novotná, and Tichá (2009) Bartolini Bussi, Canalini, and Ferri (2011) Cabassut (2011) Cifarelli, Stephan, and Pugalee (2015) Peled (2015) Silvestre and da Ponte (2015)
Effects of classroom-related conditions on problem-solving performance	Clanché and Sarrazy (1999) Gruszczyk-Kolczynska (2001)
Pupils' use of visual representations	Lemut and Dettori (1995) Novotná (1999) Ruppeldtová (2007) Novotná and Sarrazy (2009) Břehovský, Eisenmann, Ondrušová, Přibyl, and Novotná (2013)

Characteristics of problem solvers

In her presentation, Gruszczyk-Kolczynska (2001) dealt mostly with psychological factors, including emotions, as factors contributing to pupils' failure in problem solving. She emphasized that the sense of difficulty of a WP is always subjective. Her presentation included ample evidence that problem solving requires a lot of intellectual effort as well as emotion. She specifically focused on children who regarded mathematics as very difficult. By observing them during mathematics lessons, the author determined how the personality of the child, their motivation, emotional maturity, and behaviour in problem situations influenced their problem-solving behaviour. She grounded her conclusions in the idea that emotions play a guiding role for, and determined the frame of, intellectual knowledge. According to her, to be able to solve tasks successfully, a child has to possess a high level of emotional resistance against difficult situations, which is a prerequisite to the efficient learning of mathematics. The role emotional factors play in problem solving has been acknowledged by ample research (e.g. Hannula 2014; Muis et al. 2015).

Among others, Gruszczyk-Kolczynska (2001) recommended that children should not be protected from stress so that they can gradually learn to shape their emotional resistance in a safe environment, for example in games. She suggested that it is good for the child to learn to tolerate their failures, and to have hope for success in the future. They must be convinced that if they try, they will be able to do everything. Such an attitude to personal failure is very important in mathematics (consider the concept of self-efficacy in mathematics, Moddleton 2014, or growth mindset, Dweck 2017).

Much research attention has been put on the problem solving of different types of pupils (e.g. those who are talented or those with difficulties in mathematics) that has led to the realization that such pupils use different comprehension strategies. For example, Voyer (2011) found that pupils with weaker arithmetic skills construct other representations and benefit more from explanation information contained in the WP rather than the situational one. Pongsakdi et al. (2020) distinguished pupils according to their skills in text comprehension and in arithmetic. They found out, among other things, that for easy items, skills in arithmetic were more important, while for more challenging items, both skills were essential.

Four SEMT presentations investigated types of pupils in relation to WP solving. Navarra's (1995) presentation documented how pupils strong in mathematics integrated the linguistic and mathematical information present in the WP, while pupils weak in mathematics tended to use what he called 'unconsciously partisan reading', in which they selected 'friendly parts' of the text (which they had meaning for) and neglected other parts.

Peled (2015) investigated, among other things, how weak pupils behaved in a group that was working on modelling tasks—that is, WPs grounded in reality, which did not lead to a neat direct solution. She confirmed her expectation that, as such pupils were no less familiar with realistic situations than other children, they could contribute some realistic considerations to the group work. However, she also described a case of a weak pupil who substantially contributed to the group success by providing them with a suitable graphic model and later showed elements of enhanced meta-knowledge and control (the lack of which is, according to the author, a sign of weaker pupils). Peled pointed out that an ability in mathematics might not predetermine pupils' success in modelling tasks:

> As it turns out, children who have difficulties in class might exhibit good modelling performance, while students with a good mathematical background might prove to have difficulties in modelling. (p. 30)

Finally, Polotskaia et al. (2019) worked with pupils who failed their final elementary school examination, and showed that their failure was at least partially connected with their insufficient meta-cognitive knowledge. Due to their lack of knowledge about relationships and their visual representations, these pupils often drifted towards calculations in their solving process, and their attention was drawn to numerical values per se.

Navarra's (1997) presentation differs from the above, as he used WP solving (or rather, a meta-cognitive task involving WP solving) as a means of dividing pupils into groups according to the quality of control when solving a WP, both on the cognitive and meta-cognitive levels. Pupils were told that a girl could not understand why she was not successful when solving WPs. Then they were presented with nine WPs with one or two pieces of data missing and asked to help the girl to solve the problems. Navarra analysed their written explanations from the point of view of the meta-cognitive goal, considering whether the pupil observed the task to help the imaginary girl and the cognitive goal (how well they solved the problems). Navarra suggested this as a way of assessing pupils' problem-solving skills.

Conditions for harder and easier problems

Since the early days of research on WPs, it has been acknowledged that some variables made the problem more difficult. Attempts have been made to classify variables according to their nature. For example, Hembree (1992) distinguished variables affecting the difficulty of WPs as follows: *context* (abstract, concrete, factual, hypothetical, familiar and unfamiliar, imaginative, personalized), *functional arrangement* (variables related to the composition of problems: length, readability, action (static versus dynamic, involving movement or change), placement of question, order of data (proper or mixed), use of extraneous data, use of keywords, use of unfamiliar terminology, etc.), and *format* (the way of presenting a problem: written, oral, involving manipulation, drawn, telegraphic, i.e. in short phrases, with a picture, etc.).

Looking more closely at the linguistic verbal factors not directly related to arithmetic, Daroczy et al. (2015) distinguished three components: *the linguistic complexity* (variables related to structure such as sentence length and vocabulary level, and semantics such as verbal cues and object relations), *the numerical complexity of the arithmetic problem* (such as the type of number, number magnitude, the required operation, the position of the unknown, etc.), and *the relation between the two complexities.*

In an attempt to find a way of formulating WPs of increasing difficulty, in his presentation, Semadeni (1995) presented a model which included three criteria: the *mathematical structure* of the WP, the difficulty of the *context*, and the difficulty of the *format.* The last ranged from stories enacted by the teacher, stories shown in pictures, etc., to problems read by children. However, according to Semadeni, the category of format includes other variables such as the use of concrete material versus drawing pictures versus arithmetic operations on numbers only, or the proper versus mixed order of data—that is, the variables included in the category of functional arrangement by Hembree (1992).

Semadeni's model may seem to be linear, but he warned that the mutual relations between the criteria are much more intricate and that 'some kind of spiral approach is indispensable' (p. 29). Semadeni also made a number of suggestions for teachers on how to teach WPs. For example, by simulating solutions (through some objects or drawing), by setting pupils a problem-posing activity, or by using problems with missing, surplus, or contradictory data (i.e. WPs with some complicating variables, as, indeed, problems with such data have been shown to be more difficult for pupils; see Hembree 1992).

The above meta-studies confirm the viability of some parts of Semadeni's model. Some SEMT presentations contribute to this aim, too. For example, Hershkovitz's (2009) presentation focused on the inner structure of WPs—namely, on one-step and two-step additive and multiplicative WPs. It was based on the framework of intuition and schema, which Hershkovitz did not see as contradictory but rather as complementary factors of problem solving. She emphasized that they made a closed circuit: 'The more schemata a person acquires, the more intuition he has' (p. 40). Additive and multiplicative schema are the building blocks of all WPs which can be solved with the four arithmetic operations (addition, subtraction, multiplication, division). They become 'slots' in the compound schema for more complex problems (Hershkovitz 2009). She reported that pupils who used the additive or multiplicative schema succeeded in solving two-step problems better than those who did not apply the schema.

Soresi and Zucco (1995) presented on the effect of the context of the WPs (e.g. a WP on running a motorbike versus a WP on an estate agency) and language consistency. In WPs with consistent language, the keywords refer to the correct operation. In contrast, in WPs with inconsistent language, they refer to an operation opposite to the one required for the solution. For example, 'One estate agency in Milan made in one year 500 new contracts; it made 1/3 of the contracts made by one agency in Rome. How many contracts did the agency in Rome make in one year?' (p. 163). If pupils use the wrong operation suggested by the keyword(s), they commit a *reversal error* (Lewis and Mayer 1987).

Soresi and Zucco concluded that the effect of context on pupils' success in solving WPs is less evident while the language consistency effect is significant. This is in accord with many other studies (see Hembree 1992; Daroczy et al. 2015; Vondrová 2020; Vondrová et al. 2019). Soresi and Zucco further found that a familiar context lessened the difficulty of the WP caused by the inconsistent language, probably helping pupils to make a situation model. Thus, the unfamiliar context functioned here as a complicating variable. On the contrary, Vondrová (2020) documented how the presence of another complicating variable (namely, the inclusion of an irrelevant number in the WP) helped pupils to avoid a keyword strategy in an inconsistent language WP, and thus prevented them from committing a reversal error.

Context in combination with the order of numerical data was targeted in one of the more recent studies by (Vondrová and Novotná 2017). In the *proper order* WPs, the numerical data appear in the sequence needed to solve the problem. If this is not the case, such WPs are called *mixed order* ones (Hembree 1992). An example is: 'Mirka got a new jigsaw puzzle for Christmas. She solved it in three days. The second day she placed 8 pieces more than the first day, the third day by 11 pieces fewer than the second day. The first day, she correctly placed 25 pieces. How many pieces did Mirka place the third day?' (Vondrová and Novotná 2017, p. 443).

Vondrová and Novotná confirmed the existence of a small effect for the WP context (jigsaw puzzle versus chemical substances) and a significant effect for the order of the data—the mixed order WP was more difficult than the proper order one. Unlike Soresi and Zucco (1995), Vondrová and Novotná (2017) found that the unfamiliar context *lessened* the mixed order of data effect, rather than making the WP even more difficult. This opened an intriguing question: 'Could an unfamiliar context used in combination with the mixed order of data cause that pupils were alerted to possible sources of problems and were more attentive when reading the text?' (p. 447).

The consistency of language is among the most researched linguistic variables, while others get limited attention (Daroczy et al. 2015). Bruna and Havlíčková (2017) targeted the variable that is, to the best of our knowledge, original. It was suggested by a linguist in the project team (Vondrová et al. 2019) and called 'message level'. It is a term that builds on the idea that from the point of view of pragmatics a word problem is a message and as such is structured into layers, or levels, hence message levels (Bruna and Havlíčková 2017, p. 102).

We will illustrate this new concept by the WPs used in the study:

> Homework from mathematics is on page twenty-six in the workbook. That is, exactly seven pages past the half of the workbook. How many pages are there in the workbook?
>
> Pupils learnt that homework from mathematics is on page twenty-six in the workbook.
>
> Fanda said: 'That is exactly seven pages past the half of the workbook'. How many pages are there in the workbook?

In the second problem, there is an additional level present that encapsulates the original message by expressing the perspective of a protagonist (namely, pupils). The authors held conflicting hypotheses:

> On one hand, it is reasonable to assume that removing such a level would simplify the task for the pupils. On the other hand, although the level does not provide pupils with crucial information it often contains narrative elements that may motivate pupils to solve the problem. (p. 102)

It transpired that the second guess was more valid. The authors speculated that the reason might be that the WP was not demanding enough for Grade 4 pupils, and, thus, the required abstraction away from the protagonist's perspective did not present a hurdle for them.

Two presentations (Vondrová and Novotná 2017; Bruna and Havlíčková 2017) presented preliminary results of a project aimed at understanding the influence of selected variables on WP difficulty for pupils in Grades 3 to 9. The final results were published in a monograph (Vondrová et al. 2019). The results presented in the two SEMT presentations were confirmed on a larger scale. The authors (Vondrová et al. 2019) stated that the influence of the context could not be limited to a simple conclusion that the unfamiliar context is always more difficult than the familiar one, and that the unfamiliar context sometimes helped to solve a difficult WP more effectively. This was shown for WPs that included a complicating variable, such as the order of data, or the presence of an irrelevant number or of the numbers functioning as cues. However, it was only the case for some WPs; in others the unfamiliar context made the WP a bit more difficult, or did not change the difficulty at all. More refined research is needed here. The authors hypothesized that when pupils spotted an unfamiliar context, they might become alerted to a possible difficulty in the WP, or might get more interested in the problem and read it more carefully. On the other hand, when seeing a WP in a known context, they might have been drawn to a superficial solving strategy. Thus, the authors concluded, a cognitively easy WP may be as inappropriate as a cognitively demanding WP, which may lead to a cognitive overload.

In terms of the other variables presented in the two SEMT presentations, Vondrová et al. (2019) confirmed that the influence of the order of data on the difficulty of a WP was substantial. On the other hand, the results regarding the influence of the number of message levels are mixed. The number of message levels made some WPs more difficult, but it helped in some other WPs. Thus, the WPs included another variable not considered by the researchers, which influenced the pupils' success.

Effects of instructional methods on problem-solving performance

As already stated, successful WP solving requires both cognitive and meta-cognitive skills. Heuristics belong among the cognitive ones. They are:

> cognitive strategies that help to solve a task for which one does not have a clear answer or ready-made solution method. More specifically, heuristics are search strategies for problem analysis and transformation that do not guarantee, but significantly increase the probability of, finding a correct solution because they induce an active and systematic approach to the task. (Verschaffel et al. 2020, p. 5)

From the heuristics mentioned in Verschaffel et al.'s meta-analysis, the SEMT presenters paid attention to two: a) teaching pupils to deeply analyse the text, and b) leading pupils towards making graphic representations. Meta-cognitive or self-regulatory strategies are strategies that learners apply prior to, during, and after the execution of a cognitive task, to regulate one's own thinking and learning (Verschaffel et al. 2020, p. 5).

SEMT presenters have only addressed the development of meta-cognitive skills, as a way to enhance pupils' ability to solve WPs, to a limited extent. However, studies on problem posing can also be seen as part of developing meta-cognitive skills, as, when posing problems, pupils are required to consider the WP structure from a meta-cognitive point of view—they have to take on the position of an imaginary solver, and anticipate their solving strategies. The same applies to using modelling problems rather than traditional textbook WPs. Solving modelling problems requires the development of meta-cognitive strategies such as planning one's own solution, monitoring one's own progress, or evaluating the outcomes.

Deep analysis of the WP text

Solving WPs is closely connected to pupils' ability to read with understanding (e.g. Capraro, Capraro, and Rupley 2011; Pongsakdi et al. 2020; Verschaffel et al. 2020). Some SEMT presenters suggested activities which should, in their professional opinion, lead to a deeper analysis of the text of WP and, thus, to better problem-solving skills. Rather than showing that it is indeed the case, they presented these activities as an educational implication from their research.

For example, Bednářová (1999) presented the written solutions of nearly 800 pupils who solved WPs that contained information in conflict with ordinary life experience. For example:

> Fill in the missing text: Mike's father and grandfather are 100 years old together. Mike's father is 54, so Mike's grandfather is ... than Mike's father. (p. 38)

Bednářová focused on the questions of if and how pupils used their everyday experience to resolve the conflict between daily life and the mathematical structure of WPs. She proposed that such problems could prevent pupils from adopting inappropriate, impulsive approaches to problem solving by providing them with a deeper understanding of what the problem is all about and teach them not to rely on superficial clues. Navarra's (1995) technique may also be seen as a way towards careful reading. He presented pupils with a WP and with it three paraphrases. Their task was to find which one was incorrect and to justify why, which requested a deep analysis of the texts.

Bartolini Bussi, Canalini, and Ferri (2011) in their SEMT presentation introduced the Chinese method of *problems with variation*, which, in fact, calls for pupils' careful analysis of WPs. An illustration is in Table 2.4.2 below.

Pupils are encouraged to find the strategy for solving the example problem, and adapt it for the variants. To understand the similarities and differences between the variants, they are led to study the text of the WP carefully. They can also be asked to create another triplet or quartet of WPs in a similar manner.

Bartolini Bussi et al.'s presentation was an exploratory one. It presented two teachers' implementation of problems with variation in their lessons and points to some difficulties they had to overcome: for example, they had to accept that separate WPs were not used, but rather a system of WPs made the invention of general solving schemes possible. Overall, the presentation reported some promising results in terms of pupils' ability to see the mathematical structure of the WPs and 'read' its visualization through the segment representation. Further, they seemed to become better problem solvers than other pupils of the same age based on their success in solving a demanding WP set in the national assessment test.

Table 2.4.2 Beijing Education Science Research Institute and Beijing Instruction Research Center for Basic Education (1996), vol. 4, p. 88.

First, solve the nine problems below. Then explain why they have been arranged in rows and columns in this way, finding relationships.

(1) **In the river there are 45** white ducks and 30 black ducks. All together, how many ducks are there?

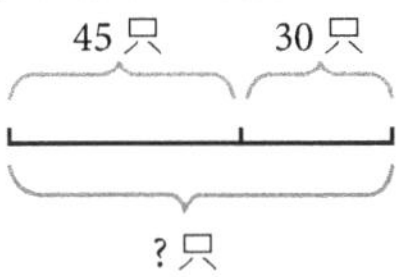

(2) **In the river there are white** ducks and black ducks. All together there are 75 ducks. 45 are white ducks. How many black ducks are there?

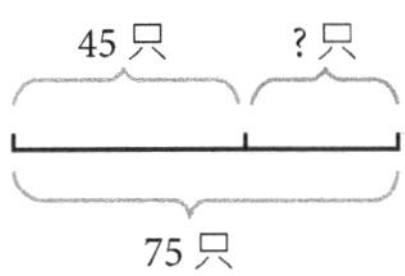

(3) **In the river there are white** ducks and black ducks. All together there are 75 ducks. 30 are black ducks. How many white ducks are there?

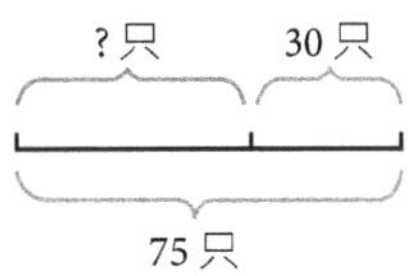

(1) **In the river there is** a group of ducks. 30 ducks swim away. 45 ducks are still there. How many ducks are in the group (at the beginning)?

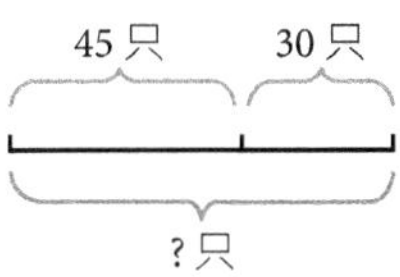

(2) **In the river there are** 75 ducks. Some ducks swim away. There are still 45 ducks. How many ducks have swum away?

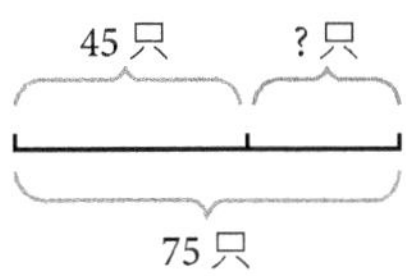

(3) **In the river there are** 75 ducks. 30 ducks swim away. How many ducks are still there?

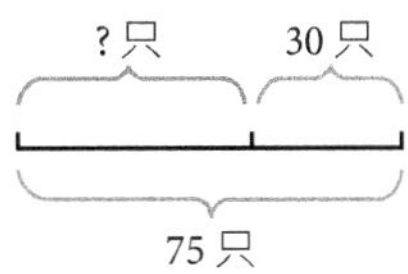

(1) **In the river there are** 30 black ducks. White ducks are 15 more than black ducks (black ducks are 15 less than white ducks). How many white ducks are there?

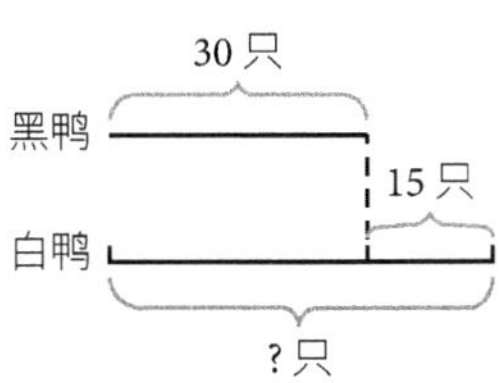

(2) **In the river there are** 30 black ducks and 45 white ducks. How many white ducks more than black ducks (How many black ducks fewer than white ducks)?

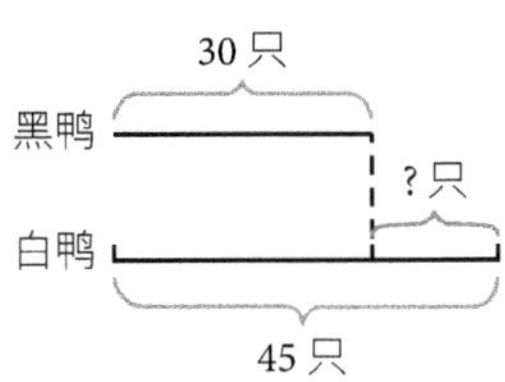

(3) **In the river there are** 45 white ducks. Black ducks are 15 fewer than white ducks (white ducks are 15 more than black ducks). How many black ducks are there?

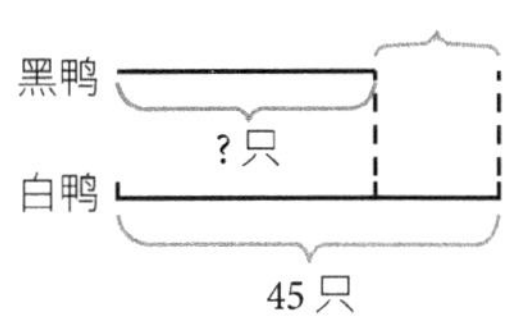

Mediating role of representations

According to Goldin (2014), mathematical representations are:

> visible or tangible productions—such as diagrams, number lines, graphs, arrangements of concrete objects or manipulatives, physical models, mathematical expressions, formulae and equations, or depictions on the screen of a computer or calculator—that encode, stand for, or embody mathematical ideas or relationships. (p. 409)

Some SEMT presenters confirmed that pupils do not generally use pictorial representation as a solving strategy for WPs (e.g. Břehovský et al. 2013; Novotná 1999), which is unfortunate, as the assumption supported by research is that:

> graphical representations may particularly help in the initial phases of the solution process of a word problem wherein the situational and mathematical model are constructed, by providing an additional relevant source of information. (Verschaffel et al. 2020, p. 6)

For example, Ott (2020) showed that teaching that combines the self-generation of graphical representations and reflection processes on them in class seemed to be positive for the development of graphical representation competency, in that the pupils from the experimental group in the study paid more attention to mathematically appropriate structures than the control group.

In their meta-analysis of studies, Verschaffel et al. (2020) reported positive outcomes of intervention studies for pupil-generated representations as well as the ones given by the teacher, and summarized some prerequisites to be fulfilled. One of them is the nature and function of the representations. Similarly, based on pupils' solutions to WPs, two SEMT presentations concluded that only some representations help the pupils to solve the problem—for example, the ones correctly showing the relations between the objects from the WP (Lemut and Dettori 1995; Ruppeldtová 2007).

Ruppeldtová (2007) investigated a specific type of additive WPs in which all the numbers are in the role of operators. For example: 'Brano is 13 cm taller than Adam. Cyril is 4 cm shorter than Brano. Compare the height of Cyril and Adam and say how many centimetres Cyril is taller or shorter than Adam'. Ruppeldtová investigated, among other things, what kind of spontaneous visualization pupils made when solving the WPs. For example, most pupils used line segments of different lengths (possibly with arrows representing operator orientation) for operators and vertical or horizontal lines for states hidden in the WP (such as the missing height of the pupils in the example above).

Using the example of a segment representation, Novotná and Sarrazy (2009) highlighted the need to take into account not only the mathematical but also the psychological perspective when dealing with graphical representations. Two different, but correct, representations may lead to different understandings of the situation described in the WP. The presenters supported their claim with examples of pupils' work.

Some SEMT presentations investigated whether the use of graphical representation is a learnable skill. Novotná (1997) presented pupils with a WP accompanied by a segment model and investigated whether they were able to use this newly presented model independently for solving a WP problem with a different context or modified in some other fashion. She did not say so in her conclusions, but the segment model was seen as a mediator between the situation and mathematical model of the WP.

Silvestre and da Ponte (2015) presented on a simple teaching experiment with a group of pupils to develop their ability to solve proportional WPs. Pupils were asked to explore the multiplicative nature of direct proportion relationships and solve both problems involving direct proportion relationships and pseudo-proportional problems. An important characteristic of teaching was that pupils were asked to work simultaneously with different representations (tables, graphs, ratios represented as fractions, ratios represented as divisions). No algorithm was taught. By analysing pupils' solutions prior to, and after, the teaching unit, the presenters concluded that the pupils enhanced their understanding of multiplicative strategies and their sense of co-variation within, and invariance between, variables, and, despite being taught no algorithm, they developed more sophisticated strategies for solving proportional WPs.

Břehovský et al.'s (2013) presentation focuses on the spontaneous use of some heuristic strategies by 11-to-12-year-old pupils. The authors concluded that those who used graphical representations solved the problem successfully. The authors later published a presentation (Eisenmann et al. 2015) focusing on their intervention research with 12-to-18-year-old pupils, in which they determined that teaching focused on problem solving, based on the use of heuristic strategies, led to a considerable improvement in pupils' creativity and in their attitude to problem solving.

Finally, consider also Bartolini Bussi, Canalini, and Ferri's (2011) presentation, above, in which the use of the segment representation contributed to pupils' becoming better problem solvers.

Development of meta-cognitive skills

Some SEMT presenters provided modest evidence of the positive influence of teaching involving the explicit development of meta-cognitive skills on the ability to solve WPs that had been established by research (Verschaffel et al. 2020).

For example, Szetela (1991) based his teaching intervention on the idea that traditional teaching in mathematics placed too much emphasis on algorithmic processes (which according to him leads to number crunching) and too little on 'thinking processes' (p. 58). Four teachers and their classes participated in the study, in which the pupils were engaged in written and oral verbalization when solving WPs with a view that such verbalization might prevent them from rushing prematurely to an answer which makes little or no sense. Promising results concerning pupils' problem-solving skills were found. Szetela's work is also important because he worked with several practising teachers and came to the conclusion that verbalization is a technique which is easily integrated into the teachers' lessons and which they found suitable for developing pupils' ability to solve WPs with understanding.

Leppäaho and Ahtee (2005) presented on their small intervention study in lessons on mathematics, mother tongue, natural sciences, and art and crafts in Grade 6. So-called *solving maps* were used to support the pupils in looking for the route towards the solution and in checking the phases used in their attempts to solve the problem. The maps also required them to write down their ideas during the solving process, as the potential of writing for promoting mathematical understanding had been acknowledged (e.g. Morgan 1998). The solving map contained the data from the WP, accompanied, possibly, by a drawing, and all the pupil's attempts to solve the problem, including the ones that were not successful. The control group did not work with the maps, and it transpired that the experimental group pupils improved more than the control group in their scores on numerical calculations and in the quality of their explanations (including drawings). The presenters pointed to other factors contributing to the success of their intervention, such as pupils working in groups, and emphasized that the mere invitation to make solving maps led pupils to realize that it was permissible to make mistakes and try different routes.

Problem posing

Problem posing is an activity with multiple educational merits (Silver 1994; Singer Ellerton and Cai 2013). It can improve pupils' problem-solving skills, attitudes towards, and confidence in, mathematics, and creativity. The process of problem posing can be considered a meta-cognitive activity, as the process of reflection is required when considering imagined mathematical activity (Schoenfeld 1994). Problem posing has been addressed by some SEMT presenters too (consider also Semadeni's (1995) recommendation for teachers mentioned above). Except for two presentations, (Bureš, Novotná, and Tichá 2009 and Bartolini Bussi, Canalini, and Ferri 2011—in the latter, problem posing was used as one of the activities in the teaching experiment), problem posing was used as a means to get an insight into pupils' understanding of WPs, as the problem-posing task provides 'a window through which to view students' mathematical thinking' (Silver 1994, p. 25).

The goal of Cifarelli, Stephan, and Pugalee's (2015) presentation was to provide an insight into eight-to-nine-year-old pupils' reasoning when posing problems aimed at their peers and their teacher. Both free and semi-structured problem-solving tasks were used. The imagined audience of the posed WP made a difference for the pupils. The numbers in the WP aimed at the teacher were far bigger than the numbers in the WP for their peers. As only the analysis of pupils' written work was used, the presenters could not account for possible reasons.

To investigate how pupils use real-life knowledge in mathematics, Bartoncová (1999) asked pupils to complete an incomplete WP and to pose a similar problem based on the WP she gave them. She noted that some children did not use their real-life experience and considered only the mathematical part. They might have played what was termed 'the word problem game' (Verschaffel et al. 2000) and used for explaining pupils' apparent suspension of common sense when solving WPs.

By observing Grade 1 and 2 pupils in problem posing, Kaslová (1999) distinguished several types of problem posing and described what difficulties they presented to the pupils. She hinted at the educational value of problem solving by showing how overcoming these difficulties helped pupils to understand the operation of addition better.

Finally, Bureš, Novotná, and Tichá (2009) investigated whether pupils were able to recognize that some WPs had the same mathematical model. In the first part of a teaching experiment organized in one class, the teacher assisted pupils to realize what a mathematical model of one WP was. Then she asked them to use this knowledge and suggest a different WP with the same model. It transpired that pupils were not generally able to distinguish all the aspects of the mathematical model—for example, the mere presence of the same operations in the posed problem was considered to be enough for the equality of the models. The presenters noted that some pupils only noticed surface features of the WP and did not get into the inner structure of the WP, which prevented them from correctly applying the idea of the mathematical model.

Redesigning WP solving as solving modelling tasks

Verschaffel et al. 2020 argued that word problems can be seen as a specific type of mathematical modelling problem:

> According to this point of view, while there always will remain a difference between a word problem being presented and solved in a school setting and an authentic real-life modelling problem with which one is confronted outside school, word problems, if appropriately designed and handled in a mathematics educational context, can act as accurate and valuable 'simulacra' of authentic mathematical modelling problems one could encounter in real life. (p. 2)

Verschaffel et al. (2010) noted that WPs fail as vehicles for encouraging pupils' disposition towards mathematical modelling and applied problem solving and suggest that modelling tasks are needed rather than traditional WPs to reach this goal. Modelling tasks are tasks that describe authentic situations and involve the analysis, organization, and mathematization of these situations (Peled 2015). A modelling situation involves a genuine problem that one might encounter, and for which one might be curious to find a solution, rather than an artificial problem.

Moreover, in her SEMT presentation, Peled (2015) noted that there had been a shift in the modelling perspective. In the early days, modelling tasks were seen as applications of knowledge learnt at school with the understanding that their solution would enhance mathematical knowledge and increase pupils' motivation:

It was only later realized that this was not enough. That is, that school mathematics should not only aim towards a rich mathematical toolbox, and be satisfied with a wide range of applications. Rather, it should also take care of building children's ability to cope with new unfamiliar and complex situations, where the situation should be analysed and decisions would have to be made about its mathematization. This recognition meant that an important goal of school mathematics should be to develop the organization and mathematization abilities that are termed modelling competencies in the modelling literature (p. 31).

Verschaffel et al. (2010) pointed to a body of research in which WP solving is changed into the task of modelling, and summarized prerequisites for their successful implementation (e.g. the use of more realistic and challenging tasks than traditional textbook problems, creating a classroom climate of inquiry, or a focus on demonstrating, experiencing, articulating, and discussing what modelling is all about).

Only two SEMT presentations explicitly dealt with modelling. Peled's (2015) presentation draws on the results of her studies with pupils from Grade 1 and Grades 5 to 7, in which modelling tasks were implemented. In response to the intervention, the ability to use tables improved in the case of young pupils, while the older pupils acquired more advanced modelling competencies (e.g. readiness to analyse relatively complex problems, the ability to construct a mathematical model, and the employment of self-control and self-evaluation). Peled's conclusion in terms of pupils with different abilities has already been mentioned above. As for modelling as such, Peled concluded:

> In the case of modelling, the development of modelling competencies does not seem to come at the expense of mathematical knowledge. On the contrary, in the case of first graders they developed a mathematical tool, tables. In the case of sixth and seventh graders they strengthened their knowledge of fractions and re-invented percentages. (p. 39)

The only other SEMT presentation on modelling is that by Cabassut (2011). Through the analysis of French textbooks and official documents, he determined that modelling is presented in them as knowledge to be taught. Thus, in theory, French pupils have a chance to be taught modelling processes.

Effects of classroom-related conditions on problem-solving performance

In this broad category of studies, Hembree (1992) includes, among others, whether pupils worked in groups or individually, whether parents provided any help, or whether any special material was used in the study. Two SEMT presentations fell within this category.

Gruszczyk-Kolczynska (2001) noted that the difficulty of a mathematical task is dependent on the conditions in which it is to be solved. For example, when the task is set among children of the same age or at home, it may seem easy, as the child has someone to ask for help. If the same task is to be solved by the child at the blackboard or in a test, its subjective difficulty may increase to such an extent that it surpasses the child's ability to solve it.

One presentation focused on cultural factors: Clanché and Sarrazy (1999) adopted an anthropo-didactic framework. They took into account both the school's didactic dimension (which is grounded in its focus on knowledge acquisition) and its anthropological dimension (the school requires that 'its members practise some language play specific of the school situation, that cannot be taught' (p. 41)). The presentation included an extended extract from a lesson in a school in the bush in New Caledonia, which documented how cultural aspects have to be taken into account even for a WP which included a social situation (namely a game of marbles) familiar to the pupils.

Conclusion

If we limit ourselves to the SEMT proceedings only, we can say that SEMT presenters devoted much attention to the question of preventing pupils from superficial solving strategies. Some suggested solutions on the level of meta-cognition. Consider, for example, the development of meta-cognitive skills in pupils to provide written or oral verbalization when solving a WP, or to make a graphical representation for it. Others proposed formulating WPs complicated by some variable. Consider the suggestion of using WPs that include information in conflict with ordinary experience, or WPs with an unfamiliar context. Quite a lot of attention has been paid by SEMT presenters to the question of teaching the solving of WPs. They suggested, and trialled, some teaching approaches, but the conclusions of these were not fully convincing, as we deal here with symposium presentations, rather than longer journal presentations—but let us say that results were encouraging.

By consulting the results of current studies on WPs published in journals, we were able to confirm the viability of some of the suggestions and conclusions made at SEMT, and we pointed to some areas that have not been given much attention by them (such as WPs as modelling tasks, and the effects of classroom-related conditions, of meta-cognitive skills development, or of problem solvers' characteristics in their ability to solve WPs).

Overall, SEMT presentations did not differ in their scope of attention from the international mathematics education research on WPs. The SEMT presentations mostly contributed to the body of research in this area rather than opening new lines of thought. This is, perhaps, not surprising, as we dealt with presentations in which authors usually presented their preliminary findings which they later published in a different medium with a potentially bigger impact.

Acknowledgements

This chapter was financially supported by the project TL03000469 by the Technology Agency of the Czech Republic Support of the integration of mathematical, reading and language literacy in primary school pupils.

References

Bartolini Bussi, M. G., Canalini, R., and Ferri, F. (2011). Towards cultural analysis of content: Problems with variation in primary school. In J. Novotná and H. Moraová (Eds) SEMT 9–21.
Bartoncová, L. (1999). How pupils complete a problem and create new one from the point of view of their life experience. In M. Hejný and J. Novotná (Eds) SEMT 34–36.
Bednářová, S. (1999). Through the children eye. In M. Hejný and J. Novotná (Eds) SEMT 37–40.

Beijing Education Science Research Institute and Beijing Instruction Research Center for Basic Education. (1996). *Mathematics (shu xue) (in Chinese)* (Vol. 4). [translated by P. Barolini].

Břehovský, J., Eisenmann, P., Ondrušová, J., Přibyl, J., and Novotná, J. (2013). Heuristic strategies in problem solving of 11–12-year-old pupils. In J. Novotná and H. Moraová (Eds.) SEMT 75–82.

Bruna, J., and Havlíčková, R. (2017). Parameters influencing difficulty of word problems: A case of four word problems in Grade 4. In J. Novotná and H. Moraová (Eds) SEMT 100–109.

Bureš, J., Novotná, J., and Tichá, H. (2009). Fostering students' mathematical understanding by developing the culture of word problems. In J. Novotná and H. Moraová (Eds) SEMT 88–96.

Cabassut, R. (2011). The teaching of modelling in France: Relation between mathematical and extra-mathematical topics. In J. Novotná and H. Moraová (Eds) SEMT 79–87.

Capraro, R. M., Capraro, M. M., and Rupley, W. H. (2011). Reading-enhanced word problem solving: A theoretical model. *European Journal of Psychology of Education*, 27(1): 91–114. doi.org/10.1007/s10212-011-0068-3

Cifarelli, V. V., Stephan, M., and Pugalee, D. K. (2015). Examining the problem posing of third graders. In J. Novotná and H. Moraová (Eds) SEMT 91–99.

Clanché, P., and Sarrazy, B. (1999). Contribution to the study of the links existing between real daily experience and daily scholar experience as regards the teaching of mathematics: Example of an 'additional structure' within a first year group in Kanak school. In M. Hejný and J. Novotná (Eds) SEMT 41–49.

Daroczy, G., Wolska, M., Meurers, W. D., and Nuerk, H. C. (2015). Word problems: A review of linguistic and numerical factors contributing to their difficulty. *Frontiers in Psychology*, 6(348): 22–34. doi.org/10.3389/fpsyg.2015.00348

Depaepe, F., De Corte, E., and Verschaffel, L. (2010). Teachers' approaches towards word problem solving: Elaborating or restricting the problem context. *Teaching and Teacher Education*, 26(2): 152–160. doi.org/10.1016/j.tate.2009.03.016

Dweck, C. S. (2017). *Mindset. Changing the Way You Think to Fulfil your Potential.* London: Constable & Robinson Ltd.

Eisenmann, P., Novotná, J., Přibyl, J., and Břehovský, J. (2015). The development of a culture of problem solving with secondary students through heuristic strategies. *Mathematics Education Research Journal*, 27(4): 535–562. doi.org/10.1007/s13394-015-0150-2

Goldin, A. G. (2014). Mathematical representations. In S. Lerman (Ed.) *Encyclopedia of Mathematics Education*, 409–413. Dordrecht, The Netherlands: Springer Reference.

Gruszczyk-Kolczynska, E. (2001). Why the solving of mathematical problems is so difficult for children? Intellectual and emotional conditions of mathematics education. In J. Novotná and M. Hejný (Eds) SEMT 14–25.

Hannula, M. S. (2014). Affect in mathematics education. In S. Lerman (Ed.) *Encyclopedia of Mathematics Education*, 23–27. Dordrecht, The Netherlands: Springer Reference.

Hembree, R. (1992). Experiments and relational studies in problem solving: A meta-analysis. *Journal for Research in Mathematics Education*, 23(3): 242–273. doi.org/10.2307/749120

Hershkovitz, S. (2009). Intuition, schema, and problem solving. In J. Novotná and H. Moraová (Eds) SEMT 31–41.

Kaslová, M. (1999). Pupils' creation of verbal problems: Method of penetrating into the world of mathematics. In M. Hejný and J. Novotná (Eds) SEMT 146–151.

Lemut, E., and Dettori, G. (1995). Representations in problem solving activities. In M. Hejný and J. Novotná (Eds) SEMT 90–93.

Leppäaho, H., and Ahtee, M. (2005). Teaching mathematical problem solving with the solving map method at primary school. In J. Novotná (Ed.) SEMT 194–202.

Lewis, A. B., and Mayer, R. E. (1987). Students' miscomprehension of relational statements in arithmetic word problems. *Journal of Educational Psychology*, 79(4): 363–371.

Moddleton, J. A. (2014). Motivation in mathematics learning. In S. Lerman (Ed.) *Encyclopedia of Mathematics Education*, 460–463. Dordrecht, The Netherlands: Springer Reference.

Morgan, C. (1998). *Writing Mathematically: The Discourse of Investigation.* London: Falmer Press.

Muis, K. R., Psaradellis, C., Lajoie, S. P., Di Leo, I., and Chevrier, M. (2015). The role of epistemic emotions in mathematics problem solving. *Contemporary Educational Psychology*, 42: 172–185. doi.org/10.1016/j.cedpsych.2015.06.003

Navarra, G. (1995). The interpretative difficulties of texts of standard problems. Some aspects of an experimental research with children 10 years old. In M. Hejný and J. Novotná (Eds) SEMT 106–109.

Navarra, G. (1997). On some aspects of the assessment of pupils involved in the solution of non standard meta-cognitive word problems. In M. Hejný and J. Novotná (Eds) SEMT 116–119.

Nesher, P., and Teubal, E. (1975). Verbal cues as an interfering factor in verbal problem solving. *Educational Studies in Mathematics*, 6(1): 11–51.

Novotná, J. (1997). Using geometrical models and interviews as diagnostic tools to determine students' misunderstandings in mathematics. In M. Hejný and J. Novotná (Eds) SEMT 61–67

Novotná, J. (1999). Do students of the 3rd to 6th grades use the everyday life schemes when solving word problems? In M. Hejný and J. Novotná (Eds) SEMT 160–164.

Novotná, J., and Sarrazy, B. (2009). Mathematical and psychological modelling in mathematics education. In J. Novotná and H. Moraová (Eds) SEMT 190–197.

Ott, B. (2020). Learner-generated graphic representations for word problems: An intervention and evaluation study in Grade 3. *Educational Studies in Mathematics*, 105(1): 91–113. doi.org/10.1007/s10649-020-09978-9

Páchová, A., and Vondrová, N. (2020). The effect of semantic cues on the difficulty of word problems and the interplay with other complicating variables. *Research in Mathematics Education*, 23(1): 85–102.

Peled, I. (2015). Modelling tasks: What develops? In J. Novotná and H. Moraová (Eds) SEMT 29–40.

Polotskaia, E., Savard, A., Calvacante, A., and Fellus, O. (2019). Middle school students' difficulties in problem solving and their roots in elementary education: What went wrong for them? In J. Novotná and H. Moraová (Eds) SEMT 315–324.

Pongsakdi, N., Kajamies, A., Veermans, K., Lertola, K., Vauras, M., and Lehtinen, E. (2020). What makes mathematical word problem solving challenging? Exploring the roles of word problem characteristics, text comprehension, and arithmetic skills. *ZDM—Mathematics Education*, 52(1): 33–44. doi.org/10.1007/s11858-019-01118-9

Reusser, K. (1990). Understanding word arithmetic problem. Linguistic and situational factors. Paper presented at the Annual Meeting of the American Educational Research Association. Boston, MA, 16–20 April 1990.

Ruppeldtová, J. (2007). Types of word problem visualisation in pupils' solutions. In J. Novotná and H. Moraová (Eds) SEMT 233–241.

Schoenfeld, A. H. (1994). Reflections on doing and teaching mathematics. In A. Schoenfeld (Ed.) *Mathematical Thinking and Problem Solving*, 53–70. Hillsdale, NJ: Erlbaum.

Semadeni, Z. (1995). Developing children's understanding of verbal arithmetical problems. In M. Hejný and J. Novotná (Eds) SEMT 27–32.

Silver, E. A. (1994). On mathematical problem posing. *For the Learning of Mathematics*, 14(1): 19–28.

Silvestre, A. I., and da Ponte, J. P. (2015). Missing value problems: Making sense of proportional strategies during an exploratory teaching unit. In J. Novotná and H. Moraová (Eds) SEMT 292–301.

Singer, F. M., Ellerton, N., and Cai, J. F. (2013). Problem-posing research in mathematics education: New questions and directions. *Educational Studies in Mathematics*, 83(1): 1–7. doi.org/10.1007/s10649-013-9478-2

Soresi, S., and Zucco, G. (1995). Effects of text characteristics on mathematical problem solving in Italian children. In M. Hejný and J. Novotná (Eds) SEMT 162–165.

Sweller, J. (2010). Element interactivity and intrinsic, extraneous, and germane cognitive load. *Educational Psychology Review*, 22(2): 123–138. doi.org/10.1007/s10648-010-9128-5

Szetela, W. (1991). A study of verbalization in problem solving in Grades 4 and 5. In J. Novotná (Ed.) SEMT 58–61.

Verschaffel, L., Schukajlow, S., Star, J., and Van Dooren, W. (2020). Word problems in mathematics education: A survey. *ZDM—Mathematics Education*, 52: 1–16. doi.org/10.1007/s11858-020-01130-4

Verschaffel, L., Greer, B., and De Corte, E. (2000). *Making Sense of Word Problems*. Lisse, The Netherlands: Swets & Zeitlinger Publishers.

Verschaffel, L., Van Dooren, W., Greer, B., and Mukhopadhyay, S. (2010). Reconceptualising word problems as exercises in mathematical modelling. *Journal für Mathematik-Didaktik*, 31: 9–29. doi.org/10.1007/s13138-010-0007-x

Vondrová, N. (2020). The effect of an irrelevant number and language consistency in a word problem on pupils' achievement and reasoning. *International Journal of Mathematical Education in Science and Technology*, advance online placement. doi.org/10.1080/0020739x.2020.1782497

Vondrová, N., Havlíčková, R., Hirschová, M., Chvál, M., Novotná, J., Páchová, A., Smetáčková, I., Šmejkalová, M., and Tůmová, V. (2019). Matematická slovní úloha: mezi matematikou, jazykem a psychologií. [*Mathematical Word Problems: Between Mathematics, Language and Psychology.*] Prague: Karolinum.

Vondrová, N., and Novotná, J. (2017). The influence of context and order of numerical data on the difficulty of word problems for grade 6 pupils. In J. Novotná and H. Moraová (Eds) SEMT 440–449.

Voyer, D. (2011). Performance in mathematical problem solving as a function of comprehension and arithmetic skills. *International Journal of Science and Mathematics Education*, 9(5): 1073–1092. doi.org/10.1007/s10763-010-9239-y

SECTION 3
ENHANCING THE TEACHING OF MATHEMATICS

Brian Doig
Deakin University, Australia

This section focuses on different approaches to teaching mathematics. Such approaches are clearly many and varied, and the authors of the chapters in this section span a range of these contexts. Interestingly, over the 30 years of SEMT, a large number of presentations have raised various issues with regard to teaching approaches.

This aspect of SEMT is exhaustingly examined in the first chapter, where the authors have traced presentations across the entire 30 years of SEMT. A retrospective examination of SEMT proceedings was undertaken, and the presentations coded by their focus on teaching. Further, presentations were then sub-divided into categories such as What do teachers think and believe about teaching? and How do teachers teach? Tables of these analyses help the reader to see how trends have developed over time.

Thus, having painted a remarkably informative picture of trends over 30 years, the authors address their own journeys through this field of research. While this is beyond the purview of SEMT, it places the SEMT presentations in an international perspective.

The second chapter is a detailed examination of Lesson Study and variation theory, with which the authors have experimented and researched for many years. While learning from Confucian Heritage Cultures (CHC), they explain how it is possible to make cultural transpositions of pedagogy to other cultures, such as that in Italy. It may be thought that, as Japanese Lesson Study has been adapted for use in many Western nations, this is a trivial item for discussion. However, Italian primary teachers have a singularly different pedagogy to most other Western teachers, which creates a need for a more extreme cultural transposition. The authors provide sufficient details and explanations of their experiences to enable the reader to both understand the difficulties in cultural transposition and note ways in which this transposition can be achieved.

In the third chapter of this section, a detailed definition and explanation of what, and how, visualization can improve teaching, and thus learning, is provided. The subjects of the research were pre-service teachers, and the chapter focuses on examining their competencies in a range of visualization strategies that have been examined in previous studies. The author refers to 32 SEMT presentations, addressing visualization, and draws conclusions on how visualization could help teachers to enrich their range of teaching strategies. These multiple ways of using visualization, the author argues, provide a usefulness that is undisputed; the professional development of teachers' competencies in this area of expertise would enable them to use visualization in various aspects of the teaching process to complement their other teaching strategies.

The final chapter in this section, on developing pre-service teachers' competencies in teaching mathematics, is related to the education of pre-service primary teachers. Due to the inertia of the school system, the authors focus on pre-service teacher education and argue that it is essential that new trends are part of their training. Six groups of papers have been selected from the SEMT proceedings for review and commentary. These articles all deal with issues that are pertinent to all those who are responsible for guiding pre-service education into their professional life.

3.1
Elementary mathematics teaching: Relevance, conceptualizations, and research

Jana Višňovská
The University of Queensland, Australia
José Luis Cortina
Universidad Pedagógica Nacional, México
Pamela Vale
Rhodes University, South Africa

Introduction

Teachers and teaching have been recognized as central to shaping learning in classroom situations, and to reforming efforts in educational change (Biesta et al. 2020; Llinares and Krainer 2006; McLaughlin 1987; Spillane 2000). Yet teaching has not always been central to research in mathematics education, and even today, the ways in which teaching is being considered, or included in different lines of mathematics education research and thought, vary broadly.

The research journeys of the authors of this chapter are grounded in the interventionist design research methodology, and in a commitment to the 'research and development' ethos. When we design, trial, modify, and study the means of supporting students' mathematical learning in classrooms, teaching is our research tool. Through teaching the mathematical activities in which students engage, we proactively shape the purposes for which they engage, the kinds of questions they come to ask, the kinds of tools they come to use and reason with, and the forms of reasoning that become salient through classroom discussions. Doing so allows us to elicit and study the kinds of *learning* that are possible, but rarely occur, in classrooms in general. In other words, we study the kinds of learning that emerge in response to highly innovative and well-resourced teaching. To understand the learning, then, requires that we understand the intricacies and complexities of the teaching that made it happen. As a result of those research experiences, we are accustomed to always view learning in relation to the teaching conditions and circumstances within which it took place.

Of course, attention to teaching is present within most theoretical and methodological endeavours that involve classrooms or learners. The differences appear in what counts as teaching and how centrally it is positioned within what is studied. For example, researchers conducting constructivist teaching experiments, with individuals or small groups of learners, study mathematical *development* that is considered to be *natural*, if the conditions for it are adequate (Steffe and Thompson 2000). Here,

Jana Višňovská, José Luis Cortina and Pamela Vale, *Elementary mathematics teaching: Relevance, conceptualizations, and research*. In: *Elementary Mathematics Teaching*. Edited by: Brian Doig, Jarmila Novotná, Berinderjeet Kaur and David K. Pugalee, Oxford University Press. © Oxford University Press (2024). DOI: 10.1093/oso/9780192869647.003.0009 the *teaching* that is included in the explanations of such development typically entails assessing students' current understandings and presenting tasks to create cognitive conflict. A particular instructional context is established, in which students are asked to do and express what makes sense to them, instead of trying to determine what answer might be acceptable to the teacher.

When insights from this research on cognitive development are used as a basis for textbooks and pedagogical approaches, they tend to carry the same view of how students need to be supported. This view consists primarily of creating an instructional environment in which students' actions and communication are driven by what makes sense to them, and where students' current understandings need to be assessed frequently, so that appropriate tasks can be introduced to support them. Such a view of teaching imposes a heavy burden on teachers, making successful adoption of the imagined teaching difficult. Teachers need to learn the framework within which to assess students' mathematical understandings in a way similar to that of the researchers, and are expected to keep track and respond to the individual development process of every student in their classroom. In some ways, it is imagined that teachers will improve their teaching once they stop being teachers and instead become researchers of their students' individual development. An alternative path, we suggest, involves viewing and supporting teachers as thoughtful professionals in the practice of teaching (Biesta et al. 2020; Sfard and Cobb 2014; Wenger 1998).

It is beyond question that multiple lines of inquiry—with and without the inclusion of teaching—have led to enriching and intriguing findings about the nature of mathematical learning. There is considerable value in this diversity of inquiry; however, when it comes to the 'development' part of 'research and development' projects, where the aim is to engender a real change in teaching and learning contexts, a tendency to focus research attention solely on learning comes at a cost. The mathematics education field once assumed that, if undeniable student learning success is demonstrated by researchers, the teaching programmes responsible for this learning will be widely and unproblematically adopted in schools (Artigue 2008). However, the absence of insights into teaching practices and decision rationales within which the learning was previously generated puts the educators, who attempt to reproduce such learning in other classrooms, at a significant disadvantage. The awareness that an innovation 'worked' in some settings, where it resulted in a specific set of learning outcomes, does not provide support for making the reasoned adaptations needed if it is to also result in similar learning in a new classroom setting (Cobb and Jackson 2008).

Writing this chapter presented us with an opportunity to take a retrospective look at how the researchers who contributed to SEMT included, conceptualized, and researched mathematics teaching over the years. We first introduce our methodology and proceed to summarize the trends we identified *across* the analysed contributions, and also *within* each of the coded categories. We pay particular attention to studies we categorized as *Studies of Teaching Practice*, where teaching itself becomes the focus of research. Against this analysis, we reflect on our own two SEMT research presentations, both categorized as Studies of Teaching Practice, and discuss how our outlook on research on teaching has shifted between 2003 and 2017. Given the complexities involved in understanding and sharing practices, we celebrate, and make a case for, maintaining SEMT workshops as a unique conference format, allowing sustained time and attention to encounter and explore each other's research approaches, and to foster the SEMT community.

Methodology

We analysed the SEMT proceedings, using as our main criteria the presence of teaching in the published presentations, including plenary addresses, workshops, and all other presentations.[1] We sought to develop a coding scheme that could account not only for the presence or absence of teaching in the presentations, but also for the role that teaching played. Using the proceedings of a single conference (Novotná 2005), we first identified presentations that included some teaching in them. If the presence or absence of teaching was not apparent from the presentation's title and abstract, we read parts of the presentation, usually the introduction and the methodology section. In some cases, we read the whole presentation as contained in the proceedings.

In following our goals for this chapter, we noted some differences in the roles that teaching played in the proceedings, and in how it was framed. We then formulated tentative coding categories and tested them. Two of the authors independently coded the 2005 proceedings. We then compared our results and found we had a 60% agreement in the coding. We discussed the discrepancies, further specified the categories, and agreed that a new category needed to be included. Next, the same two authors independently coded the Novotná and Moraová (2007) proceedings with agreement of over 90%. Upon reconciling the remaining differences, the coding scheme was used by the second author to consistently code all 15 proceedings. We describe the resulting coding categories and how we saw them as extending the presence of mathematics teaching in the research endeavours.

We first made a distinction between presentations that did not include teaching in any way, and those that did:

No Teaching — The presentations do not include teaching. They might be solely concerned with student mathematical learning or report the results of clinical interviews or paper-and-pencil assessments.

The presentations judged as including teaching were further differentiated. The most rudimentary code was assigned to presentations that did not study teaching but acknowledged its importance.

Teaching Implied — The presentations do not focus on the study of teaching. Teaching is acknowledged, or implied, as having played a part in the researched phenomena. Tests of instructional innovations, in which viability of the innovation is assessed by comparing the outcomes of a 'treatment' and a 'control' group of students, would be coded here.

In the next two categories, the research focus itself shifts beyond learning. Teaching is recognized as central to the learning endeavour. Perhaps more importantly, teachers

[1] Publications in languages other than English in the initial proceedings were not included in our analysis.

are recognized as intellectual actors who shape what happens in the classrooms, and so research focuses on how their knowledge, beliefs, or points of view influence their teaching.

Aspects That Influence Teaching	The presentations study aspects and traits that the authors consider to be influential for mathematics teaching, such as teachers' mathematical knowledge for teaching.
What Teachers Think–Believe	The presentations document teachers' views of their teaching, or issues directly relevant in their practice, such as a new national curriculum. The studies typically involve surveys or interviews.

In the following two categories, teaching as a social practice is foregrounded. Presentations aim to provide a window into teaching practices, or sometimes do so inadvertently. They do not provide a conceptualization for a theoretical account of those practices.

How Teachers Teach	The presentations are based on observations of teachers in their classrooms. They provide a descriptive account of the teaching practice.
Reflections or Accounts of Own Teaching	The presentations are written by teachers themselves to communicate discoveries or insights obtained in their teaching practice.

Finally, the last category captures presentations in which the practices of teaching and planning for classroom instruction are themselves the object of conceptualizing and theorizing. The function and/or importance of some elements of the practice that are studied are justified with respect to a bigger, more complex endeavour of supporting the learning of students in the environment of a mathematics classroom.

Studies of Teaching Practice	The presentations report studies of teaching practice. Observations of, and artefacts from, planning or teaching are analysed and form a basis for theorizing.

SEMT findings

The analysis of proportions of contributions to SEMT symposia that in some manner included teaching, and those that did not, suggests that, over the past 30 years, inclusion of teaching has gradually become an expectation in elementary mathematics education research (see Figure 3.1.1). While very few SEMT presentations could be characterized as including any teaching in the first decade of SEMT, in the last decade the majority of the presentations included teaching as relevant to the reported study and some researched teaching practices specifically.

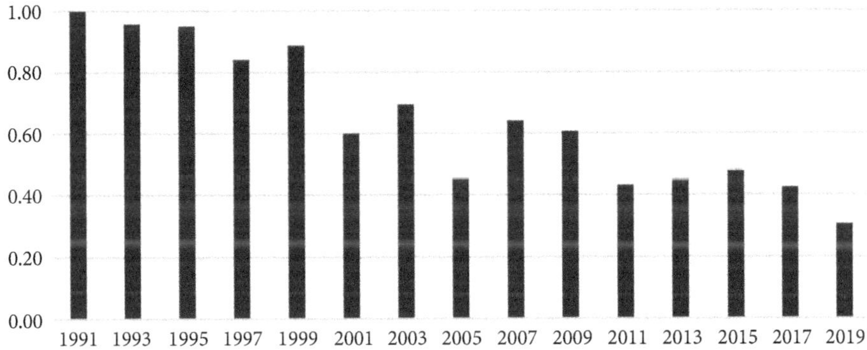

Figure 3.1.1 Proportion of SEMT contributions coded as 'No Teaching' across the proceedings

In Figure 3.1.1, we chose to represent the proportion of presentations that *did not include teaching*, in order to highlight that through the 1990s, elementary mathematics education researchers routinely omitted mathematics teaching from reports on mathematics learning. This is especially surprising, given that our data come from research symposia where the title delineated the focus as elementary mathematics *Teaching*. It seems very likely to us that the meaning of 'mathematics teaching' was, in the 1990s, synonymous with mathematics education broadly, and—contrary to our and others' contemporary interpretations—did not position mathematics *teaching* as the focus for research interest.

A decrease in the proportion of presentations that did not include teaching became more prominent in the early 2000s. The largest boost to this trend came from the *Teaching Implied* contributions, a category with the most coded presentations across the 2000s symposia (see Table 3.1.1). The only exception to this pattern was SEMT 2011, where the symposium theme foregrounded the construct of mathematical knowledge for teaching, thus boosting the number of presentations coded under *Aspects That Influence Teaching.*

The exploratory analysis of the proportions of the *No Teaching* presentations gave us a reason to examine contribution patterns in relation to the three decades of SEMT (1991–2000, 2001–2010, 2011–2020), revealing a trend of increase in the numbers of contributions across the decades in all coded categories. A singular exception was the *How Teachers Teach* category, where the most recent decade included fewer presentations (12) than the 2000s (13). While the number of total symposium contributions also increased across the decades, the absolute numbers of the *No Teaching* presentations (not just their proportions) showed a decreasing trend.

Teaching Implied

The number of contributions coded as *Teaching Implied* undeniably grew through the three SEMT decades (see Table 3.1.2). Within all the presentations coded as *Teaching*, this was the first category with a double-digit number of contributions, and the one

Table 3.1.1 Frequencies of SEMT presentations by coding category by symposium

Coding Category	1991	1993	1995	1997	1999	2001	2003	2005	2007	2009	2011	2013	2015	2017	2019	Total
Teaching Implied	0	1	2	0	1	6	10	16	10	8	7	17	15	14	24	131
Aspects That Influence Teaching	0	1	1	1	0	6	2	3	3	4	15	3	3	7	14	63
What Teachers Think–Believe	0	0	0	0	1	1	0	1	0	2	1	1	1	5	1	14
How Teachers Teach	0	0	0	3	3	2	1	6	1	3	4	2	1	3	2	31
Reflections/Accounts of Own Teaching	0	0	0	0	1	1	0	2	0	0	3	3	1	0	0	11
Studies of Teaching Practices	0	0	0	1	0	0	2	3	2	1	4	3	3	2	4	25
Total: TEACHING	0	2	3	5	6	16	15	31	16	18	34	29	24	31	45	275
Total: NO TEACHING	22	44	57	27	48	24	34	26	29	28	26	24	22	23	20	454

Table 3.1.2 Frequencies of SEMT presentations by coding category by decade

Coding Category	Decade 1 (1991–2000)	Decade 2 (2001–2010)	Decade 3 (2011–2020)	Total
Teaching Implied	4	50	77	131
Aspects That Influence Teaching	3	18	42	63
What Teachers Think–Believe	1	4	9	14
How Teachers Teach	6	13	12	31
Reflections/Accounts of Own Teaching	1	3	7	11
Studies of Teaching Practices	1	8	16	25
Total: TEACHING	16	96	163	275
Total: NO TEACHING	198	141	115	454
TOTAL (contributions coded)	214	237	278	729

that continued to include most of the *Teaching* contributions across most years (see Table 3.1.1). We argue that this was the case, at least partially, because of the changes in writing expectations for the presentations that would previously have fallen within the *No Teaching* category.

We illustrate this by comparing an example of a *Teaching Implied* presentation from the second SEMT decade with a similar presentation published in the 1990s, and coded as *No Teaching.* The presentation entitled *Mental organs' work in the number expression solving processes (One approach to teaching number expressions)*, by Skalická (1997), is an example of a *No Teaching* contribution. Even though the title suggested that the presentation would address a specific approach to teaching number expressions, the approach was not addressed or clarified in the presentation. Instead, the presentation provided a detailed analysis of children's written calculations accompanied by a theorized account of their thinking, positing the conjectured cognitive structures that generated it. A focus on student learning, and use of a psychological perspective, were some of the shared characteristics of many of the 1990s *No Teaching* presentations. Not only did these presentations not discuss the teaching involved, but the authors did not always recognize that this was the case, and used the word 'teaching' rather freely.

Littler and Benson's (2005) presentation entitled *Patterns leading to generalisations*, coded as *Teaching Implied*, was similar to Skalická (1997) in several ways. It focused on students' responses to number pattern tasks. A description of several tasks was followed by an analysis and discussion of the performance of children on these tasks. However, unlike the *No Teaching* presentations, Littler and Benson included a long section titled *The Classroom Effect* (p. 204), in which they outlined some of the key teaching strategies that needed to be in place so that similar student solutions would be expected to emerge in a classroom. While the authors did not study the teaching—indeed their focus was on types of tasks and on students—they clearly recognized that a specific form of teaching was implicated in the student responses analysed.

More recent presentations in this category included quasi-experimental designs with a control group, and other forms of evaluation of intervention projects. For example, Marchini and Vighi (2009) explored, using a quasi-experimental design, whether introducing isometries in elementary grades can support children's geometric understanding. The 'treatment' (p. 172) that the authors describe in detail is a teaching strategy, and the authors aim to establish its comparative effectiveness in early grades. More recently, Prusak and Hershkowitz (2019) reported on an analysis of one classroom activity used as a part of a classroom design experiment consisting of 28 classroom sessions. Children's performance and engagement were analysed both quantitatively and qualitatively. The principles followed in the design of the task were made explicit, and the teaching approach that aimed at promoting problem solving and conceptual learning was alluded to throughout the presentation.

The *No Teaching* presentations did not disappear. They remained relatively frequent among poster presentations and as specific topical analyses. In 2019, such presentations reported on analyses of tools and resources (e.g. digital, games, texts; 8), products of student work (e.g. test outcomes, mistakes, solutions, homework, or individual interviews; 9), or were related to teacher education (3). Only two (out of 21) plenary addresses in the last decade did not include teaching, one in 2011 (Wittmann, *Early mathematical education: A plea for mathematically founded conceptions*) and one in 2013 (Krauthausen, *Digital tools in elementary mathematics education*). Overall, it appears that the view of the relevance of teaching to research endeavours of the SEMT community became established, and the inclusion of teaching became a norm across a variety of research methodologies and foci.

Aspects That Influence Teaching

A total of 63 presentations fell in this category, the second most frequent count among the presentations that included teaching. Decade by decade, the number of these SEMT contributions more than doubled (see Table 3.1.2). While authors did not study teaching as a practice, teaching was regarded as important, under the assumption that certain teacher traits and insights significantly influenced the nature, quality, and results of teaching. Many of the presentations referred to what Ball and Bass (2003) came to call *mathematical knowledge for teaching* (MKT). For instance, the unusually high count of these presentations in 2011 (15) corresponds to the conference theme that focused on MKT.[2] Even before the term was coined, the topic was present in mathematics education debates, including at SEMT. For instance, Tall's (2001) plenary addressed the mathematics that teachers needed to know in order to 'participate actively in the long-term mathematical development of their children' (p. 26).

[2] The reason for a similar spike in 2019, with 14 presentations, is less clear to us.

What Teachers Think and Believe

The presentations in this category acknowledged the importance of teaching, but did not study it directly. The focus was instead on teachers' beliefs and points of view, which were related to, or assumed to affect, their teaching. These presentations were less frequent, with one plenary address and 13 other presentations. Again, a pattern of considerable increase in these contributions is observable across the three decades (see Table 3.1.2). Illustrative of this category, Thanheiser and Philipp (2017) presented a study in which they surveyed 23 prospective elementary teachers. They documented and analysed the prospective teachers' reactions to a vignette, in which a classroom teacher dismissed a student's solution where the student used a culturally based non-standard algorithm. In the presentation, the authors focused on the prospective teachers' lack of acknowledgement of the source of the algorithm, and argued why teachers of mathematics need to learn to focus on, and then proactively include, the students' culturally based mathematical practices in the classroom activities and discussions.

How Teachers Teach

The 31 presentations that were coded in this category cut across all SEMT presentation formats. Interestingly, this was the only teaching category where the most presentations appeared during the second decade (13), not in the most recent one (12). The authors of these presentations typically approached teaching in an exploratory and descriptive fashion. The observations that were carried out in mathematics classrooms were aimed at identifying what made a difference to students' learning. The presentation by Cheeseman (2009) is a representative example. This author reported findings from a study that focused on the practice of four teachers of mathematics. The teachers had been identified as highly effective in working with young children. Classroom observations, video recordings of their lessons, and subsequent teacher interviews were used to identify the behaviours characteristic of, and responsible for, the teachers' effectiveness.

Reflections and Accounts of Own Teaching

While this category comprised only 11 presentations (see Tables 3.1.1 and 3.1.2), these were noticeably different from presentations coded elsewhere. The authors, who acted as classroom teachers, offered descriptions, accounts, or reflections of what happened while they were teaching, thus going beyond descriptive qualities of presentations in the previous category and uncovering some of the reasoning involved in a teacher's decision making. For example, Kloboučková (2011) analysed an intervention she conducted in her fourth-grade classroom, which focused on teaching geometry through cube building. Although the author's teaching practice was not the main focus of the presentation, much of it became accessible to the reader through the explanations of her planning, her selection of geometry tasks, and the changes she made to the intervention over time. Transcripts of the teacher's interactions with students were also included in the presentation, providing the reader with additional images of the teaching involved.

Across the first two decades of SEMT, these presentations provided opportunities for engagement in mathematics education debate for local (Czech and Slovak) mathematics teachers and doctoral candidates,[3] often during their initial engagement with the research in the field. In the last decade, the pool of authors in this category broadened with respect to their background (Poland, Italy, USA), and included established Czech and international researchers (see, e.g., Pytlak 2013; Stephan and McManus 2013). We suggest that this might indicate an increase in value given—from the research perspective—to the insights into teaching practices that reflective accounts of teaching make available.

Studies of Teaching Practices

As Tables 3.1.1 and 3.1.2 show, we coded a total of 25 presentations in this category. The great majority of these (16) were reported in the third decade. We propose that this trend reflects the advances in conceptualizing and operationalizing the complexity of classroom events, and the related increased attention to the study of 'lived' teaching practices in mathematics classrooms. Two of the 25 presentations were plenary addresses (Kaur 2019; Sarrazy and Novotná 2005), and 22 other presentations, including two contributions by the first two authors of this chapter. Four of the presentations (e.g. Moraová and Novotná 2017) focused on student teachers, and five (e.g. Shabtay 2019) reported on a broader study that involved a teacher professional development (PD) component.

The conceptualizations of teaching practices and how these were studied varied. Some of the presentations were framed through Brousseau's *Theory of Didactical Situations* (Brousseau 1997). For instance, Sarrazy and Novotná's (2005) plenary introduced the notion of *didactical contract* as a specific 'theoretical frame for the analysis of phenomena of teaching mathematics' (p. 33). Some presentations related to an emergent perspective, or theorizing (Cobb and Yackel 1996). For example, Symeonidis, Tatsis, and Kaldrymidou (2015) explored the notions of social and socio-mathematical norms that were established in one fifth-grade classroom, where students were expected to solve complex mathematical tasks collaboratively. Both the notions of didactical contract and the norms and classroom practices arising from these are considered valuable beyond their analytical potential for researchers' understanding of classroom learning. They also envision, as part of a workable teaching perspective, how one can shape the learning in the mathematics classroom. The extent to which such a workable teaching perspective could become a reality, and under what conditions of support, frames questions we find worth pursuing.

[3] Including the first author, Višňovská 1999), as she first engaged with mathematics education as a research field.

Additional presentations in this category could be characterized as focusing on specific routines within a teaching practice that was itself not overtly theorized but where authors claimed intellectual affiliation with a specific perspective on learning, or otherwise justified the routine as pedagogically sound. These included 'task selection' (Chopin and Roiné 2013), 'use of prediction questions' (Kasmer and Billings 2011), 'building on students' understandings' (Askew et al. 1999) 'elicitation patterns in classroom talk' (Back 2005), and 'formulating lesson plans' (Moraová and Novotná 2017).

The teaching story across the three decades of SEMT

Our analysis of SEMT presentations illustrates how the importance of understanding teaching increased in the research on mathematics education over the past three decades. We argue that this mirrors an emerging recognition, in the mathematics education research community, that understanding how mathematics teaching 'works', how it changes, and how it needs to be supported; are all essential if the advances in research on mathematical learning are to become workable in classrooms.

With SEMT commencing shortly after the publication of the 1989 National Council of Teachers of Mathematics Standards in the USA, the three decades were shaped by calls for reform and innovation in mathematics education, and by honest efforts to move away from learning mathematics by rote towards learning mathematics with conceptual understanding—*on a large scale*. However, even within those efforts, teaching was initially treated as being relatively simple by the researchers. The teachers were asked to *implement* the products of research in their classrooms, as if teaching itself was captured in these products and was, at least partially, external to, or independent from, the teacher. At times, teachers were asked to create specific conditions in their classrooms (e.g. include group work, create a 'culture of inclusion', or adopt a specific belief or perspective) for the innovations to 'work'. In some sense, rather than being offered resources that would work for, and with, the teacher, the teachers were being asked to work for the resources, often under conditions of limited or no support.

The lack of impact that such approaches tend to have in classrooms contributed to the rise of conceptualizations of teaching that acknowledge its complexity. Some researchers turned to investigate the paradigmatic cases of teachers and researchers who happened to teach mathematics productively, attempting to understand what these teachers had, or knew (traits and knowledge necessary to teach well), that other teachers typically did not have, or did not know (i.e. Aspects That Influence Teaching). Others started to look closely at the classrooms to understand what teachers were actually doing—and how successful they were at promoting mathematical learning (i.e. How Teachers Teach, Reflections on Own Teaching). While no overt research agreement yet exists on what productive teaching practices are or what they need to be, researching teaching as a complex practice became increasingly common (i.e. Studies of Teaching Practices).

Our teaching story

The interventionist path that we came to follow as design researchers also led us to increasingly turn our attention to teaching. In the early 2000s the first and second authors undertook research apprenticeships as doctoral students in the research team of Cobb and McClain, at Vanderbilt University. At the time, the active project explorations of the team made the transition from classroom design experiments, including the second author's dissertation study on ratios as units of measure (Cortina 2006), to design experiments aimed at supporting teacher PD, including Višňovská's dissertation on PD in middle school statistics (Višňovská 2009).

Classroom design experiments as windows into ambitious and equitable teaching

Our first joint SEMT presentation (Višňovská and Cortina 2003) was in a somewhat typical short format, in which we used data from a classroom design experiment to bring attention to, and explain, a specific theorized issue. In this case, we aimed to illustrate and explain how one feature of a complex teaching practice (i.e. a way in which instruction was organized) enabled the researchers who acted as teachers in the classroom to systematically advance students' mathematical reasoning. We illustrated how diversity of students' reasoning can be viewed and used by the teacher as a resource in whole-class discussions.

In hindsight, other issues were at play in the 2003 presentation, which we did not stress overtly—possibly because we did not fully appreciate their importance to establishing and maintaining the teaching practice we found productive. For instance, the classroom discussion excerpts we provided in the presentation highlighted the key role of the teacher's focus on *quantitative* explanations. In other words, an expectation was formed in the classroom that students would reason with numbers as quantities, rather than using numbers as merely things to be manipulated in calculations (Thompson et al. 1994). The difference in the nature of the mathematical discourse in which students were expected to engage *made the difference*: when talking about calculations with 'bare numbers' (e.g. the rule of three), students contributed to classroom discussion rather homogeneously and were all highly uncertain about both the correctness of, and justification for, their contributions. In contrast, once students accepted that they were expected to talk about the quantities that the numbers in the tasks represented (e.g. that we can expect to spend half the fuel when travelling half the distance) the new focus readily allowed them to generate various ideas to contribute, and provided sense-making tools in following the ideas of others.

Undeniably, we aimed to bring attention to teaching that helped us to make the students' learning happen. However, the omission of the discussion of the importance of quantity in our presentation illustrates that we were not yet orientated towards generating and sharing the means by which such teaching practice could be gradually and thoughtfully reinvented by others. In a sense, we were still *connecting the dots* as we conceptualized our own teaching, and, as we made the transition from researching the learning that was possible in classrooms, to researching the teaching that was possible, and under what circumstances such teaching could take shape.

Towards designing resources that work for and with teachers

A lot had changed in our research in the lead-up to our second SEMT contribution. Aiming to generate opportunities for quantitative conceptual discourse in classrooms, we developed and tested in under-resourced classrooms an instructional sequence for teaching fractions that is prominently based on length measurement (Cortina, Višňovská, and Zúñiga 2014; Vale and Graven 2018). We have argued that instructional design theories should guide us in designing resources for classroom teaching and teacher learning, instead of aiming to support students' learning directly (Cobb, Zhao, and Višňovská 2008; Višňovská and Cortina 2018).

Our second SEMT presentation, *Learning to support all students' fraction understanding* (Višňovská and Cortina 2017), is a result of explorations into designing for teaching and teachers. In it, we discussed how one Mexican teacher 'came to base her instructional decisions in students' reasoning broadly, and in the reasoning of students for whom the ideas were proving the most challenging, in particular' (pp. 430–431). By this point, we had seen multiple teachers who, while they collaborated with us over the long term (typically over a year or longer), came to use our designed resources in ways that resulted in changes in both their teaching and their students' learning. Our research interests refocused on the process and the means of supporting teachers to make the *transition* towards increasingly ambitious and equitable mathematics teaching. In conducting what could be called teacher PD design experiments (Cobb et al. 2003; Cobb, Zhao, and Dean 2009), our role was to both design and research the circumstances and means of support under which powerful teacher learning and teaching would take place. Our research questions included: what helps the teacher include the new resource as part of their practice? What does not help and needs to be further re-designed with the teacher in mind?

Elsewhere, we highlighted that mathematical coherence of curriculum resources, which cannot yet be taken for granted, is of high importance in making teaching mathematics in classrooms manageable (Višňovská and Cortina 2003). In the 2017 SEMT presentation, we discussed several important ways in which the instructional sequence, and the teacher's initial experiences with it in her classroom, facilitated the teacher's learning and teaching. Specifically, the early experiences with the sequence suggested to the teacher that it was possible to hold whole-class mathematical conversations *within her classroom*. In other words, it allowed the teacher to experience how all her students were capable of actively listening and contributing ideas. Second, the teacher's classroom experiences allowed us to have meaningful conversations with her about two contrasting ways in which she could proceed in using the sequence: to faithfully enact each instructional activity, on the one hand, or to use the sequence rationale in designing various targeted activities as she progressively supported her students' use of specific forms of mathematical reasoning. Finally, the sequence allowed us to co-design, with the teacher, specific activities that she then used to proactively support how her students reasoned in the classroom. As a result, the teacher then proceeded to making independent instructional decisions about subsequent activities for her classroom, which were informed by the sequence rationale.

Rather than trying to explain or defend how we conceptualize teaching at this point, we offer the summary above as an illustration of the set of concerns and considerations directly related to teaching—rather than learning—that we came to believe are crucial to making advances in students' dispositions towards and participation in mathematics. Multiple avenues worth exploration and insight are open in this direction of educational research.

No resource can be a miracle pill

We continue our work under the conjecture that the resources can be developed via classroom and PD design experimentation so that they would work for, and with, the classroom teacher. At the same time, we are fully aware that these resources would not be suited for implementation in most schools and school systems, as these currently operate.

We indeed meet many teachers who express excitement over both the resources and the learning outcomes of students from classrooms that they recognize as very similar to theirs when these are presented to them in workshop settings, and yet they never engage with these in their own classrooms. Similarly, under conditions of a short collaboration, teachers are willing to try aspects of the resources, but are not confident to keep trying without more pronounced support. For example, two South African fourth-grade teachers worked with the third author for four weeks, using the instructional sequence on fractions (Cortina, Višňovská, and Zúñiga 2014) in three weekly afterschool mathematics club sessions. They reported that their students 'enjoyed it very much' and that they themselves were pleasantly surprised by the mathematical ideas that students shared, and that both they and the students were able to hold a conversation about mathematical meanings. One of the teachers started asking students to explain how they thought ('Why do you say this?') and reported including such questions subsequently in her regular teaching. Yet, neither teacher was keen to try the same activities in their regular classrooms, fearing the lesson would turn into 'chaos'. Like the Mexican teacher, they both identified and valued the positive changes in their students' interactions and mathematical reasoning. Unlike the Mexican teacher, they were not afforded the time needed to reinvent the instructional sequence rationale so that these could become part of their own teaching repertoire. For a group of Australian third- and fourth-grade teachers who collaborated with the first author, the order of mathematical ideas in the designed sequence did not match the order expected by the national curriculum and assessment. Unsurprisingly, while these teachers, too, expressed appreciation for their students' learning, only one of them continued collaborating with us in the subsequent years.

Indeed, the designed resources are tools for change, and so change is to be expected. But it needs to come as a solution to a problem, not as a problem to a previously stable situation. If we are to change how mathematics is taught on a large scale, supporting teachers to teach in ways that today appear unimaginable has to come as a solution to a system-level problem. This is a big expectation, because instead of pursuing teachers' greater compliance, support will need to be directed towards teachers' continuing opportunities to develop and refine complex practices.

Researchers within the mathematics education design research community know this can be done—we repeatedly see similar profound changes happen. Students who did not desire to speak in their mathematics classroom choose to talk once the classroom is organized in ways that make it possible for them to have ideas, and to share these safely (Cobb 1999; Višňovská and Cortina 2003). Teachers who feared losing face, and the little control they had over their teaching, remain with the research teams once they see the changes and possibilities in how their classrooms interact, and once they become aware of, and are shielded from, their institutional realities (Višňovská and Cobb 2013; Višňovská and Cortina 2018). It is up to us to find out how to bring the systems that resist productive change in teacher support, to come to see such change as a solution to their problem.

A comment on sharing research and workshops

In 2019 the three of us collaboratively shared our journeys related to the fraction sequence in the form of a presentation entitled *Building learning opportunities in classrooms of disadvantage: Rethinking the learning trajectories* (Višňovská, Cortina, and Vale 2019). The presentation format provided us with opportunities to share our approach to research in considerable detail, in three 90-minute sessions. By allowing us to generate, within the space of the sessions, a range of varied shared experiences, the presentation format encouraged substantial conversations with, and among, participants, whose work came from similar, and different, research traditions.

As a community, we are only rarely afforded the luxury of deeply engaging with others' research, and that of having other researchers engage with ours, beyond what is possible to accomplish by reading academic publications. Unfortunately, the diminishing scope of the research practice rationales that typically make it to our publications (where 80-page research reports seem to be conclusively an artefact of the past) does not facilitate our understanding of each other's aims and approaches, and makes publication in the competitive, and theoretically diverse, environment yet more difficult. We would like to express our hope that SEMT presentations will continue functioning as a platform for embracing and critiquing theoretical and methodological approaches, and for making meaningful cross-communication a possibility and a norm.

Conclusion

In this chapter, we have shared findings from the analysis of teaching and the role it has played in research across the three decades of SEMT, followed by discussion of how our own research thoughts changed along similar lines. Our aim was to provide an argument for the need for the centrality of teaching practices in mathematics education research, and illustrate through the historical evolutions in SEMT research the broader awareness of this need in our community.

The quest for understanding and supporting ambitious and equitable teaching practices has only just started, and there is much yet to be understood and organized into workable perspectives and frameworks that could orientate the work of teaching and that of teacher support. In the past 30 years, we have made the transition from very little interest in teaching as such, to the recognition that teaching matters, and is indeed complex. We hope this path will continue towards teaching being not a context for, or a factor within, learning, but instead a researchable and learnable practice in its own right. Perhaps one day soon the *Teaching Sciences* will be born as a capable response to the *Learning Sciences.*

References

Artigue, M. (2008). La didactique des mathématiques face aux défis de l'enseignement des mathématiques. In G. Gueudet and Y. Matheron (Eds) *Actes du séminaire national de didactique des mathématiques 2007*, 14–39. Paris, France: IREM de Paris, Association pour la Recherche en Didactique des Mathématiques (ARDM).

Askew, M., Brown, M., Denvir, H., and Rhodes, V. (1999). Numeracy in early years: Building on children's understanding or delivering the curriculum? In M. Hejný and J. Novotná (Eds) SEMT 120–124.

Back, J. (2005). Small shifts, big changes: Elicitation patterns in classroom talk. In J. Novotná (Ed.) SEMT 71–78.

Ball, D., and Bass, H. (2003). Toward a practice-based theory of mathematical knowledge for teaching. In B. Davis and E. Simmt (Eds) Proceedings of the 2002 Annual Meeting of the Canadian Mathematics Education Study Group, 3–14. Edmonton, Canada: CMESG/GCEDM.

Biesta, G., Takayama, K., Kettle, M., and Heimans, S. (2020). Teacher education between principle, politics, and practice: A statement from the new editors of the Asia-Pacific Journal of Teacher Education. *Asia-Pacific Journal of Teacher Education*, 48(5): 455–459. doi.org/10.1080/1359866X.2020.1818485

Brousseau, G. (1997). *Theory of Didactical Situations in Mathematics.* Dordrecht, The Netherlands: Kluwer.

Cheeseman, J. (2009). Challenging children to think: Teacher behaviours that stimulate children to examine their mathematical thinking. In J. Novotná and H. Moraová (Eds) SEMT 11–23.

Chopin, M.-P., and Roiné, C. (2013). The anthropological and didactical constraints of the selection of pupils' tasks in elementary teaching of mathematics: An introduction to the concepts of didactic 'blindness' and 'visibility'. In J. Novotná and H. Moraová (Eds) SEMT 83–90.

Cobb, P. (1999). Individual and collective mathematical development: The case of statistical data analysis. *Mathematical Thinking and Learning*, 1(1): 5–43.

Cobb, P., and Jackson, K. (2008). The consequences of experimentalism in formulating recommendations for policy and practice in mathematics education. *Educational Researcher*, 37(9): 573–581. doi.org/10.3102/0013189X08327826

Cobb, P., McClain, K., Lamberg, T., and Dean, C. (2003). Situating teachers' instructional practices in the institutional setting of the school and school district. *Educational Researcher*, 32(6): 13–24.

Cobb, P., and Yackel, E. (1996). Constructivist, emergent, and sociocultural perspectives in the context of developmental research. *Educational Psychologist*, 31: 175–190. doi.org/10.1080/00461520.1996.9653265

Cobb, P., Zhao, Q., and Dean, C. (2009). Conducting design experiments to support teachers' learning: A reflection from the field. *The Journal of the Learning Sciences*, 18: 165–199. doi.org/10.1080/10508400902797933

Cobb, P., Zhao, Q., and Višňovská, J. (2008). Learning from and adapting the theory of realistic mathematics education. *Education et Didactique*, 2(1): 55–73.
Cortina, J. L. (2006). Instructional design in ratio (Unpublished doctoral dissertation). Vanderbilt University, Nashville, TN, USA.
Cortina, J. L., Višňovská, J., and Zúñiga, C. (2014). Unit fractions in the context of proportionality: Supporting students' reasoning about the inverse order relationship. *Mathematics Education Research Journal*, 26: 79–99.
Kasmer, L., and Billings, E. (2011). Prediction questions in the elementary classroom. In J. Novotná and H. Moraová (Eds) SEMT 189–197.
Kaur, B. (2019). Opportunity to learn: Reason and communicate mathematically in the elementary class. In J. Novotná and H. Moraová (Eds) SEMT 9–28.
Kloboučková, J. (2011). Cube building in primary school. In J. Novotná and H. Moraová (Eds) SEMT 206–214.
Krauthausen, G. (2013). Early mathematical education: A plea for mathematically founded conceptions. In J. Novotná and H. Moraová (Eds) SEMT 33–44.
Littler, G. H., and Benson, D. A. (2005). Patterns leading to generalisations. In J. Novotná (Ed.) SEMT 202–211.
Llinares, S., and Krainer, K. (2006). Mathematics (student) teachers and teacher educators as learners. In A. Gutiérrez and P. Boero (Eds) *Handbook of Research on the Psychology of Mathematics Education: Past, Present and Future*, 429–459. Rotterdam, The Netherlands: Sense Publishers.
Marchini, C., and Vighi, P. (2009). Can we develop geometrical understanding by focusing to isometries? A teaching experiment by the means of geometrical artefacts. In J. Novotná and H. Moraová (Eds) SEMT 169–176.
McLaughlin, M. (1987). Learning from experience: Lessons from policy implementation. *Educational Evaluation and Policy Analysis*, 9(2): 171–178.
Moraová, H., and Novotná, J. (2017). Higher order thinking skills in CLIL lesson plans of pre-service teachers. In J. Novotná and H. Moraová (Eds) SEMT 336–345.
Novotná, J. (2005). Proceedings of the International Symposium on Elementary Mathematics Teaching. Prague, Czech Republic: Charles University.
Novotná, J., and Moraová, H. (2007). Approaches to teaching mathematics at the elementary level. In J. Novotná and H. Moraová (Eds) SEMT Prague, Czech Republic: Charles University.
Novotná, J., and Moraová, H. (Eds) (2007). Approaches to teaching mathematics at the elementary level. Prague, Czech Republic: Charles University.
Prusak, N., and Hershkowitz, R. (2019). Task designed for promoting the concept of area in a problem-solving environment: The story of one activity. In J. Novotná and H. Moraová (Eds) SEMT 325–334.
Pytlak, M. (2013). Playing with three-dimensional geometry in the 3rd class of primary school. In J. Novotná and H. Moraová (Eds) SEMT 264–273.
Sarrazy, B., and Novotná, J. (2005). Didactical contract: Theoretical frame for the analysis of phenomena of teaching mathematics. In J. Novotná and H. Moraová (Eds) SEMT 33–59.
Sfard, A., and Cobb, P. (2014). Research in mathematics education: What can it teach us about human learning? In R. K. Sawyer (Ed.) *The Cambridge Handbook of the Learning Sciences*, 545–564. Cambridge: Cambridge University Press.
Shabtay, G. (2019). The link between classroom discourse and students agency. In J. Novotná and H. Moraová (Eds) SEMT 401–408.
Skalická, L. (1997). Mental organs' work in number expression solving processes (One approach to teaching number expressions). In M. Hejný and J. Novotná (Eds) SEMT 61–67.
Spillane, J. P. (2000). Cognition and policy implementation: District policy-makers and the reform of mathematics education. *Cognition and Instruction*, 18: 141–179.
Steffe, L. P., and Thompson, P. W. (2000). Teaching experiment methodology: Underlying principles and essential elements. In A. Kelly and A. Lesh (Eds) *Handbook of Research Design in Mathematics and Science Education*, 267–307. Mahwah, NJ: Lawrence Erlbaum Associates.
Stephan, M., and McManus, G. (2013). Pre-interviewing as formative assessment. In J. Novotná and H. Moraová (Eds) SEMT 306–313.

Symeonidis, N., Tatsis, K., and Kaldrymidou, M. (2015). Norms in decision-making and in shaping attitudes, emotions and performance. In J. Novotná and H. Moraová (Eds) SEMT 322–331.

Tall, D. (2001). What mathematics is needed by teachers of young children? In J. Novotná and M. Hejný (Eds) SEMT 26–36.

Thanheiser, E., and Philipp, R. (2017). Do prospective elementary teachers (in the United States) notice cultural aspects of mathematics in a teaching scenario? In J. Novotná and H. Moraová (Eds) SEMT 403–411.

Thompson, A. G., Philipp, R. A., Thompson, P. W., and Boyd, B. A. (1994). Calculational and conceptual orientations in teaching mathematics. In A. Coxford (Ed.) *1994 Yearbook of the National Council of Teachers of Mathematics*, 79–92. Reston, VA: National Council of Teachers of Mathematics.

Vale, P., and Graven, M. (2018). Facilitating conceptual engagement with fractions through suspending the use of mathematical terminology. In E. Bergqvist, M. Österholm, C. Grandberg, and L. Sumpter (Eds) Proceedings of the 42nd Conference of the International Group for the Psychology of Mathematics Education, vol. 4, 331–338. Umeå, Sweden: PME.

Višňovská, J. (1999). Children and the discovery of the mathematical rules. In M. Hejný and J. Novotná (Eds) SEMT 175–179.

Višňovská, J. (2009). Supporting mathematics teachers' learning: Building on current instructional practices to achieve a professional development agenda (Unpublished doctoral dissertation). Vanderbilt University, Nashville, TN, USA.

Višňovská, J. and Cobb, P. (2013). Classroom video in teacher professional development: Community documentational genesis perspective. *ZDM—Mathematics Education*, 45(7): 1017–1029. doi.org/10.1007/s11858-013-0523-5

Višňovská, J., and Cortina, J. L. (2003). Building on students' reasoning: Using students' preconceptions of proportional relations in developing instruction. In J. Novotná (Ed.) SEMT 157–159.

Višňovská, J., and Cortina, J. L. (2017). Learning to support all students' fraction understanding. In J. Novotná and H. Moraová (Eds) SEMT 430–440.

Višňovská, J., and Cortina, J. L. (2018). Resourcing teachers in transition to plan for interactions with students' ideas. In L. Fan, L. Trouche, C. Qi, S. Rezat, and J. Višňovská (Eds) *Research on Mathematics Textbooks and Teachers' Resources: Advances and Issues*, 277–295. Cham, Switzerland: Springer.

Višňovská, J., Cortina, J. L., and Vale, P. (2019). Building learning opportunities in classrooms of disadvantage: Rethinking the learning trajectories. In J. Novotná and H. Moraová (Eds) SEMT 452–454.

Wenger, E. (1998). *Communities of Practice: Learning, Meaning, and Identity*. Cambridge: Cambridge University Press.

Wittmann, E. C. (2011). Early mathematical education: A plea for mathematically founded conceptions. In J. Novotná and H. Moraová (Eds) SEMT 43–55.

3.2

From variation pedagogy and Lesson Study to cultural transposition

Maria Bartolini Bussi
Department of Education and Humanities, University of Modena and Reggio Emilia, Italy
Alessandro Ramploud
Department of Mathematics, University of Pisa, Italy

Introduction

This paper is organized around three key phrases: variation pedagogy, Lesson Study, and cultural transposition. The first two keywords explicitly refer to research traditions developed in Asia. The third is an original construct developed in the West, in Italy, to cope with the deep cultural differences between Eastern and Western research traditions. This situation suggested that we ask two research questions:

1. To what extent do classroom teaching practices (e.g. variation pedagogy, Lesson Study) depend on the culture of the region where they were used for the first time?
2. Does the difficulty depend on a 'simple' translation problem (of tasks, methodologies, etc.) from one language to another?

What is variation pedagogy?

The first author came in touch with *variation pedagogy* (VP) about 15 years ago when she looked at a personal collection of Chinese textbooks in order to enrich the lessons for primary school teacher education in Reggio Emilia, where she acted for more than 20 years as both the university programme Dean and the teacher of didactics of mathematics. In the textbooks, a particular feature of the arithmetic problems is evident: from the beginning, students are not given a single problem, but rather a pair or even a triplet of word problems in the same context. For instance, the following image is taken from a booklet used in the first semester of the first grade. The image is aimed at introducing the pair of numbers 6 and 7.

Figure 3.2.1 represents a primary school classroom where a group of students is cleaning together with their teacher. From the image, one can extract different word problems (that are not stated in the text, as the lesson is done at a very early stage

Maria Bartolini Bussi and Alessandro Ramploud, *From variation pedagogy and Lesson Study to cultural transposition*. In: *Elementary Mathematics Teaching*. Edited by: Brian Doig, Jarmila Novotná, Berinderjeet Kaur and David K. Pugalee, Oxford University Press. © Oxford University Press (2024). DOI: 10.1093/oso/9780192869647.003.0010

Figure 3.2.1 The introduction of the numbers 6 and 7 in the Chinese textbook (Mathematics Textbook Developer Group for Elementary School 2005, p. 42)

of learning to read and write in Chinese). We assume that the students are invited to produce different texts orally:

There are 6 students and 1 teacher (that is 7 characters):

$$(7 = 6 + 1)$$

There are 6 characters within the classroom and 1 girl coming in:

$$(7 = 6 + 1)$$

There are 6 student desks and 1 teacher's desk:

$$(6 + 1 = 7)$$

There are 6 student chairs close to the student desks and 1 is carried by a boy:

$$(6 + 1 = 7)$$

And so on.

The textbook (121 pages) is for the first school semester that usually lasts from the beginning of the school (September) to the Chinese New Year celebration (between 21 January and 19 February).

The position in the booklet of these problems, on p. 42 (around 1/3 of the total length), suggests that the lesson is given in around November or December of the first grade, when the early numbers are introduced, not one by one but in pairs.

A few pages later (p. 56), a problem about dinosaurs is given, where the same poster suggests different arithmetic operations represented by the sheets prepared and handed to the students.

The link between addition and subtraction is highlighted. The above image suggests from the very beginning the shift from:

$$a + b = c \ \ to \ \ c - a = b \ \ and \ \ c - b = a$$

This strict link was traditional in China. Siu (2004) reported that the ancient book by Yang Hui, published in 1274, read: *Whenever there is an addition there is a subtraction* (p. 164) (see Figure 3.2.2 below).

Figure 3.2.2 The dinosaur problem (Mathematics Textbook Developer Group for Elementary School 2005, p. 56)

In another textbook, for the second grade, the following system of nine problems about addition and subtraction is given (Bartolini Bussi, Canalini, and Ferri 2011). The students are required not only to solve the problems, but also to understand why the problems have been arranged by rows and columns in a table. One can analyse the system either by rows or by columns. Graphic schemes give additional information about the solutions and suggest analogies. In the first line, the first problem (with the grey background) suggests the following addition:

$$45 + 30 =$$

While the second and third problems suggest the following subtractions:

$$75 - 45 =$$
$$75 - 30 =$$

If one reads the system of nine problems by columns, it is evident that, even if the situation and the numbers are changed slightly, each column suggests the same operation. The context, 'On the river there are ducks', is the same (see Table 3.2.1 below).

The relationship with the semantic analysis of additive problems (Moser, and Romberg) is evident, as different lines refer to *combine*, *change*, *compare*, and *equalize*. Yet a difference appears. The Chinese approach is not a way to *classify* arithmetic problems, but rather to consider a problem and its variations as a component of a *whole system* aiming at the construction of a complex meaning: the same problem situation (ducks in a river) is likely to generate in each row three different word problems depending on the choice of data and the unknowns in the text. The situation might be interpreted as if an observer is looking at the river from a bridge and sees a group of ducks: the text of each problem arises from the focus—that is, the available data—and requires one to look for the unknown.

This system (that is proposed for the end of the second grade) aims at constructing the complex meaning of addition and subtraction, highlighting also the connection between them. Similar sets of problems about multiplication and division are given in all the grades. During dialogues with our Chinese colleague Sun Xu Hua, she highlighted the fact that in the Western approach semantic analysis aims at classifying problems (hence separating them into classes), while in the Chinese one the aim is to join them in the same whole.

Drawing on the analysis of several textbooks for the early grades of primary school, we observed that gradually the attitude towards interpreting the same situation as the source of different arithmetic problems was encouraged. This intuition, arising from the analysis of early grades textbooks, was later confirmed in a more general way, when the contact of the authors with the Chinese community became more systematic. For instance, in the China National Presentation at ICME 2011 (Chinese Delegation 2008) the characteristics of the problem solving of Chinese students were described as follows:

Table 3.2.1 Beijing Education Science Research Institute and Beijing Instruction Research Center for Basic Education (1996), vol. 4, p. 88.

First solve the nine problems below. Then explain why they have been arranged in rows and columns in this way, finding relationships.

(1) **In the river there are** 45 white ducks and 30 black ducks. All together, how many ducks are there?

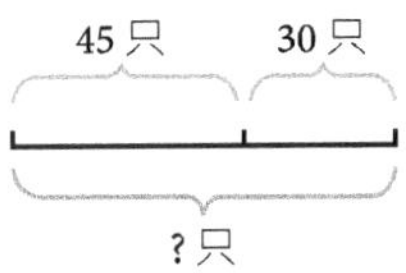

(2) **In the river there are white** ducks and black ducks. All together there are 75 ducks. 45 are white ducks. How many black ducks are there?

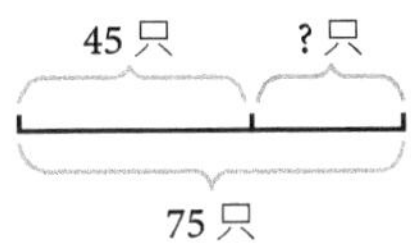

(3) **In the river there are white** ducks and black ducks. All together there are 75 ducks. 30 are black ducks. How many white ducks are there?

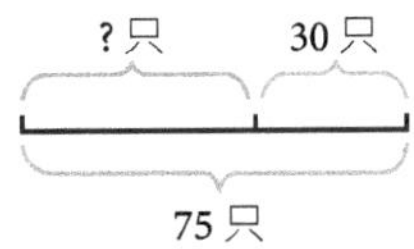

(1) **In the river there is** a group of ducks. 30 ducks swim away. 45 ducks are still there. How many ducks are in the group (at the beginning)?

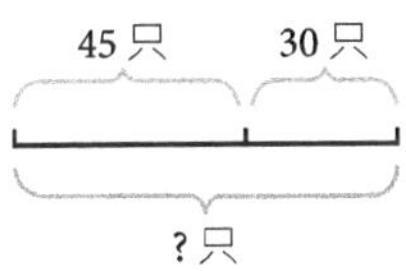

(2) **In the river there are** 75 ducks. Some ducks swim away. There are still 45 ducks. How many ducks have swum away?

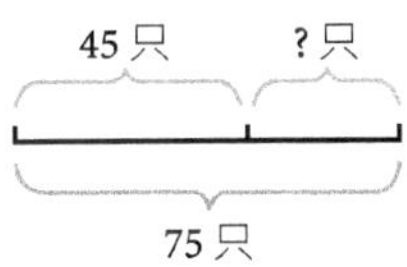

(3) **In the river there are** 75 ducks. 30 ducks swim away. How many ducks are still there?

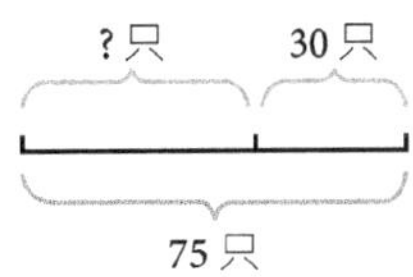

(1) **In the river there are** 30 black ducks. There are 15 more white ducks than black ducks (there are 15 fewer black ducks than white ducks). How many white ducks are there?

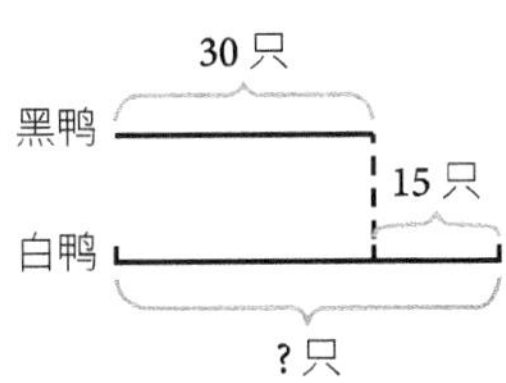

(2) **In the river there are** 30 black ducks and 45 white ducks. How many white ducks more than black ducks (How many black ducks fewer than white ducks)?

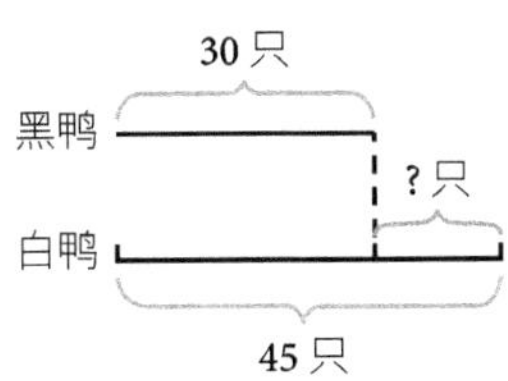

(3) **In the river there are** 45 white ducks. Black ducks are 15 fewer than white ducks (white ducks are 15 more than black ducks). How many black ducks are there?

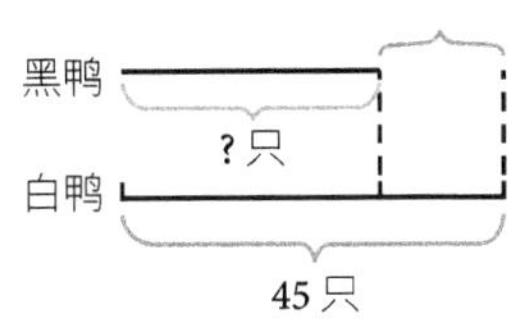

> Students usually need to complete three different activities during problem solving: extending the problem to different variation, using different methods to solve the problem and applying the solutions to different situations.
> The basic strategy to solve a new difficult problem is by reducing it to a solved or easier problem. This is called 'the reducing strategy'. In order to use a reducing strategy, students must remember some typical samples in each area of mathematics contents. These typical samples will become step stones during problem solving.
>
> Students' style of problem solving is somewhat similar to the work of a 'pure mathematician': formal deduction, theorem-based reasoning, multi-steps complex symbolic operation, rich of mathematics knowledge, but no realistic context. (pp. 27–28)

From the beginning, we named these systems of problems *bianshi* (变式). The root of this name in Chinese was reported by Leung in a small but intense seminar on VP, held in Oxford in 2018. Leung is a special witness, as he is Chinese from Hong Kong, lived in Canada and worked with Marton, who is well known in the West for his theory of learning (Huang and Li 2017).

According to Leung, there were different possible choices when in Hong Kong a group of scholars (including Chinese ones and Marton) chose a name for VP hinting at the *Book of Changes* (*I Ching*), which is one of the milestones of ancient Chinese thought. According to Leung (personal communication), the decision opened a new dimension to the theory that has not yet been fully explored. This choice for the word *change* (*bian*, the first character above) happened to be consistent with Sun's, who, at that time, was studying in Hong Kong.

VP was actually a traditional method in ancient China and is still one of the features of classroom teaching. This is well known by Chinese teachers. Once we met in Italy a respected Chinese teacher who was visiting her daughter. When we interviewed her about *bianshi* problems, she was surprised and answered that it was the *only* method for problem solving and asked whether in Italy we had different methods. The focus on algebraic thinking was acknowledged as one of the advantages for Chinese students, which can explain, at least partly, the high performances of them in the Programme for International Student Assessment (PISA) (Cai and Knuth 2011).

Bianshi problems have the advantage of highlighting the connections between different occurrences of word problems. The idea of a mathematical meaning as a system, with rich connections with others and not limited to a single statement (definition), is consistent with the view of professional mathematicians. Moreover, connectedness was one of the keywords of the Chinese style of mathematics teaching used by Ma (2010) in her famous book, where she compared the US style of teaching with that of the Chinese.

The early experiments in Italy

A small group of Italian mathematics education researchers and primary school teachers started to study the possibility of exploiting the Chinese approach to variation problems to improve the quality of learning of Italian students. Primary school teachers had noticed the good performance of Chinese students who had migrated to

Italy with their families, in spite of the different languages (Chinese versus Italian). The long-term experiments about addition and multiplication reported at SEMT 2011 (Bartolini Bussi, Canalini, and Ferri 2011) were part of these early studies.

The experiments focused on some open questions to be addressed in future studies—for instance, the way of introducing schemes to solve problems and the features of the linguistic structure of variation problems. The word problems in Table 3.2.1 have a very dry and concise linguistic structure.

We aimed at reproducing the structure of the problems of the Chinese textbooks. In the duck problems we translated the task from Chinese literally, maintaining the language structure of the original. In the beginning, this choice was, in some sense, natural and instinctive, and not guided by a theoretical framework. Yet, working with teachers and students, we became aware that this concise structure fostered the shift to the arithmetic and algebraic structure, and to the interpretation and production of schemes. In these texts the narrative component was very limited. This feature seemed to clash with the common attitude in Western teaching of word problems (Verschaffel et al. 2009): word problems often have a lot of narrative enrichment that is expected to make them closer to students' experience (e.g. in the duck problem, one could read that *it was a sunny day and that the ducks were happy to swim in the sun*).

Our experiments raised the issue: does a richer narrative text help or inhibit the arithmetic or algebraic solution? From a cognitive point of view, is the interpretation of a narrative text more demanding or more helpful compared with a dry and concise text? Does this apply in the same way to high and low achievers, to students of Italian extraction, or to students from other cultures with a more limited knowledge of Italian?

In an Italian second-grade classroom, the students were not yet familiar with graphic schemes to solve additive word problems, but they observed that the very structure of the text suggested the shift from the verbal text to the scheme of the table and were able to pair each text with the correct scheme (Bartolini Bussi et al. 2011).

In another second-grade classroom (Maffoni, personal communication, 2013), the students themselves produced similar schemes, with a gradual shift from realistic images (too long to be drawn) to symbolic images (one circle, black or white, for each duck) to points and rectangles and segments to represent the whole set of either white or black ducks in regular arrays. See Figure 3.2.3 below.

In the previously mentioned meeting on VP held in Oxford in 2018, a presentation on *conceptual variation* was given by Debbie Morgan (personal communication, 2018), Director Primary Mathematics at the National Centre for Excellence in Teaching Mathematics, who is involved in exchanges between British teachers from London and Chinese teachers from Shanghai. Conceptual variation concerns the situation when the teacher considers how the concept might be represented in order to draw out its structure in mathematics. The whole table of the duck problem is a good example of conceptual variation. The first author of this chapter, in the discussion after Morgan's presentation, gave the example of Table 3.2.1 as a paradigmatic case of conceptual variation. Another example of conceptual variation was given by Morgan herself, in the form of a set of tasks to be solved using concrete material (sticks). A Chinese teacher, in London, reported this example during a joint meeting. The set of tasks is the following:

Arrange 21 sticks into squares. How many squares can you make, and how many are remaining?

Arrange 21 sticks into 'greater than' (>) symbols. How many symbols can you make, and how many are remaining?

Arrange 21 sticks into hexagons. How many hexagons can you make, and how many are remaining?

And so on. After some cases the following task is given:

With your 21 sticks now choose a shape or group and repeat till you can make no more. How many shapes (groups) have you made? How many sticks make each shape (group)? How many remaining? Use the following sentence to explain what you have made:
I have _____ groups of _____ and _____remaining.
True or false?
With 21 sticks I have 3 groups of 6 and 3 remaining.
With 21 sticks I have 4 groups of 5 and 1 remaining.
With 21 sticks I have 5 groups of 4 and 1 remaining.
Write each case as an equation. What does each number in the equation represent?

Then, other examples are discussed, and students are asked to write their solution on the blackboard, and to produce new examples by themselves. At the end a summary is given:

What did we learn?

- Partition a number into a maximum number of groups with a remainder.
- Express this as an equation.
- Generalize that if we maximize the number of groups, the remainder is always smaller than the size of the group.

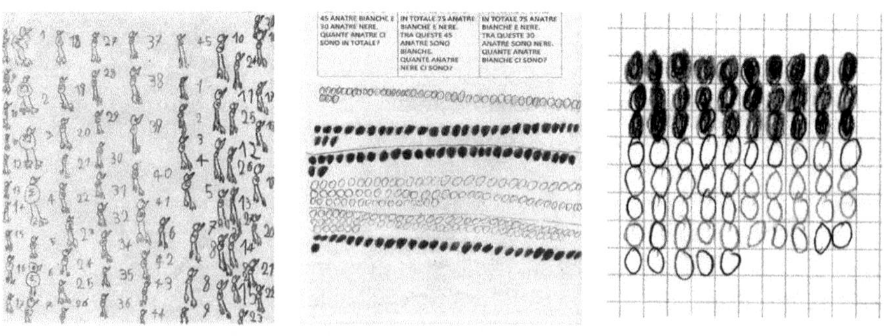

Figure 3.2.3 Representations of different levels of student solutions of the duck problems (see Table 3.2.1) (Photographs by the author)

The production of different solutions is encouraged. What is typical is the insistence on the same linguistic structure in the solution of all the tasks, which for Western researchers is supposed to be boring. The structure hints towards the suggested equation scheme:

$$\ldots = \ldots x \ldots + \ldots$$

This may be used with young students also.

The Italian experiments on variation pedagogy were later merged with experiments on Lesson Study (LS).

What is LS?

The origin of LS in Asia

During our travels, we happened to visit several Chinese classrooms and observe the implementation of VP. A good example of VP is in Chapter 11 of the ICMI STUDY 2023 (Bartolini Bussi and Sun 2018). Some dozens of participants in the study conference had been invited to observe a first-grade classroom in a lesson about addition with regrouping. The paradigmatic operation given in the lesson was:

$$24 + 9 = \ldots$$

Different solutions are given by small groups of students, drawing on boxes of cakes (each full box with ten cakes, but some boxes with fewer cakes) to be regrouped in different ways. The teacher encouraged newer and newer solutions with different ways of *making ten*—that is, ways of reconstructing tens by regrouping the cakes in the boxes (see Figure 3.2.4).

The classroom observation was introduced to a large number of participants in the study conference (some dozens) by giving them the lesson plan in advance, that is the careful design of the lesson, was discussed in advance.

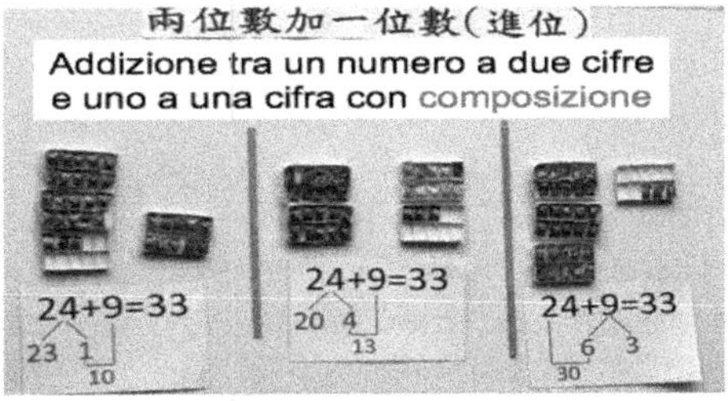

Figure 3.2.4 Representations of different student solutions of the cakes problem (Photographs by the author)

The use of a lesson plan reminds us that Japanese LS (for differences between the Chinese and the Japanese models of LS, see Bartolini Bussi et al. 2017; Fujii, this volume; Doig and Groves, this volume).

A Japanese LS cycle:

> consists in preparation, actual class and class review sessions ... This process begins with finding and selecting materials relevant to the purpose, and is then followed by refining the class design based on the actual needs of the students and tying all this information together into a lesson plan. The significance of LS is that all of these processes are performed in collaboration with other teachers. A class ... is then taught based on the teaching plan devised. The class is observed by many teachers, who are sometimes joined by university instructors and supervisors from the board of education, and a review session is held for all observers after the class. (Baba 2007, p. 2, cited in Bartolini Bussi and Funghi 2019).

LS is a model for mathematics teacher education and development used all over the world (Quaresma et al. 2018). It was initiated in Japan in the late nineteenth century (Isoda et al. 2007) and extended with adaptation in many other countries such as Australia, China, North and South America, and Europe, leading to the constitution in 2007 of the World Association of Lesson Studies (walsnet.org).

In Italy teachers rarely use a very structured way of lesson planning. The Italian scholastic context, in fact, is characterized by *inclusiveness* (Bartolini Bussi et al. 2017), meaning that the same class can include children with learning difficulties, children with physical and mental disabilities, children that are not Italian speakers, and so on, and a *child-centred vision of teaching* in which teachers tend to believe that each child has his or her own specific characteristics, and needs individualized support. Moreover, in Italy teachers are free to choose their own teaching methods and the sequencing of the topics to be treated. Italian primary teachers usually spend many hours in the same class and teach the same class for more than one year, and this makes them feel less anxious about the short-term effects of their teaching, and encourages them to take care of, and to observe, long-term processes.

LS, instead, reflects a completely different way of conceiving both the planning and the management of the lesson. The version of lesson plan we proposed to Italian teachers (for more details, see Bartolini Bussi et al. 2017) consisted in a table to plan a lesson of exactly 60 minutes, requiring teachers to have explicit objectives and prerequisites for the lesson and to describe, and justify, for each phase of the lesson, the activity to be carried out, its educational aims, the time required for it, materials needed, and the focus of observation.

Our research group implemented several hundred LS cycles in Italy with both prospective and in-service teachers. Together with Funghi, we studied the effects of LS on the teachers' cultural beliefs (Bartolini Bussi and Funghi 2019).

What is cultural transposition (CT)?

The two cases in the previous sections (VP and LS) suggest, from different perspectives, the importance of cultural roots. The former concerns the epistemological analysis of mathematical meaning (that in the SEMT 2011 presentation was mentioned as *the cultural analysis of content* (Guala and Boero 2017)) and the latter, the methodological aspect of design and implementation of a lesson.

Is it possible to read these examples without considering the cultural elements that cross them? Are we faced with a simple translation problem (tasks, methodologies, etc.), or is there something more complex to keep under control? To try to answer these questions, it is necessary, first of all, to explain some lines of development that have characterized the research in mathematics education in recent years, starting from a cultural frame.

Mathematical enculturation (Bishop 1988) represents one of the first essential references in recognizing the crucial role played by culture in mathematics education research. Bishop identified different human activities (like counting, locating, measuring, designing, playing, or building a model) which can be seen as common ground in producing mathematical signs that are always locally developed and culturally determined. Moreover, Barton's studies (e.g. Barton 2007; Barton and Frank 2001), by showing how different languages and cultures embody and express different mathematical ideas, reinforced the importance of the cultural perspective in studying mathematics education phenomena in all their complexity.

In parallel with these approaches, although with crucial differences, D'Ambrosio (2001) posed the ethical issue at the core of mathematics education, following an ethno-mathematics line of research, by studying and exploring the mathematical activity of different 'cultural groups'.

Indeed, even if the changing and dynamic nature of cultures makes it difficult to recognize the boundaries and limits between one culture and another, according to Monticelli (2012), Lotman's sociological studies have pointed out that there are some elements (like languages, myths, knowledge, customs, etc.) that can help us to recognize different cultures; mathematics, as a cultural expression of human beings, is one of these elements. As pointed out by Radford (1997):

> [One of the main ideas of the work is] that mathematical knowledge is more than merely concomitant with its cultural environment and that the configuration and the content of mathematical knowledge is properly and intimately defined by the culture in which it develops and in which it is subsumed. (p. 32)

For example, red and black sticks for positive and negative numbers, respectively, the use of which, in the Chinese practice of calculations, is documented from the second century BC, is a beautiful case of mathematics as part of a particular cultural heritage, as the artefact of the sticks was imported from non-mathematical practices like, for example, the *I Ching* divinatory arts. The two types of sticks offered operational possibilities, and thus mathematical assumptions and meanings, very different from those realized by the alphabetical numbers or the numerical segments used in the arithmetic of the Greek culture, which is at the roots of Western mathematics. For example, while zero was an aberration for the Greek episteme and actually appeared in the official Western mathematics only at the end of the seventeenth century, it is in this Chinese calculation practice a form of 'being' in all respects, which is the empty space obtained by removing the black and red sticks in case of balance. The balance principle between *Yin* and *Yang* flows in all Chinese cultural milieu, as we have just seen in mathematics, but also in poetry (Radford 1997), painting, and the *I Ching*, one of the founding books of Chinese thought. Indeed, according to Jullien (2006), the structure of the book itself, in which every hexagram has its complementary hexagram, deeply embodies the balance principle.

So different cultures produce different mathematical meanings and, of course, they also generate different mathematics education school practices (Bartolini Bussi, and Sun 2018). The meeting between different mathematics education school practices is the core of the CT perspective (Mellone et al. 2019; Mellone, Ramploud, and Carotenuto 2020) and, in particular, is perceived as a potential space of reflection and awareness development for researchers, educators, and teachers. The idea of CT is not to perform a comparative study or a slavish import–export of mathematics education methodologies and tools between different cultures, but rather to open a dialogue between two different cultures in which every thought, meeting the other culture, questions its own 'unthought' (Jullien 2006). Indeed, every cultural paradigm, understood as different levels in which the culture is stratified, is situated on a series of implicit assumptions that appear unthinkable (Bartolini Bussi and Funghi 2019) within the same paradigm. In this sense, rethinking one's own unthinkable means moving into a different paradigm to be able to carefully analyse what is given as absolutely self-evident in our culturally specific paradigm.

In order to look at foreign mathematics education practices from the perspective of CT, it is essential to activate a process of *deconstruction*—that is, an analysis of the different levels in which culture is stratified. Derrida (1996) conceives deconstruction as an exercise of 'radical criticism' on cultural stratifications, without any attempt to reach a conceptual or an abstract dimension; rather, his approach seeks specifically to analyse the process of abstraction and conceptualization of Western thought:

> radical critique, namely a procedure ready to undertake its self-critique. This critique wants itself to be in principle and explicitly open to its own transformation, re-evaluation, self-reinterpretation. (p. 110)

In this sense, the didactic practices of mathematics as a cultural product can be understood and given as different levels of stratification. For example, the different ways to read fractions in Italy (from numerator to denominator) and in China (from denominator to numerator) are varying cultural stratifications (Bartolini Bussi, Baccaglini-Frank, and Ramploud 2014). However, they have important consequences on learning.

The deconstruction process of a mathematics education practice is a process implemented by a group of researchers in mathematics education who, coming into contact with teaching practices of other regions, begin to activate a deconstruction of the different cultural levels in which that teaching practice is stratified through the lens of their cultural beliefs and values. In the light of this deconstructive process, the researchers try also to make the 'foreign' teaching practice compatible with their own cultural beliefs, inaugurating other interpretative keys of the didactic practice of their own cultural context. Building on this experience, the researchers design and implement professional development and induce and propose a rethinking of teaching practices. We believe, in fact, that the introduction of an education practice from the CT perspective, allowing the teachers to come into contact with different values and beliefs on teaching and learning mathematics, can support them in becoming aware of their unthoughts—that is, some of the 'invisible' cultural beliefs about teaching and learning that they have absorbed and re-elaborated within their own culture.

Conclusion

Now let's go back to our initial questions.

In the first, we asked ourselves whether it is possible to read these examples without considering the cultural element that runs through them. Obviously it is not possible, because, as we have seen, different cultures produce different mathematical meanings and, of course, they also generate different mathematics education school practices (Bartolini Bussi and Sun 2018).

If we move on to the second question—that is, if we are faced with a simple translation problem (of tasks, methodologies, etc.), or if something more complex is hidden to keep the situation under control—we could say that simple translation completely hides the deep meanings of what we face. It is therefore necessary to carry out a CT in which translation exclusively plays the role of opening a deconstruction process, in which researchers and teachers give meaning to their research and teaching practices. Jullien (2006) writes:

> This is not about comparative philosophy, about paralleling different conceptions, but about a philosophical dialogue in which every thought, when coming towards the other, questions itself about its own unthought. (translated from Jullien, 2006, p. 8)

In our opinion, from the point of view of CT, Jullien's sentence could be paraphrased in this way: we believe that it is not a question of comparative didactics of mathematics, of the parallelization of different conceptions (also for the differences in scholastics systems), but we would like to try to open a transpositive dialogue between the different didactics of mathematics, which, in meeting each other, question their own unthought.

Acknowledgements

We acknowledge the collaboration of many primary school mathematics teachers, of many students, and of many prospective teachers in our education programme, who have taken part with enthusiasm in the different phases of our study. We thank also the publishers of Chinese textbooks who put at disposal different series of primary school textbooks.

References

Baba, T. (2007). How is Lesson Study implemented? In M. Isoda, M. Stephens, Y. Ohara, and T. Miyakawa (Eds) *Japanese Lesson Study in Mathematics: Its Impact, Diversity and Potential for Educational Improvement*, 2–7. Singapore: World Scientific Publishing.

Bartolini Bussi, M. G., Funghi, S., and Ramploud, A. (2020), Mathematics teachers' cultural beliefs: The case of Lesson Study. In O. Chapman and Potari (Eds) *Handbook of Mathematics Teacher Education*. 2nd edition, vol. 1, 132–154. Rotterdam, The Netherlands: Sense Publishers.

Bartolini Bussi, M. G., and Sun, X. H. (2018). *Building the Foundation. Whole Numbers in the Primary Grades. The 23rd ICMI Study*. Cham, Switzerland: Springer Nature.

Bartolini Bussi, M. G., Canalini, R., and Ferri, F. (2011). Towards cultural analysis of content: Problems with variations in primary school. In J. Novotná and H. Moroavá (Eds) SEMT 9–21.

Bartolini Bussi, M. G., and Funghi, S. (2019). Lesson Study in primary pre-service teachers' education: Influences on beliefs about lesson planning and conduction. In M. Graven, H. Venkat, A. Essien, and P. Vale (Eds) Proceedings of the 43rd Conference of the International Group for the Psychology of Mathematics Education, vol. 2, 81–88. Pretoria, South Africa: PME.

Bartolini Bussi, M. G., Baccaglini-Frank, A., and Ramploud, A. (2014). Intercultural dialogue and the geography and history of thought. *For the Learning of Mathematics*, 34(1): 31–33.

Bartolini Bussi, M. G., Bertolini, C., Ramploud, A., and Sun, X. (2017). Cultural transposition of Chinese Lesson Study to Italy. An exploratory study on fractions in a fourth-grade classroom. *International Journal for Lesson and Learning Studies*, 6(4): 380–395.

Barton, B. (2007). *The Language of Mathematics: Telling Mathematical Tales*, vol. 44. New York: Springer Science & Business Media.

Barton, B., and Frank, R. (2001). Mathematical ideas and indigenous languages. In B. Atweh, H. Forgasz, and B. Nebres (Eds) *Socio-Cultural Research on Mathematics Education: An International Perspective*, 135–150. London: Lawrence Erlbaum Associates.

Beijing Education Science Research Institute and Beijing Instruction Research Center for Basic Education. (1996). *Mathematics (shu xue) (in Chinese)*, vol. 4. (P. Barolini trans).

Bishop, A. (1988). *Mathematical Enculturation: A Cultural Perspective on Mathematics Education*. Dordrecht, The Netherlands: Kluwer Academic Publishers.

Cai, J., and Knuth, E. (Eds) (2011). *Early Algebraization: A Global Dialogue from Multiple Perspectives*. Berlin–Heidelberg: Springer Science and Business Media.

Carpenter, T., Moser, M., and Romberg, T. (Eds) (1982). *Addition and Subtraction: A Cognitive Approach*. Hillsdale, NJ: Lawrence Erlbaum Associates.

Chinese Delegation (2008). *Mathematics Education in China: Tradition and Reality – China National Presentation*, ICME 11, ICME: Chinese Delegation.

D'Ambrosio, U. (2001). *Ethnomathematics*. Rotterdam/Taipei: Sense Publishers.

Derrida, J. (1996). *Spectres of Marx: The State of the Debt, the Work of Mourning and the New International* [translated by P. Kamuf]. New York: Routledge.

Guala, E., and Boero, P. (2017). Cultural analysis of mathematical content in teacher education: The case of elementary arithmetic theorems. *Educational Studies in Mathematics*, 96(2): 207–227.

Huang, R., and Li, Y. (2017). *Teaching and Learning Mathematics through Variation. Confucian Heritage Meets Western Theories*. Rotterdam, The Netherlands: Sense Publishers.

Isoda, M., Stephens, M., Ohara, Y., and Miyakawa, T. (Eds) (2007). *Japanese Lesson Study in Mathematics: Its Impact, Diversity and Potential for Educational Improvement*. Singapore: World Scientific Publishing Company.

Jullien, F. (2006). *Si parler va sans dire. Du logos et d'autres ressources*. Paris: Le Seuil.

Ma, L. (2010). *Knowing and Teaching Elementary Mathematics: Teachers' Understanding of Fundamental Mathematics in China and the United States*. New York: Routledge (Original work published 1999).

Mathematics Textbook Developer Group for Elementary School. (2005). *Mathematics, First Grade (in Chinese)*. Beijing, China: The People's Education Press.

Mellone, M., Ramploud, A., and Carotenuto, G. (2020). Cultural transformation experience of the El'konin-Davydov course. *Educational Studies in Mathematics*, 106: 379–396

Mellone, M., Ramploud, A., Martignone, F., and Di Paola, B. (2019). Cultural Transposition: Italian didactic experiences inspired by Chinese and Russian perspectives on Whole Number Arithmetic (WNA). *Special Issue ZDM*, 51(1): 199–212.

Monticelli, D. (2012). Challenging identity: Lotman's 'translation of the untranslatable' and Derrida's différance. *Σημειωτκή-Sign Systems Studies*, 40(3–4): 319–339.

Quaresma, M., Winsløw, C., Clivaz, S., Da Ponte, J. P., Ní Shúilleabháin, A., and Takahashi, A. (Eds) (2018). *Mathematics Lesson Study Around the World*. New York: Springer International Publishing.

Radford, L. (1997). On psychology, historical epistemology, and the teaching of mathematics: Towards a socio-cultural history of mathematics. *For the Learning of Mathematics*, 17(1): 26–33.

Siu, M. K. (2004). Official curriculum in mathematics in ancient China: How did candidates study for the examination? In L. Fan, N. Y. Wong, J. Cai, and S. Li (Eds) *How Chinese Learn Mathematics: Perspectives from Insiders*, 157–185. Singapore: World Scientific.

Verschaffel, L., Greer, B., van Dooren, W., and Mukhopadhyay, S. (Eds) (2009). *Words and Worlds: Modelling Verbal Descriptions of Situations*. Rotterdam, The Netherlands: Sense Publishers.

3.3

Visualization as a tool in teaching: Pre-service teachers' competencies in visualization

Jasmina Milinković
University of Belgrade, Serbia

Introduction

In some cultures, a person who appears not to comprehend some thing would not be asked, 'Shall I explain it to you?' Instead the question would be: 'Shall I make a drawing of it for you?' It seems that we believe that drawing something makes us see things easily. From birth, the visual sense is a primary tool for learning about world. The importance of seeing does not diminish in the following years of growth of young children. The question is whether the school appreciates the importance of this visualization in the mathematical world to which we introduce pupils. How can a teacher help pupils to 'see' abstract mathematical concepts? How do we know what is a good visual representation of an arithmetic procedure? Do we know how to present a problem visually? From 'the socio-cultural aspect, it is almost a commonplace to state that we live in a world where information is transmitted mostly in visual wrappings, and technologies support and encourage communication which is essentially visual' (Arcavi 2003, p. 26).

Depending on one's professional position, we find somewhat different perspectives on what visualization is. Presmeg (2006) cited the psychological perspective of Piaget and Inhelder that:

> when a person creates a spatial arrangement (including a mathematical inscription) there is a visual image in the person's mind, guiding this creation. Thus visualisation is taken to include processes of constructing and transforming both visual mental imagery and all of the inscriptions of a spatial nature that may be implicated in doing mathematics. (p. 207)

Clearly the focus is on visual constructs as an aid in the formation of mental concepts. But if we focus on mathematical reasoning in problem solving and problem posing, then one's attention should be shifted. Along these lines, Arcavi (2003) explains:

> Visualization is the ability, the process, and the product, of creation, interpretation, use of and reflection upon pictures, images, diagrams, in our minds, on paper or

Jasmina Milinković, *Visualization as a tool in teaching: Pre-service teachers' competencies in visualization*. In: *Elementary Mathematics Teaching*. Edited by: Brian Doig, Jarmila Novotná, Berinderjeet Kaur and David K. Pugalee, Oxford University Press. © Oxford University Press (2024). DOI: 10.1093/oso/9780192869647.003.0011

> with technological tools, with the purpose of depicting and communicating information, thinking about and developing previously unknown ideas and advancing understandings. (p. 26)

Nowadays, visualization is considered to be an important tool in teaching. It is a way of communicating ideas between teachers and children. When one knows how to transform knowledge into forms comprehensible to learners, it becomes an element of pedagogical content knowledge (Livingston and Borko 1990). According to Shulman, the ways of representing and formulating the subject that make it comprehensible to others are one of the key teachers' competencies related to pedagogical content knowledge. Shulman (1986) stated:

> Since there are no single most powerful forms of representation of ideas, the teacher must have at hand a veritable [arsenal] of alternative forms of representation, some of which derive from research whereas others originate in the wisdom of practice. (p. 9)

The ways of representing the subject are considered to be an element of subject matter knowledge for teachers of mathematics as well as for pupils (Ball, Thames, and Phelps 2008). School practice shows that 'precisely the mathematical terminology and symbolism tends to represent an obstacle for some pupils in understanding mathematics' (Roubíček 2007, p. 231). Roubíček observed that semiotic analysis can shed light on problems in communication between the teacher and the pupil, since the problems tend to be caused by different mental representations of the teacher and the pupil's non-conventional writing forms. Visual representations are often seen as a mediator between children's meta-language and mathematics language. They are external tools used for representing ideas, such as drawings, tables, and graphs. This is why, currently, the issue of visualization is one of the key areas in research in mathematics education.

That visualization is useful is considered to be an intuitively undisputable claim, but historically this was not always the case. There is a long history of debates about the role of visualization in mathematics. In their study, Brating and Pejlare investigated how, historically, some mathematicians supported the role of visual thinking in mathematics (Brating and Pejlare 2008): for example, going back to the seventeenth century, we find a debate between Thomas Hobbes and John Wallis regarding visualizations in mathematics. Could reasoning about realistic objects as models of abstract mathematical objects be sufficient proof? Or, are only mathematical definitions acceptable? Two centuries later, Felix Klein attempted to construct an abridged system of mathematics solely based on naïve intuition but 'Klein had a problem with visual thinking versus the formal development of mathematics in the wake of Weierstrass' function' (Brating and Pejlare 2008, p. 346). It was argued that the individual interacts with a mathematical visualization in a way that is better or worse depending on previous knowledge and the context. This dynamic is sometimes overlooked.

Kilpatrick, Swafford, and Findell (2001) define five features—conceptual understanding, procedural fluency, strategic competence, adaptive reasoning, and productive disposition—as elements of 'mathematical proficiency'. Visualization may be one of the cross-cutting competencies among those features. Particularly in the initial teaching of mathematics, in Grades 1 to 4, visualization is considered to be a critical skill, since pictorial representations facilitate the introduction of new concepts, methods of computation, and problem-solving strategy. Iconic representations serve as a link between the real world and abstract mathematical concepts (Abrams 2001). Ball (1990) reported on a research finding that pre-service teachers, although successful in calculation, were not able to produce good pictorial representations for a problem involving division of a fraction by a fraction. Looking at preparatory programmes for elementary and secondary teachers, one can see a string of reasons for such a weakness. Student teachers only occasionally see professors using visual representations in their mathematics classes. Therefore, the idea of using visual aids does not become appealing to them. Further, unfamiliarity with different representations means that they are not readily available for novice teachers to use.

Arcavi (2003) highlights three functions of visual representations: as a support and illustration of symbolic representations, as a tool for resolving conflict between intuition and symbolic solution, and as a tool to re-engage and recover conceptual understanding. He suggested that 'seeing the thing may sharpen our understanding or serve as a springboard for questions which we were not able to formulate before' (p. 216). Extending the idea of representation as a medium for cognition, Arcavi identified visual representation as a cognitive technology aid for thinking, learning, and problem-solving activities in technology-driven communication.

When mathematicians speak about different representations, they usually think about different ways we can represent a problem. A concept close in meaning to representations is that of schematizations. It refers to the presentation of a certain problem on different levels of abstraction, or thinking about the problem in different paradigms. Freudenthal (1983) stated that historical phases in the development of mathematics signified that knowledge achieved by discovery at one moment had been transformed by schematization (or coding) into new skills and/or understandings on a higher level of abstraction. Early phases of development of mathematical ideas were characterized by visual representations of concepts. A leap forward was the introduction of numerals (symbolic representation of numbers). Schematization should not be viewed as a historical necessity but as part of humans' psychological development in understanding their surroundings (Freudenthal 1983). Bruner proposed a theory of representations in which understanding a representation is a final product of processing and coding information. Bruner's theory of three types of representation (active, iconic, and symbolic) provides a flow chart of progressive schematization and formalization that occurs in the learning process (Bruner 1960). We consider visualization a synonym for an iconic representation.

Visualization in the formation of concepts

Mathematical reasoning ensures the ability to recognize patterns, existing in various contexts on different levels of abstraction, and to transform it into a mathematical form. Effective mathematical thinking involves understanding the relationships among different representations of 'the same concept as well as the structural similarities (and differences) among representational systems' (Goldin and Shteingold 2001, p. 9). Regardless, we need to recognize certain difficulties in engaging in visualization of mathematical concepts, from a cultural, to a cognitive, to a sociological perspective. First, it is a culturally unappreciated method, and therefore there is little space in the school curriculum for presenting such an approach, particularly in higher levels of school (Sfard 1991). But if we go down to earlier school years, things are somewhat different. Perhaps we can claim that all primary mathematics is 'visible'. Somewhere along the line of the school years, the effort to visualize new concepts or procedures diminishes, either because there is a lack of practice (habit) or because it is actually considered not useful. Learning mathematics is perceived as a gradual process, through which meaning is built up by meeting carefully selected problem situations and by drawing on diverse systems of representation and devices, since mathematical objects are not directly accessible to the senses. Bruner's theory of representations underlined the value of visual models as the main path between action and symbolic representations (Bruner 1960).

One can observe the effects of Bruner's theory in a school curriculum's recommendation for using manipulatives and visual models. For example, in the Serbian curriculum, the arithmetic operation of multiplication is primarily modelled as a realistic situation of n elements in m places (e.g. m plates, each with n pieces of cake), m x n diagrams, or set representation, whereas the product is the number of elements of the union of disjoint equivalent sets. Barmby, Harries, and Suggate (2009) examined children's ability to recognize common representations of multiplication. In doing so, they set out a theoretical framework for how children use representations, highlighting that recognition of representations is one of the skills required. We draw on empirical evidence that suggests that this skill of recognition is distinct, and that it is affected by how the representation is composed. Representations of multiplication used in the study were: the number line and string of beads representations, the array collection of m x n stamps, apples, and (sets of) strawberries (Barmby, Harries, and Suggate 2009). The Year 6 children were most able to recognize all of the representations. The authors stress the importance of developing representations of mathematical concepts by moving 'from a process view of a concept to a structural view, for example multiplication as a process of repeated groupings to a static representation such as the array' (Barmby et al. 2009). Milinković (2012) reported several representations, for multiplication, produced by pre-service teachers. Figure 3.3.1 shows nine representations of multiplication discussed, but not all of which were equally frequent in teaching practice in primary grades.

Some mathematical concepts, formulae, or procedures are best understood if represented visually (Figure 3.3.2). For example, coming to a formula for calculating the volume of a cube is best intuitively discovered by pupils if they can see if they can manipulate a three-dimensional model of the formula and/or by studying a well-chosen illustration imitating manipulation.

Not all formulae are so easily discernable, and for some it is not so easy to find good visualizations which will prompt discovery. It is necessary to take into consideration what we want to visualize and for whom. For example, the educated mathematician has no problem communicating through visualizations, nor the student who has grasped the role these visualizations play in mathematics. For a person who is not familiar with the relevant mathematical theory, these visualizations may mean something completely different.

Krátká (2005) focuses on the role of geometrical picture of fundamental geometrical objects. She reflects on why we can consider drawn representations as obstacles

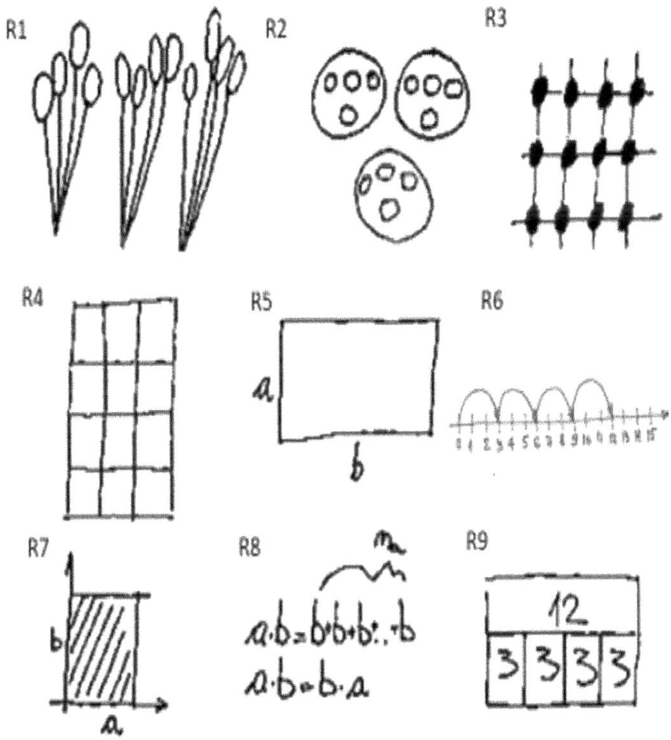

Figure 3.3.1 Teachers' representations of multiplicative situations (Milinković 2012, p. 212)

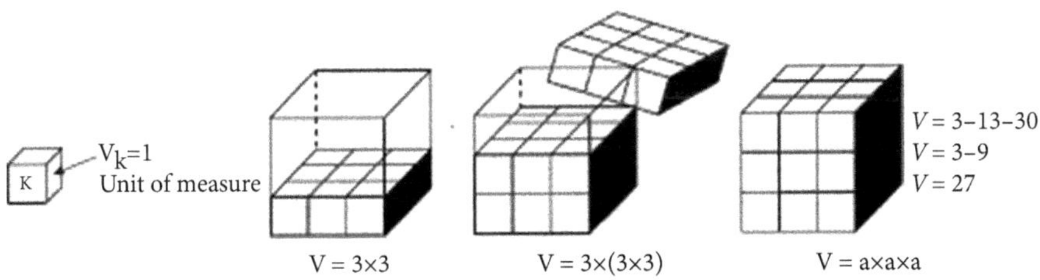

Figure 3.3.2 Volume: coming to a formula *via* visualization (Milinković 2021, p. 89)

from Brousseau's point of view, presenting the historical development of the conceptions of a point and a straight line and its parts. She identified the phenomena of influence of pictorial representations and demonstrated the process of detachment from a picture.

> When we use Brousseau's theory of obstacles, we should find in addition to these stages several obstacles, which a child must overcome. It seems that the drawing of geometrical pictures, respectively the handling geometrical objects by way of geometrical pictures, could play the role of an (epistemological) obstacle. (p. 180)

And further:

> When knowledge went from experiences, the origin of geometry is in practical problems of daily life. From practical problems toward Euclidian geometry is really a long intellectual way, which was got over just with geometrical pictures. (p. 182)

The obstacle is a knowledge, which can produce incorrect answers when used in a new context. This knowledge must be familiar to its owner and must be used successfully many times. These requirements can fill a geometrical picture as a model of a point or a straight line. The children acquaint with fundamental geometrical objects through pictures. Krátká concluded that the pictures of fundamental geometrical objects may help children to work with them, but 'they can prevent them creating better knowledge' (Krátká 2005, p. 186).

Several studies have examined the formation of geometrical concepts (area, perimeter, cube, cuboid, symmetry, etc.) and the role of visual representations. Jirotková, Vighi, and Zemanová (2019) described the results of research focused on the misconceptions about the relationship between perimeter and area, conducted in primary schools with 10-to-11-year-old children. They found that pupils' com-municative skills could be insufficient to articulate their ideas in a comprehensible way. One of the reasons could be that the visual perception of the area is dominant and overrides visualization of the perimeter. The study confirmed that forbidding the use of a rule put the pupils in a really difficult situation.

Prusak and Hershkowitz (2019) presented a task designed for promoting the concept of area in a problem-solving environment aimed at instilling inquiry learning and argumentative norms, relying on shared visual representations. Their main focus was to investigate if, and in what way, principled design is effective for instilling a problem-solving culture and for promoting conceptual learning.

Kloboučková and Vondrová (2019) presented a study aiming at the description of concept development in mathematics. Their question was, 'How do pupils get to know that shapes which do not appear congruent can have the same area, by building on one another's knowledge in a classroom situation?' While individual pupils discovered at least two ways of folding a square into congruent parts, the authors hypothesized that the class as a whole, by discussing and observing one another's work, reached the generic model of the concept. All pupils divided a square into four congruent squares and congruent triangles and visually convinced themselves that there is an infinite number of ways to obtain them.

Mathematical science deals with abstract objects and entities quite different from the real world, and for that reason historically it did not rely on visual representations. But mathematics can also be seen as a science of patterns that can be represented on various levels of abstraction, from real to visual or mental. Some fields of mathematics, such as geometry or statistics, are at least partially predisposed to using visual representations. Regardless of such an obvious fact, for a long time a traditional approach to teaching even these fields at secondary school or university level did not substantially rely on visualization. Illustrative is Tennant's argument (1986) that a diagram is only a heuristic to prompt certain trains of inference that has no proper place in the proof. For him (Tennant 1986) 'the proof system is sound if every sentence deducible from a set of premises is a logical consequence thereof. It is complete if every logical consequence of a set of premises is deducible there from' (p. 303). Tennant (1986) understands proof as a syntactic object consisting only of a finite array of sentences. Such a perspective is undermined by other opinions. For example, Thurston (1994) noted that 'when the mathematical idea is clear, the formal set-up is usually unnecessary and redundant' (Thurston 1994, cited in Bayers 2007, p. 255). Thurston (1994) remarks that mathematicians are making effort to find ways for pupils to understand and think about mathematics. Along this line, Bayers states that 'formal mathematical reasoning is blind to the dimension of the mathematical idea because the formalism of mathematics is downstream from the idea. The mathematical practice involves almost taking formalism for granted' (Bayers 2007, p. 255). Bayers elaborated on this observation, writing: 'when we think about the origins and nature of mathematical thought we must move upstream, past the formal proofs, past the axioms, into the realm of the mathematical idea' (p. 255). In this line of argument, the term 'mathematics' refers to what has sometimes been called 'informal mathematics', as opposed to 'formal mathematics'. Bayers (2007) points out that working from the perspective of ideas enables the development of an argument on the relationship between logic and mathematics, and contrasts Platonism with formalism, and discovery with invention.

Given their previous experience, mathematics teachers may be prone to use only symbolic (analytical) representation, considering this more pedagogically appropriate. Giaquinto (1994) argued that visual thinking can be a means of discovery in geometry but only in restricted cases in elementary analysis. Brating and Pejlare (2008) provided empirical evidence on a problem of drawing 'correct' mathematical conclusions on the basis of visualizations in a specific context. They concluded that 'mathematical concepts may be difficult to understand in any other way than in mathematically well-defined contexts' (p. 2). Some studies do not support the use of visual representations: for example the studies of de Bock et al., 2003, and Siegler and Thompson, 2014.

Bicknell and Young-Loveridge (2015) presented a New Zealand study with five-year-old children that focused on the use of multiplication and division word problems to build number knowledge. Using a design research methodology, researchers and the class teachers developed word problems using familiar contexts and materials with groups of two, five, and ten. Children were assessed before and after a series of targeted lessons and showed improvement in basic number knowledge including subitizing, counting, multiplication, division, and place value. Although familiarity with larger numbers increased, most children did not always make links between number knowledge and representations such as the ten-frame and the number line. We suggest that using problems involving numbers beyond 20 and multiple representations could help extend children's mathematical reasoning beyond common expectations.

Larsson (2013) presented a study examining 24 fifth-grade students' use of multiplicative thinking and whether this is related to their understanding of commutativity and their choice of form of representation (equal groups, array, not corresponding). Larsson reported that students who use multiplicative thinking show understanding of commutativity to a greater extent, as well as the ability to use multiple forms of representation for multiplication compared with students who use additive thinking. The findings suggest that repeated addition of equal groups is not enough to construct and practice multiplicative thinking, and that multiple forms of representation can enhance understanding of commutativity and distributivity.

> When the findings of this study is considered in comparison to other work on different representations [e.g. Barmby et al. 2009] it is not possible to draw conclusions of what forms of representation would best serve to enhance the students' possibilities to understand properties of multiplication. (Larsson 2013, p. 186)

Dabić and Milinković's (2015) presentation focused on pupils' understanding of representations for the commutative property of multiplication (produced by student teachers). Their exploratory study on the use of visual representations as means of communication between pre-service teachers and pupils investigated how well teachers transmit the idea of the commutative law through pictures. A sample of 36 prospective teachers was enrolled in the first part of the study. In the second part of the study, data were collected from a sample of 200 pupils from Serbia. The data analysis of individual interviews with children highlighted discrepancies between what teachers meant to convey through pictures, and how children interpreted this. The authors perceived difficulties in visual communication between teachers and pupils. The combined area and array model was the most preferred representation. The most successful inferences were also drawn from that model. Prospective teachers in the study actually did not distinguish which of those representations were suitable for representing the commutative property. For example, they did not recognize the Cartesian product as a fitting representation for visual presentation of this property. The results of this study have shown that children had significant difficulties in interpreting representations produced by teachers. They tended to assign meanings unlike the ones attempted by teachers. Contrary to earlier assumptions (Skemp 1989; Anghileri 2000; Barmby et al. 2009), Dabić and Milinković (2015) found that the array representation was the least comprehensible for children. The results led them to the conclusion that more abstract models are better means of visual communication between teachers and students in the case of the commutative law of multiplication.

The presentation by Navarra and Giacomin (2005) focused on some aspects of a study about the distributive law carried out with 8-to-11-year-old pupils within the theoretical framework of early algebra. Their hypothesis was that a path leading to understanding the law should start from an analysis of the ways in which perception leads to different mental images that influence the choice of either the (a+b)xc or the (axc)+(bxc) representation. They argued that understanding these dynamics is a fundamental step for both constructing a meaningful approach to the law and favouring its gradual conceptualization.

Gagatsis et al.'s (2003) presentation was on an investigation of the ability of sixth-grade primary school students to translate representations (verbal, tabular, graphical, and symbolic forms) of algebraic relations from one form to another, and identified their preferred types of representation and translation. For the analysis and processing of the data, Gras's implicative statistical analysis was used. The results provided supported the assumption that 'equivalent' mathematical representations constitute different entities for the children and thus each one needs special attention during instruction. Findings also supported the idea that different systems of mathematical representation must be interrelated in teaching, and translations between them should be encouraged.

Taking the Japanese abacus—the soroban—with its Earth and Heaven beads as its metaphor, Doig and Groves (2015) examined the Earth beads necessary to reach the goal of a Heavenly bead—that is, an effective structured problem-solving mathematics lesson.

Way (2005) using two stages of content analysis identified the issues that a cohort of Australian primary pre-service teachers perceived to be important or problematic, particularly regarding the learning and teaching of fractions with young children, in response to reading and in-class experiences. It was found that reflection on the nature of understanding provoked the students to evaluate their own understanding of fraction concepts and to contemplate the pedagogy required to develop children's conceptual understanding.

Figure 3.3.3, below, shows an example of a successful representation and an example of a misleading representation of the addition of fractions with the same denominator. Lipovec and Podgorošek (2017) presented the results of research conducted with 1272 students, aged between 10 and 12 years. Students were asked to provide a drawing depicting numerical expressions such as 3/5 of 15 and 2^3. They analysed the drawings based on three aspects: the adequacy of the drawing to embody the required concept, the conceptual or procedural orientation of the drawing, and the correctness of the result of the numerical expression. It turns out that approximately 60% of students represented the required concept adequately. Procedurally orientated drawings prevailed for both concepts, and predominantly for exponentiation. Additionally, procedurally orientated drawings were more prone to arithmetic mistakes than conceptual ones.

Kolar, Janežič, and Čadež (2015) analysed the understanding of primary student teachers, mathematics student teachers, and eighth-grade primary school students of the following aspect of fractions: the importance of the fixed whole when comparing

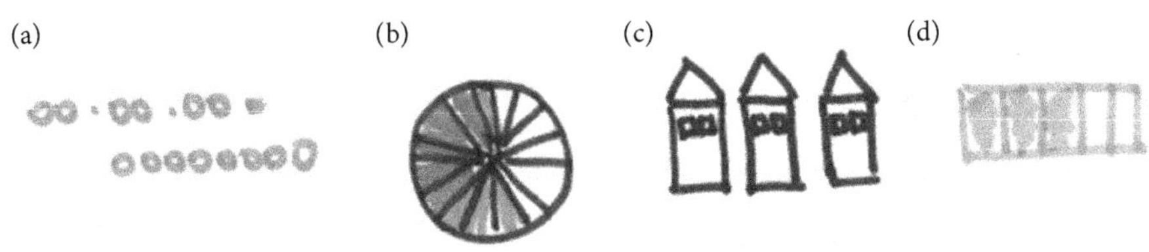

Figure 3.3.3 Conceptual and procedural drawings (Lipovec & Podgoršek, 2017, p. 322)

fractions and representations of fractions greater than one. The comparison of results among these groups gives an insight into the most typical misconceptions in relation to this topic, including a good or bad choice of visual representation.

Tichá (2003) presented a study of the possibilities of utilizing students' translations between modes of representation as a tool for diagnosis in her study of images, level of understanding, misunderstandings, and obstacles for understanding of the notion of fractions. It dealt with students' formulations of word problems for a given arithmetical calculation or for the given visual representation. The translation to a word problem was used as a technique that can help us in answering questions about *why* students have serious difficulties with fractions, *where* the understanding and problem-solving obstacles are, and *how* to replace misunderstandings. Concrete examples of problems formulated by students and demonstrations of misunderstandings were the basis of her SEMT presentation. She concluded that the research indicated that, since some misconceptions, observed in the fifth grade and in the eighth grade and among students at higher secondary schools and universities, confirmed that some misconceptions, that obviously originate in the stage of creation of pre-conceptions, have persisted. So, she inferred that these are in part teacher-induced misunderstandings.

Visualization in problem solving and problem posing

A training course in the didactics of mathematics for elementary school teachers usually includes developing future teachers' competencies in problem posing and problem solving. One of the elements considered in this part of the course is the representations used in problem posing and problem solving. The idea that visual representation is a tool among mathematical reasoning tools is well documented in the literature (Janvier 1987; Cobb, Yackel, and Wood 1992; Michalewicz and Fogel 2000, Cuoco and Curcio 2001; Woleck 2001). 'Real-world meanings can be acted out, modeled with objects, and drawn with simplified math[ematical] drawings' (Fuson 2004, p. 118).

The presentation by Roubíček (2007) explained how communication in the teaching of mathematics could be accomplished using various languages as semiotic representation systems. He stressed that their proper acquirement plays an important role, not only in communication, but also in cognitive processes. Roubíček stated that functional communication in the class represents an important precondition of an effective teaching process. It is apparent that it is never possible to arrive at complete agreement in communication between the teacher and the pupil, as their knowledge of mathematics is different. Roubíček argued that, in some class situations, the pupils communicate on a given mathematical problem more easily among themselves than with the teacher. While the teacher attends to the formal aspect of the communication, the pupils often use unconventional expressions. Perhaps this is too vague from the teacher's point of view, despite being understandable by the pupils.

Ruppeldtová's presentation (2007) addressed several cognitive phenomena he had identified through analysing the solutions, made up by pre-service preschool and elementary teachers, when creating word problems with operators of a conceptual character. Special attention was paid to those questions and solving strategies which led to a word problem transformed into problems of a syllogistic type. A different focus was presented by Venuto and Hart (2017) in their discussion of a teaching experiment in a first-grade classroom in a large metropolitan school district in the United States, which looked at the benefits of adding written discourse within the context of eight conceptually based lessons grounded in the Cognitively Guided Instruction (CGI) approach. Examples of the students' written explanations from before and after the teaching unit were shared. They demonstrated the change in the children's ability to explain their mathematical reasoning in writing combined with drawing, suggesting that adding writing to conceptually based lessons improves children's ability to communicate their thinking. See, for example, Figure 3.3.4 below.

It has been argued, also, that visualization is important in problem solving (Friedlander and Tabach 2001; Goldin and Shteingold 2001). One way to help students to become confident in using visualization in problem solving is to confront them with problems posed in a visual form. Researchers emphasize the importance of problem visualization as a link between external representations and mental images. Using several representations of problems provides models for their use in the solution process (Friedlander and Tabach 2001). Wittmann (2005) supported the idea of using visualization as a mediating tool preceding symbolic representations. Similarly, Singer, Ellerton, and Cai (2015) pointed out that the task format underlines sequences of transfers from external to internal representation.

In her presentation, Highfield (2013) described the mathematical engagement of 31 children, aged between three and seven, as they participated in a 12-week programme

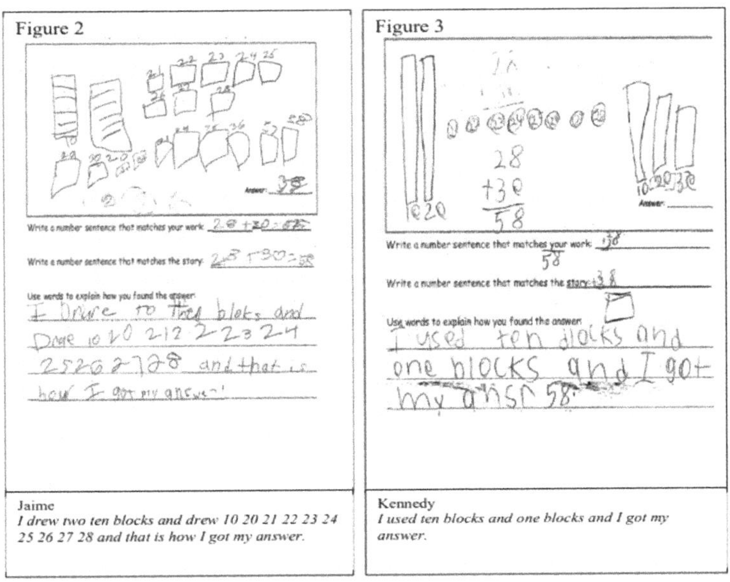

Figure 3.3.4 Children's solutions in words and drawings (Venuto and Hart 2017, p. 425)

using simple robots. The study examined the children's mathematical learning and play in two contexts: a prior-to-school setting and Grade 1 in a primary school. Data collected from videotaped experiences, semi-structured interviews, and children's representations highlighted the rich mathematical learning in number, space, and measurement. The children engaged in a cycle of programming the robotic toys that promoted reflective problem solving. Jirotková and Slezáková's (2013) presentation showed how pupils' 'stay' in the mathematical learning environment and can become the source of their experience with the role of numbers, the operations of addition and subtraction, and relations between numbers. It also showed how this 'stay' stimulates their further development. The researchers' attention focused on recording the processes of the 5-to-7-year-old pupils, and the development of their written recording in structuring relevant information using a chart.

Csachová and Jurečková (2019) focused their presentation on the success of 10-and 11-year-old pupils in problem solving in the Slovak nationwide testing, and found that pupils have had difficulties with some problems because they were not able to read data from figures in the questions. On the other hand, some selected problems were easy for pupils because figures (e.g. illustrations) made the problem-solving process easier.

There is no agreement among researchers or among mathematicians regarding the benefits of diverse representations in teaching. While some researchers oppose it, others make strong arguments for the use of them. For example, Cobb (1988), Erlwanger (1973), and Lawler's (1981) research showed that students have difficulties in recognizing the relationship of different solutions of identical problems given in different contexts (cited by Hiebert and Carpenter 1992). There is evidence that teachers have difficulty in creating visual representations, and distinguishing good ones from poor ones (Milinković 2012; Dabić and Milinković 2015). Gökçek and Azak (2015) examined sixth-grade students' problem-solving strategies when they encountered non-routine problems in class. They reported that students used strategies involving guessing and checking, drawing a model or diagram, and writing an equation in solving problems. Piñeiro and colleagues (2019) reported on the design of an instrument to explore prospective primary school teachers' mathematical problem-solving knowledge, which included items assessing their ability or recognizing a problem-solving strategy (looking for a pattern, guess and check, building a table, drawing a diagram).

Several researchers have called for the use of multiple representations in teaching (Greer 1992; Nunes 1992; Arcavi 2003; Sfard 2003). Arcavi (2003) considered visual representations as a cognitive technology tool for thinking, learning, and problem solving. He recognized three functions of visual representations: as a support and illustration of symbolic representations; as a tool for resolving conflict between intuition and symbolic solution; and as a tool to reorganize and recuperate conceptual understanding. It was suggested that visualization sharpens our understanding and serves as a springboard for questions that we would not pose otherwise (Milinković 2015). Visualization is linked to the process of mathematical modelling (Greer 1992; Nunes 1992): 'Understanding several different situations involving the same invariant leads the abstraction and generalization of the core concept (the invariant), and to the enrichment of the concept by extending the set of situations to which it applies' (Nunes 1992, pp. 571–572).

How does one come up with a conclusion to the question of how much is the sum of *n* odd numbers? How would ancient Greeks solve this problem? Pythagoreans visualized this situation. Pythagoras expressed his ideas by using words and symbols rather than numbers. His ideas about the universe were expressed with a series of symbolic visual representations. Pythagoreans believed that everything starts with a monad, a representation for number one, with a point and a circle around it. Thus, they would start with the monad, and continue the pattern as shown in Figure 3.3.5.

One way to generate all odd numbers from 1 to (2n–1) is visualized with the dot grid made of n^2 dots (Figure 3.3.6). The first dot in the upper-left corner represents 1 (stands alone). The next three dots are linked, then five, etc. In that way, we have generated a sequence of n consecutive odd numbers from 1 to (2n–1). Finally, we get the conclusion that the sum of the n odd numbers equals n^2. This is shown in Table 3.3.1 below.

The Pythagoreans used this method of visualization as a way of coming up with a proposition (theorem) and showing the validity of various propositions written with symbols, building Pythagorean number theory. Historically, once mathematical induction was introduced as a way to prove these kinds of theorems, visual representations of claims were abandoned. Diagrams were essential to mathematical reasoning at least until the flourishing of algebra in the seventeenth and eighteenth centuries. Moreover, the formalist conception of proof prevailed after mathematicians developed the notion of abstract structure in the nineteenth century.

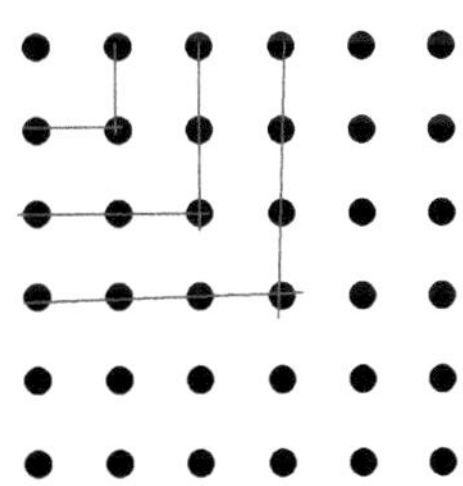

Figure 3.3.5 Visualizing the sum of *n* numbers

Table 3.3.1 The sum of n odd numbers

n	Σ
1	1
2	$1 + 3 = 4$ (= $2^{2)}$
3	$1 + 3 + 5 = 9$ (= 3^2)
	...
n	$1 + 3 + \ldots + (2n - 1) = n^2$

Using variants of this problem, Rütten and Weskamp (2019) presented their research that used a design of a combinatorial learning environment for fostering reasoning skills: diagrammatic reasoning, conjectures, and justifications, in building towers of cubes and cuboids:

> Diagrammatic reasoning synthesizes a manifold of relations that integrates the construction [and] observation of a diagram, the observation of structural relations among its parts, and the perceptual manipulation and thought-experimentation to infer new possible relations conducive to the attainment of the conceptualization of the Object of the sign-vehicle (the Object-as-it-is). (p. 364)

They point out that the aim of diagrammatic reasoning is the construction of a mathematical argument that is a warrant for the abstract structure of the mathematical object, whereas visualization mediates the emergence of diagrammatic reasoning to arrive at generalization.

López Centella's presentation (2019) explored how students in the third grade of primary education (eight to nine years) think about questions based on a generalizing problem, involving a functional relation pictorially presented. She examined the strategies, functional relations and rules, and systems of representations employed by students through the analysis of their responses to paper-based questionnaires. The findings illustrated the variety and combined use of strategies that pupils employ to express the relationship between co-varying quantities, the importance of number sizes involved in questions to foster functional thinking, and the influence of the picture used in the problem statement on their understanding and solving processes. It was observed that many students used more than one strategy in their responses, operational strategies, and counting of drawings. The fact that they combined operational strategies with counting of drawings seemed to be more related to their need for a visual support to try to find out the relationship between the quantities in play. In a previously conducted research, Blanton (2008) posed an original problem to investigate students' reasoning when confronted with the task of organizing seating of a particular number of people at a sequence of tables placed next to each other, but in accordance with rules that at a trapezium-shaped table one person can sit at each side of the table, except at the longest side, where two persons can be seated (see Figure 3.3.6). Students were asked to explain their reasoning in determining the necessary number of tables. Regarding the pictorial representation system, their drawings of the tables influenced their identification of the functional relation involved in the problem.

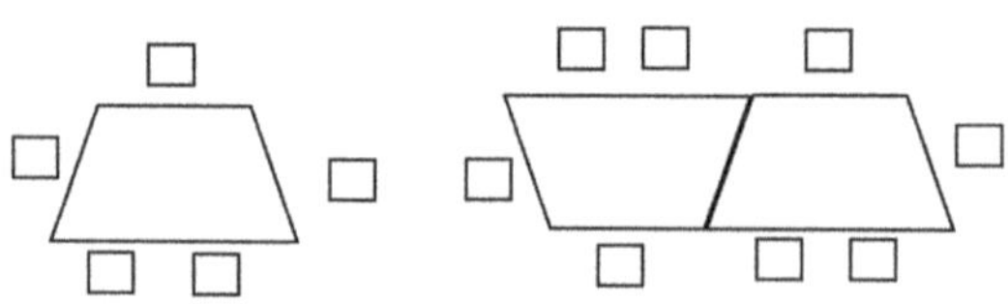

Figure 3.3.6 The table problem

López Centella's (2019) presentation described numerical, verbal, and pictorial systems of representation used in responses by eight- and nine-year-old students to a similar context. The drawings were analysed with a focus on tables and guests as elements of the drawing, and the relationship established between the elements (shape and number of guests drawn). Regarding the pictorial representation system, the drawings of the tables influenced the identification of the functional relation involved in the problem. Pupils tended to represent as many guests as they apparently could fit on each side of the tables they had drawn. It was remarked that the visual transformation of the figures needed for an appropriate joining of the tables was an obstacle for some students.

According to Barwise and Etchemendy (1991):

> Mathematicians have been aware of the value of diagrams and other visual tools both for teaching and as heuristics for mathematical discovery ... But despite the obvious importance of visual images in human cognitive activities, visual representation remains a second-class citizen in both the theory and practice of mathematics. In particular, we are all taught to look askance at proofs that make crucial use of diagrams, graphs, or other non-linguistic forms of representation, and we pass on this disdain to students [however], visual forms of representation can be important ... as legitimate elements of mathematical proofs. (cited in Arcavi 2003, p. 32)

Knowledge of representations, as was noted earlier, is particularly important in problem solving (Polya 1957; Goldin 1987). One way to help students become confident in using different representations in problem solving is to confront them with different forms of problems. Friedlander and Tabach (2001) argued that teachers' presenting problem situations in different representations encourages flexibility in students' choice of representations. They stated, 'The presentation of a problem in several representations gives legitimatization to their use in the solution process' (p. 176). Similarly, Singer, Ellerton, and Cai (2015) emphasized 'the tasks' format underlines cycles of transfer in order to internalize' (p. 147). But, in practice, teachers rarely consider different representations in problem posing as an important issue. Second, it may bring some challenge to a classroom because, before using visual representations in solving problems, children have to pass through a phase of learning to visualize. This requires the ability to translate between symbolic and visual representation.

In their presentation, Cifarelli et al. (2017) used the results of a study of students solving 'net worth' problems, a class of real-world problems that model the operations of addition and subtraction, and, thus, represent familiar mathematical situations for elementary-grade students. Using episodes drawn from classroom work, they focused in their analysis on examining the concepts involved in solving these types of problems and teachers' implementation of an instructional sequence, including visualization, on a number line, designed to support students in making connections between the underlying concepts and formal integer addition and subtraction operations.

Fritzlar (2007) presented a case study of how pre-service student teachers implemented a mathematical problem, with much potential for problem solving, into their own lessons in class. Models and representations were 20% of the issues across 14 of the discussion threads that dealt with the significance of the different ways in which fractions are modelled and represented. It appeared that the idea of working with a variety of representations for fractions made considerable impact on many students, and again they made explicit connections to building conceptual understanding.

Pugalee and Seifert (2013) studied sixth-grade students combining visual and written argumentation in problem solving. Results showed that, over time, an emphasis on writing positively influences students' mathematical communication and higher levels of reasoning (see Figure 3.3.7). The problem was as follows:

> Both Diana and Tabatha ate ¼ of a pizza. Diana argues that she ate more pizza than Tabatha. Is it possible for Diana to be right?

Student's written response:

Well as you can see the pizza could have been bigger with more pieces.

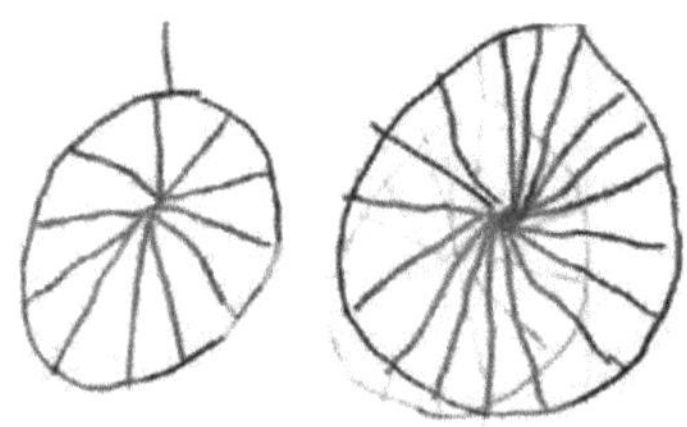

Figure 3.3.7 Responses to the fraction question (Pugalee & Seifert, 2013, p. 261)

An exemplary student's answer on the fraction task included a drawing of a small pizza divided into ten (unequal) pieces and a big pizza divided into 17 (unequal) pieces and a verbal explanation as follows: 'Well, as you can see, the pizza could have been bigger with more pieces' (Pugalee and Seifert 2013, p. 261).

Anastasiadou (2009) presented a structural equation model representing the hierarchical structure of translation among representations of a frequency concept. She aimed to contribute to the understanding of the approaches students use in solving tasks related to the frequency concept and to examine which approach is more strongly correlated to their success on such tasks. The participants were Greek sixth-grade students. Implicative statistical analysis was applied to evaluate the relation between students' approach and their ability to solve problems. Confirmatory Factor Analysis (CFA), in the form of structural equation modelling, was applied to the data. The structural model that resulted from the analysis confirmed the existence of five first-order factors relative to the frequency representations.

Divišová and Stehlíková (2011) presented their research that dealt with a certain type of geometric problem—that is, problems effectively solvable without algebraic calculations. They observed pupils' preference for using (often lengthy) algebraic solutions over (often quick) geometric ones. In particular, they studied pupils' solving strategies, hints that might lead towards a geometric solution, and whether the strategies were age related.

Guidoni et al. (2011) presented some of the main aspects of the cognitive model of 'resonance' as a tool to describe, plan, and manage teaching and learning processes in mathematics. According to the Vygotskian tradition, in this model, mathematics education is taken as the gradual development and refinement, by means of cultural mediation and social interaction, of natural strategies that are known to be at the root of mathematics. The authors showed how resonance dynamics may be triggered, through the intentional use of suitable artefacts. In particular, Guidoni and colleagues described the first steps of a possible approach to the polynomial representations of numbers in a first-grade class.

A somewhat unusual way of representing problems is visual. In that case, information is presented either exclusively in pictures, or in a combination of picture(s) and text. Figure 3.3.8, below, illustrates a reading picture problem, adapted from a third-grade mathematics textbook (Milinković & Dabić, 2020, p. 14). The picture is an integral part of the problem, otherwise the issue of attending to the same mathematical content could be posed exclusively in a textual form.

What is the mass of the pills on the left side of the scale?
Circle the weights that need to be placed on the other side of the scale to keep it in balance.

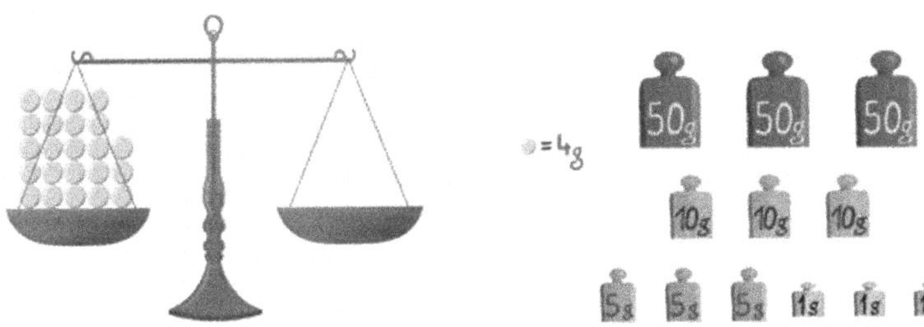

Figure 3.3.8 Reading picture, adapted from Textbook *3rd Grade Mathematics* (Milinković & Dabić, 2020, p. 14)

The same context, but with a modification of the question, could be used in the domain of fractions. Similarly, visually posed balance-scale problems could be used at various grade levels to practise arithmetic operations and solve algebraic tasks. Picturing data is a domain of school mathematics where it is not unusual, but common practice, to present information visually via graphs. Graphs such as pictograms, pie charts, bar graphs, or line graphs are alternative ways to present data visually. Often, understanding information provided in the visual format of a graph, analysis of the graph, and evaluation will enable the pupil to find an answer.

Less frequently, problems in picture format are present in other domains. The next problem, from the domain of numbers and combinatorics, is a variation of a problem from a fourth-grade workbook (Milinković 2021, p. 113). The same problem may be posed in words, but it would require many words and perhaps would sound difficult.

Bosko and Milica put together five-digit numbers by pulling the balls out of the box. What numbers they could make?

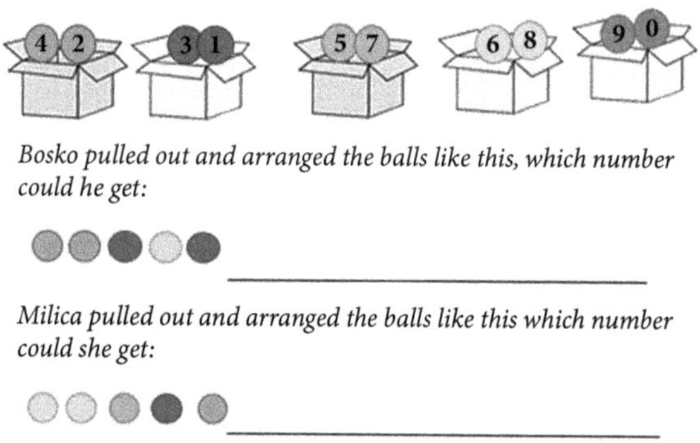

Figure 3.3.9 Problem in picture, from the *4rd Grade Mathematics textbook* (Milinković, 2021, p. 20)

An important type of problem is the one in which the final goal is to find the pattern of growth and predict the solution for the n-th iteration. Would a visual representation of the problem help or not? A good example of such a problem was designed for the US *Mathematics in Context* curriculum for middle grades (Romberg 1996). In one unit, students develop an understanding of the idea of patterns. In particular, they learn how to express a pattern with mathematical symbols (see Figure 3.3.9 below).

In the activity, they study the growth pattern of a snake with red (R) and black (B) rings. They are asked to explain the snake's skin pattern of growth, based on a visual representation, and to predict how the snake's skin would look like at the n-th iteration (see Figure 3.3.10). A drawing showed how, from one R ring, in the next step

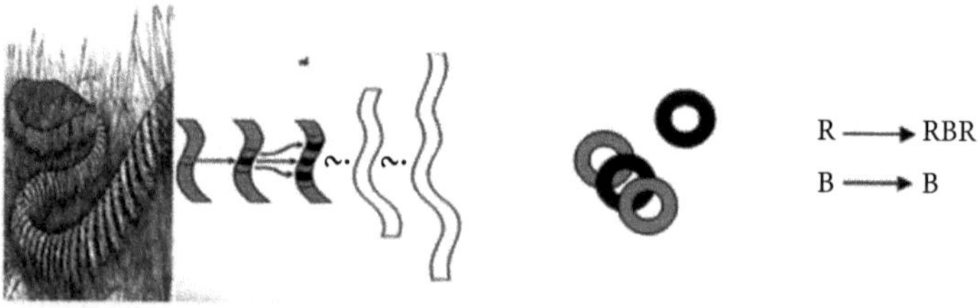

	PATTERN	number of red rings	number of black rings	number of rings
1	R	1	0	1
2	RBR	2	1	3
3	RBRBRBR	4	3	7
4	...	8	...	...
5	...	...	15	...

R rings 1, 2, 4, 8, ... 2^{n-1}

B rings 0, 1, 3 ..... 2^{n-1}-1

Total 1, 3, 7 ..... 2^{n}-1

Figure 3.3.10 Adapted from Mathematics in Context textbook series (Romberg, 1996).

in the cycle of growth, the snake's skin had a more complex RBR pattern, and in the next cycle, the pattern became even more complex. Students were expected to study the pattern of growth and predict how the snake would look at the n-th iteration of growth (Romberg, 1996).

Conclusion

This chapter addressed the role of mathematical visualization in concept formation, problem posing, and problem solving. It was argued that mathematical visualization provides cognitive accessibility to abstract mathematical concepts and problems. Traditional mathematical practice may hinder teachers in fully appreciating how useful visualization may be in mediating the passage from the real world to abstraction, as needs to occur in pupils' minds. The history of mathematics shows us that, for a long time, visualization was disregarded, in contrast to pure theoretical proof. However, there has been a strong push for visualization in problem solving, contrary to traditional notions, and in accordance with the findings of research into the psychology of mathematics learning. Accordingly, visualization is, nowadays, seen as an important pedagogical tool in teaching, as it provides a mode for reasoning.

If one looks at the examples provided here, taken from mathematical research and practice, one sees that the use of visualization is supported by evidence from the research in model situations, although there is a dependence on the context and teacher expertise.

References

Abrams, J. (2001). Teaching mathematical modeling and the skills of representation. In A. Cuoco and F. Curcio (Eds) *The Roles of Representation in School Mathematics, 2001 Yearbook*, 269–282. Reston, VA: NCTM.

Anastasiadou, S. (2009). A Structural equation model representing the hierarchical structure of translations among representations in frequency concept. In J. Novotná and H. Moraová (Eds) SEMT 57–64.

Anghileri, J. (2000). *Teaching Number Sense*. London: Continuum.

Arcavi, A. (2003). The role of visual representations in the learning of mathematics *Educational Studies in Mathematics*, 52: 215–241.

Ball, D., Thames, M., and Phelps, G. (2008). Content knowledge for teaching: What makes it special? *Journal of Teacher Education*, 59(5): 389–407.

Ball, L. (1990). The mathematical understanding that pre-service teachers bring to teacher education. *Elementary School Journal*, 90: 499–466.

Barmby, P., Harries, T., and Suggate, J. (2009). Recognizing representations of multiplication. In J. Novotná and H. Moraová (Eds) SEMT 65–72.

Barmby, P., Harries, T., Higgins, S., and Suggate, J. (2009). The array representation and primary children's understanding and reasoning in multiplication. *Educational Studies in Mathematics*, 70(3): 217–241.

Barwise, J., and Etchemendy, J. (1991). Visual information and valid reasoning. In W. Zimmermann and S. Cunningham (Eds) *Visualization in Teaching and Learning Mathematics*, 9–24. Washington, DC: Mathematical Association of America.

Bayers, W. (2007). *How Mathematicians Think.* Princeton, NJ: Princeton University Press.
Bicknell, B., and Young-Loveridge, J. (2015). Using problem solving and representations to build number knowledge. In J. Novotná and H. Moraová (Eds) SEMT 63–72.
Blanton, M. L. (2008). Algebra and the elementary classroom: Transforming thinking, transforming practice. Heinemann Educational Books.
Brating, K., and Pejlare, J. (2008). Visualizations in mathematics. *Erkenntnis*, 68(3): 345–358.
Bruner, J. (1960). *The Process of Education.* Cambridge, MA: Harvard University.
Cifarelli, V., Stephan, M., Pugalee, D., and Wang, C. (2017). Implementing instructional sequence for solving net worth problems. In J. Novotná and H. Moraová (Eds) SEMT 148–158.
Cobb, P. (1988). The tension between theories of learning and instruction in mathematics education. *Educational Psychologist*, 23: 87–103.
Cobb, P., Yackel, E., and Wood, T. (1992). A constructivist alternative to the representational view of mind in mathematics education. *Journal for Research in Mathematics Education*, 23(1): 2–33.
Couco, A., and Curcio, F. (Eds) (2001). *The Roles of Representation in School Mathematics.* NCTM Yearbook. Reston, VA: National Council of Teachers of Mathematics.
Csachová, L., and Jurečková, M. (2019). Analysis of problems with figures in primary grades mathematics. In J. Novotná and H. Moraová (Eds) SEMT 140–148.
Dabić, M., and Milinković, J. (2015). Teachers' representations of multiplication – do children understand them? In J. Novotná and H. Moraová (Eds) SEMT 99–107.
De Bock, D., Verschaffel, L., Janssens, D., Van Dooren, W., and Claes, K. (2003). Do realistic representations always have a beneficial impact on students' performance? Negative evidence from a study on modeling non-linear geometry. *Learning and Instruction*, 13(4): 441–463
Divišová, B., and Stehlíková, N. (2011). Geometric problems effectively solvable without algebraic calculations. In J. Novotná and H. Moraová (Eds) SEMT 126–134.
Doig, D., and Groves, S. (2015). Reaching for the heavenly bead of teaching through structured problem solving. In J. Novotná and H. Moraová (Eds) SEMT 117–127.
Erlwanger, S. H. (1973). Benny's conception of rules and answers in IPI mathematics. *Journal of Children's Mathematical Behavior*, 1(2): 7–26.
Freudenthal, H. (1983). *Didactical Phenomenology of Mathematical Structures.* Dordrecht, The Netherlands: Reidel.
Friedlander, A., and Tabach, M. (2001). Promoting multiple representations of Algebra. In A. Cuoco and F. Curcio (Eds) *The Roles of Representations in School Mathematics, 2001 Yearbook*, 173–186. Reston, VA: NCTM, Macmillan Publishing Company.
Fritzlar, T. (2007). Why is it so difficult to use a 'good problem' in a 'good way' for teaching? In J. Novotná and H. Moraová (Eds) SEMT 109–116.
Fuson, K. C. (2004). Pre-K to grade 2 goals and standards: Achieving 21st-century mastery for all. In D. Clements and J. Sarama (Eds) *Engaging Young Children in Mathematics: Standards for Early Childhood Education*, 105–148. Mahwah, NJ: Lawrence Erlbaum.
Gagatsis, I., Elia, A., Mouyi, A., and Evangelidou, A. (2003). Primary students' aptitude to express algebraic relations. In J. Novotná (Ed.) SEMT 72–78.
Giaquinto, M. (1994). Epistemology of visual thinking in elementary real analysis. *The University of Chicago Press Journal*, 45: 789–813.
Gökçek, T., and Azak, S. (2015). Sixth grade students' strategies for solving mathematical problems. In J. Novotná and H. Moraová (Eds) SEMT 127–137.
Goldin, G. A. (1987). Cognitive representational systems for mathematical problem solving. In C. Janvier (Ed.) Problems of Representation in the Leaching and Learning of Mathematics, 125–145. Hillsdale, NJ: Lawrence Erlbaum Associates.
Goldin, G., and Shteingold, N. (2001). System of representations and the development of mathematical concepts. In A. Cuoco and F. Curcio (Eds) *The Roles of Representation in School Mathematics*, 1–23. NCTM Yearbook. Reston, VA: NCTM.
Greer, B. (1992). Multiplication and division as models of situations. In D. Grouws (Ed.) *Those Who Understand: Knowledge Growth in Teaching of Research on Mathematics Teaching and Learning*, 276–295. Reston, VA: NCTM, Macmillan Publishing Company.

Guidoni, P., Mellone, M., Minichini, C., and Serpico, M. (2011). Towards a 'Resonance' comprehension of the polynomial representation of numbers. In J. Novotná and H. Moraová (Eds) SEMT 142–150.
Hiebert, J., and Carpenter, T. (1992). Learning and teaching with understanding. In D. A. Grouws (Ed.) *Handbook of Research in Mathematics Teaching and Learning*, 65–97. New York: Macmillan.
Highfield, K. (2013). Playing with simple robotics: A tool for mathematics learning in the early years. In J. Novotná and H. Moraová (Eds) SEMT 122–130.
Janvier, C. (Ed.) (1987). *Problems of Representation in the Teaching and Learning of Mathematics.* Hillsdale, NJ: Erlbaum.
Jirotková, D., and Slezáková, J. (2013). Didactic environment bus as a tool for development of early mathematical thinking. In J. Novotná and H. Moraová (Eds) SEMT 147–154.
Jirotková, D., Vighi, P., and Zemanová, R. (2019). Misconceptions about the relationship between perimeter and area. In J. Novotná and H. Moraová (Eds) SEMT 221–230.
Kilpatrick, J., Swafford, J., and Findell, B. (2001). *Adding It Up: Helping Children Learn Mathematics.* Washington, DC: National Academic Press.
Kloboučková, J., and Vondrová, N. (2019). Getting to know shapes of the same area – how pupils build on one another's work. In J. Novotná and H. Moraová (Eds) SEMT 231–240.
Kolar, V., Janežič, A., and Hodnik Čadež, T. (2015). Diagnosing students' difficulties in understanding the concept of fraction. In J. Novotná and H. Moraová (Eds) SEMT 233–241.
Krátká, M. (2005). A geometrical picture as an obstacle. In J. Novotná (Ed.) SEMT 179–186.
Larsson, K. (2013). Multiplicative thinking in relation to commutativity and forms of representations. In J. Novotná and H. Moraová (Eds) SEMT 179–187.
Lawler, R. W. (1981). The progressive construction of mind. *Cognitive Science*, 5: 1–30.
Lipovec, A., and Podgorošek, M. (2017). Student's visual representation of fraction and exponentiation. In J. Novotná and H. Moraová (Eds) SEMT 319–328.
Livingston, C., and Borko, H. (1990). High school mathematics review lessons: expert/novice distinctions. *Journal for Research in Mathematics Education*, 21(7): 372–387.
López Centella, E. (2019). Functional thinking of third grade students: A study from early algebra framework. In J. Novotná and H. Moraová (Eds) SEMT 251–240.
Michalewicz, Z., and Fogel, D. (2000). *How to Solve It: Modern Heuristics.* Berlin/Heidelberg: Springer.
Milinković, J., and Dabić, M. (2020). *Matematika za 3. Razred (Mathematics for 3rd Grade).* Beograd: Kreativni centar.
Milinković, J. (2012). Pre service teachers' representational preferences. In Tai-Yih Tso (Ed.), Opportunities to Learn in Mathematics Education. Proceedings of the 36th Conference of the International Group for the Psychology of Mathematics Education, vol. 3, 209–215. Taipei, Taiwan: PME (2012).
Milinkovic, J. (2015) Conceptualizing problem posing via transformation. In F. Singer, N. Ellerton, and J. Cai (Eds) (2015). *Mathematical Problem Posing from Research to Effective Practice.* New York: Springer. doi.org/10.1007/978-1-4614-6258-3_26
Milinković, J. (2021). *Matematika za 4. Razred. (Mathematics for 4th Grade).* Beograd: Kreativni centar.
Milinković, J. (2021). *Mathematics, A Textbook for 4th Grade.* Beograd: Kreativni Centar.
Milinković, J., and Dabić, M. (2020). *Matematika za 3. Razred (Mathematics for 3rd Grade).* Beograd: Kreativni centar.
Navarra, G., and Giacomin, A. (2005) The achievement of the equivalence of representations of the distributive law with young pupils. In J. Novotná and H. Moraová (Eds) SEMT 235–243.
Nunes, T. (1992). Cognitive invariants and cultural variation in mathematical concepts. *International Journal of Behavioral Development*, 14(4): 433–453.
Pelczer, I., Singer, F., and Voica, C. (2011). Between algebra and geometry: the dual nature of the number line. In M. Pytlak, T. Rowland, and E. Swoboda (Eds) Proceedings of the 7th Conference of the European Society for Research in Mathematics Education (CERME 7), 376–385. Rzeszów, Poland: University of Rzeszów.
Piaget, J., & Inhelder, B. (1971). *Mental imagery and the child.* London: Routledge & Kegan Paul.

Piñeiro, J., Chapman, O., Castro-Rodríguez, E., and Castro, E. (2019). Students' visual representation of fractions and exponentiation. In J. Novotná and H. Moraová (Eds) SEMT 305–331.
Polya, G. (1957). *How to Solve It.* Princeton, NJ: Princeton University Press.
Presmeg, N. (2006). Research on visualisation in learning and teaching mathematics. In A. Gutierrez and P. Boero (Eds) *Handbook of Research on the Psychology of Mathematics Education: Past, Present, and Future,* 205–236. Rotterdam, The Netherlands: Sense Publishers.
Prusak, N., and Hershkowitz, R. (2019). Task designed for promoting the concept of area in a problem solving environment: The story of one activity. In J. Novotná and H. Moraová (Eds) SEMT 325–334.
Pugalee, D., and Seifert, A. (2013). Including writing to extend cognitive level for mathematical tasks. In J. Novotná and H. Moraová (Eds) SEMT 257–264.
Romberg, T. A. (Ed.) (1996). *Mathematics in Contexts: A Connected Curriculum for Grade 5–8.* Chicago, IL: Encyclopaedia Britannica Educational Corporation.
Roubíček, F. (2007). Some aspects of communication in the teaching of mathematics: The semiotic viewpoint. In J. Novotná and H. Moraová (Eds) SEMT 225–232.
Ruppeldtová, J. (2005). Types of word problem visualisation in pupils' solutions. In J. Novotná and H. Moraová (Eds) SEMT 233–240.
Rütten, C., and Weskamp S. (2019). Conjectures and justifications in building towers of cubes and cuboids: Design of a combinatorial learning environments. In J. Novotná and H. Moraová (Eds) SEMT 344–353.
Sfard, A. (1991). On the dual nature of mathematical conceptions: Reflections on processes and objects as different sides of the same coin. *Educational Studies in Mathematics,* 22(1): 1–36.
Sfard, A. (2003). Balancing the unbalanceable: The NCTM Standards in light of theories of learning mathematics. In J. Kilpatrick, W.G. Martin, and D. Schifter (Eds) *A Research Companion to Principles and Standards for School Mathematics,* 353–392. Reston, VA: NCTM.
Shulman, L. S. (1986). Those who understand: Knowledge growth in teaching. *Educational Researcher,* 15(2): 4–14.
Singer, F., Ellerton, N., and Cai, J. (2015) (Eds). *Mathematical Problem Posing: From Research to Effective Practice.* New York: Springer.
Skemp, R. (1989). *Mathematics in Primary School.* London: Routledge.
Siegle, R. and Thompson, C. (2014). Numerical landmarks are useful—except when they're not. *Journal of Experimental Child Psychology,* 120, pp. 339–58.
Tennant, N. (1986). The withering away of formal semantics? Mind and language 1, 302–318. Retrieved July 2014, from https://cpb-us-w2.wpmucdn.com/u.osu.edu/dist/a/4597/files/2014/07/tennant_ml1986-w4txio.pdf
Thurston, W. (1994). On proof and progress in mathematics. *Bulletin of the American Mathematical Society,* 30(2): 161–177.
Tichá, M. (2003). Following the path of discovering fractions. In J. Novotná (Ed.) SEMT 16–25.
Venuto, N., and Hart, L. (2017). First graders write about mathematics: A teaching experiment. In J. Novotná and H. Moraová (Eds) SEMT 421–430.
Way, J. (2005). Umm ... I think we're lost! Pre-service teachers' thoughts on developing children's understanding of fractions. In J. Novotná and H. Moraová (Eds) SEMT 248–256.
Wittmann, E. (2005). Mathematics as the Science of Patterns: A guideline for developing mathematics education from Early Childhood to adulthood. Paper presented at the International Colloquium Mathematical Learning from Early Childhood to Adulthood, Mons, Belgium.
Woleck, K. (2001). Listen to their pictures: An investigation of children's mathematical drawings. In A. Cuoco and F. Curcio (Eds) *The Roles of Representation in School Mathematics: 2001 Yearbook,* 215–227. Reston, VA: National Council of Teachers of Mathematics.

3.4
Pre-service primary student teachers' development of competencies in teaching mathematics

Jana Slezáková
Charles University, Prague, The Czech Republic
Darina Jirotková
Charles University, Prague, The Czech Republic
Katerina Juzová
Charles University, Prague, The Czech Republic
Karolina Mottlová
Charles University, Prague, The Czech Republic
Tereza Vybíralová
Charles University, Prague, The Czech Republic
Renata Zemanová
University of Ostrava, Ostrava, The Czech Republic

Introduction

The primary focus of SEMT symposia is primary school mathematics and the education of primary teachers in mathematics. As some of the authors of this chapter (the first three) have been involved in pre-service primary teacher education throughout their careers and the remaining authors have embarked on this career within their doctoral studies, we chose to present this chapter on this topic. From the SEMT proceedings from 1991 to 2019, we selected 102 papers that touch on our topic. In the end, however, we decided not to include those that primarily focus on the field of preschool teacher education, or lower secondary school teachers. We also excluded papers that were too narrowly focused on one particular phenomenon and did not, in our view, contribute in any way to addressing the issue of (future) teacher education. Thus, 66 contributions were selected for our examination.

Throughout the whole 30 years of the SEMT symposia, similar issues can be seen as intertwining. None of the topics appears to have a broader impact or to have been brought to a conclusion. This is not surprising because the view of what quality mathematics teaching at primary level is has been gradually changing worldwide. The criteria for quality teaching have been shifting from an emphasis on content to an emphasis on the pupil, their individuality, and personal development, and, in the last few years, on the teacher, their competencies, pedagogical skills, and mathematical

Jana Slezáková et al. *Pre-service primary student teachers' development of competencies in teaching mathematics*. In: *Elementary Mathematics Teaching*. Edited by: Brian Doig, Jarmila Novotná, Berinderjeet Kaur and David K. Pugalee, Oxford University Press. © Oxford University Press (2024). DOI: 10.1093/oso/9780192869647.003.0012 knowledge. These changes bring dynamism to the area of pre-service and in-service teacher education.

Our approach to the review of SEMT presentations

When analysing the SEMT proceedings, we started by characterizing each selected presentation by its main idea. This enabled us to create six groups of presentations with a common theme, which we arranged from the more general to the more specific perspectives. Many presentations could not be placed in a particular group unambiguously because topics overlap considerably. For example, the issues of pre-service and in-service teacher education overlap, as well as the issues of working with primary pupils and pre-service primary teachers. Each of these groups of presentations is explored in the sections that follow.

The first group includes presentations that focus on pre-service primary teachers' attitude to mathematics and the possibilities of affecting it in such a way that this relationship becomes positive. Based on many years of experience, we consider a positive attitude towards the subject studied to be crucial for pre-service teacher students, and to have an impact on the quality of their later work with pupils. That is why this topic is presented first. The second group of presentations deals with the concept of primary school teacher education and the questions of what a primary school teacher should know in terms of mathematical content and pedagogical content in order to be a good mathematics teacher. The third idea brings together presentations focusing on pre-service teacher education programmes. The fourth and fifth groups of presentations discuss examples of partial programmes and courses in the fields of arithmetic and geometry. The sixth group of presentations presents those that deal with teaching methods suitable for the development of pre- and in-service teacher competencies.

Pre-service primary teachers' attitudes to mathematics

Twelve presentations addressed the issue of the relationship of pupils, students, and teachers to mathematics and discussed the possibilities of influencing this relationship. During the 30 years of the SEMT symposia, this issue has not undergone major changes. The presentations generally were about the negative attitude of pre-service primary teachers to mathematics, which usually has roots from their time as pupils. A negative relationship is often triggered by mathematical anxiety, or even fear. It is a common reason why students choose to study primary school teaching. They believe that they will master the mathematics needed for primary school without having to learn it. It goes without saying, and many studies confirm, that a student's relationship to a subject, not only mathematics, is shaped by their teachers at all levels. The authors of the presentations analysed look for ways to untie this Gordian knot, and improve the relationship of pre-service primary teachers to mathematics so that they can help their future pupils build a similarly good relationship.

Littler (1995) sought to improve students' attitude to mathematics by changing the teaching strategies. Based on a questionnaire, he stated that pre-service primary teachers perceive play, 'investigation', and problem solving as an important component of teaching mathematics. It is important that both teachers and children enjoy mathematics. Later, Urbańska (1991) mapped pre-service primary teachers' attitudes to mathematics using a questionnaire also. It turned out that most of these respondents had difficulty in mathematics, that they had experienced boring mathematics lessons as pupils, and that most chose to study primary school teaching because they thought they would not have to learn mathematics. This survey was compared with a survey among in-service primary teachers, where 10% of them remembered mathematics as the worst subject.

Memories of mathematics lessons, from the position of a pupil, naturally affect teachers in their teaching practice. If a teacher has not mastered didactic principles well, if they have experienced no examples of good practice, they usually come back to the teaching style and methods, to procedures and work with pupils, that they experienced while pupils at school. This experience is also reflected in their choice of mathematics textbooks. Perkkilä (2001) came to this conclusion by examining the relationship between teachers' pedagogical beliefs and their practice. The research was carried out using a questionnaire (N = 230), classroom observation, analysis of video recordings of lessons, and written lesson plans. It is interesting to note the discrepancies, or contradictions, between teachers' beliefs and their teaching practices.

Obviously, no teacher education programme can offer its students experience and examples of good practice on how to teach every mathematical topic. These courses must, with respect to the participants and their needs, be on a more general level of pedagogical content knowledge, as we argue in later sections of this chapter.

Hoskonen and Pehkonen (2005) presented the results of a study of the impact of university lecturers on pre-service primary teachers' attitudes to mathematics. They presented students' experiences with their mathematics teacher at university. Their comments included:

> Kind, caring, supportive, fair and interactive—good atmosphere in the classroom. Even though the teacher was strict, the atmosphere was good when he was fair. At other times, there was the atmosphere of fear and anxiety. Then the students did not dare to answer and the situation did not support learning anything. On the other hand, a nice teacher could also cause problems.

Many students were convinced that their learning depended on the teacher's personality, which leads to the question of who should teach pre-service teachers, which is discussed below.

Other presentations presented examples of how relationships and attitudes of pre-service primary school teachers to mathematics can be changed.

Kratochvílová (2001) described one of the ways in which a pre-service primary teacher's relationship to teaching mathematics can be influenced. The student, the subject of the research, planned and implemented a teaching experiment with a pupil that they then carefully analysed with an 'expert' (university teacher) in terms of the pupil's cognitive processes while solving the problem. It was clear how the pre-service teacher's professional enrichment in the field of teaching mathematics resulted in an increase in their professional self-confidence, and, thus, also in a qualitative shift in their relationship to mathematics.

Swoboda (2001) focused on teaching geometry in pre-service primary teacher education. The students' relationship to geometry was influenced through building students' ideas about mathematics and ways of understanding the world of mathematics. One way is the method of experimenting. Students who have decided to study the field of primary school teaching do not necessarily have to know mathematics at any higher level than their future pupils, but they should know which methods can help the students to overcome any difficulties. The author drew on Tall's (1995) reflections on cognitive growth in mathematics.

Jirotková and Zemanová (2013) presented on the phenomenon of pre-service teachers' empathy for pupils' thinking processes when solving mathematical problems, and introduced the Method of Combined Discussion that allows development of a teacher's empathy and, thus, affects a teacher's teaching style and improves the quality of the mathematics education.

Marcinek (2013) addressed a similar issue. He tried to influence the students' relationship to mathematics through interpretations of mathematical ideas—that is, activities leading to increasing sensitivity to the ideas of their peers. The impact of these activities was obvious. Students greatly improved their ability to interpret ideas and work with them.

Samková (2017) studied the change in students' relationship to mathematics in the research that she implemented for the course *Planning and Conducting Inquiry-Based Mathematics* for pre-service primary teacher education. She noted students' positive changes in their argumentation (a shift towards more effective use of counter-examples and a shift from empirical to deductive arguments), in their openness to mathematics (a shift in looking for more, or all, solutions, and a shift towards accepting more forms of recording), in their beliefs about mathematics and the teaching of mathematics (e.g. a newly formed belief that discovery in itself helps to understand and memorize a phenomenon, and also that mathematics is not just memorizing formulae and procedures).

Slezáková, Jirotková, and Kloboučká (2017) used student portfolios to find the turning points in the students' relationship to mathematics. Analysis of student portfolios, submitted for the state final examinations, showed that finding the meaning and significance of a constructivist approach to teaching in general, not only in mathematics, changed their teachers' pedagogical beliefs. The key moments of these changes were marked when theory and practice were connected and when the students gained direct experience with pupils.

Conception of pre-service primary teacher education

There were 24 presentations at SEMT that focused on the conception of pre-service primary teacher education. A major topic, addressed throughout the 30 years of SEMT symposia, is the connection between theory and practice. It is clear that the possibilities for each university, and each country, both financial and personal, are different, and affect how this issue is addressed.

Coufalová (1991) asked who should be training pre-service teachers. This question is still relevant, and the answer to it changes over time. Policy makers and other stakeholders debate this question; it is discussed by deans of faculties of education, and other faculties preparing teachers; and representatives of state bodies such as ministries of education also address it. Later, in 1993, Coufalová concluded that if a teacher is to be able to educate a creative pupil, they must be creative themselves. Hence, her requirement for teacher education was to produce a creative teacher, regardless of who is doing the training.

Vinner (2011) agreed with the idea of reducing the theoretical curriculum. In his presentation he asked for a careful reconsideration of what knowledge a teacher needs to teach mathematics in primary schools. The university teacher, when working with pre-service primary teachers, should leave out topics that are outside the pre-service teachers' zone of proximal development. He pointed out that some teachers, although they are good people and teach as best they can, do not excel in teaching and never will, and thus it is unnecessary to demand more from them than their best efforts.

The presentation by Bobik (1993) showed that knowledge of history is important for teachers, with an emphasis on the processes of phylogenesis and ontogenesis, as it helps teachers understand their pupils. We add that this knowledge not only helps teachers to understand pupils, but also helps them, and the authors of teaching materials, to design the curriculum in such a way that the individual undergoes the cognitive process in a similar order to that which underlies the development of human thought.

In his presentation, Tall (2001) suggested the importance of teachers' knowledge of, or at least their awareness of, theories that provide an overview of how mathematics develops cognitively, and how mathematical knowledge is formed in their pupils' minds. He even considered it more important than knowing the sophisticated rationale for why it works. This means that the teacher must be aware of how children 'do' mathematics in different ways and why they face conflicts and difficulties. This awareness needs to be cultivated by sharing ideas and experience during their whole teaching career. Tall points out that a teacher's competence in school mathematics, and the confidence that they are doing it well, is essential for mathematics teachers.

The *Theory of Didactic Situations* (TDS), developed by Guy Brousseau in France in the 1970s and further elaborated in recent decades, helps teachers in designing and planning teaching, and in analysing the learning process; it was presented in detail by Chopin and Novotná (2011). With the help of archival documents, testimonies, and video recordings, they showed, during their workshop, how researchers in mathematics education and primary school teachers co-operated, with the goal of gaining a deeper understanding of how pupils learn while doing mathematics, and of developing situations for the improvement of this leaning.

The presentation by Ribeiro and Carrillo (2011) focused on key areas of teacher education content. These areas included knowledge of topics and tasks that teachers find challenging in practice. From a slightly different perspective, Van Bommel's (2011) presentation described the critical moments related to the curriculum that need to be considered in pre-service primary teacher education.

Kaslová's (2001) presentation pointed out that there was insufficient development of assessment skills in pre-service teacher education. She stated that her university faculty did not prepare its students sufficiently to justify the assessment of these essential skills that are important, not only for working with pupils, but also for communicating with other teachers, and parents.

Pre-service teachers' problems and their common misconceptions in mathematics that must be worked with during their studies were the foci of Tichá's (2011) presentation. She gave examples and explored their probable causes. Her poster showed examples of students suggestions for worded tasks that should be solved using division of two vulgar fractions. Further, she illustrated students' misconceptions of both vulgar fractions and the operation of division on them, and explored probable causes of these misconceptions.

In their presentation, Hodgson, Lajoie, and Poirier (2011) addressed the issue of the professionalization of prospective teachers, with specific regard to the importance of 'knowing-to act' in classroom situations. This approach aimed at overcoming difficulties in university courses, and avoiding the decontextualization of these courses.

The mathematical knowledge needed for teaching in primary schools, from the perspective of a university pre-service teacher education course, was the focus of the presentation by Gomes, Martins, and Ribeiro (2011). They elaborated on their study, which focused on trainee teachers' shortcomings in geometry, especially the understanding of the concept of the quadrilateral, at the beginning of their training. They investigated different levels of understanding of 2D and 3D shapes, and definitions with properties on the enumeration of features.

The presentation by Fung (2011) focused on the content of the pre-service primary teacher curriculum and used three examples taken from the primary school curriculum to illustrate the form and depth of mathematical knowledge. He shared his experience with a course for teachers in which the discussion of mathematical concepts took place at two different cognitive levels—at the level of pupils and at the level of teachers. Such mathematical discourse strengthened the pre-service teachers' self-confidence and encouraged them to go into more mathematically demanding areas.

In her presentation, Bambico (2001) described the extent to which an in-service teacher's personal characteristics (their length of teaching experience and their age) are related to their performance in mathematics. Using a correlational approach, she found that the teachers who have spent more years in teaching mathematics tend to demonstrate better instructional techniques than those with less experience in teaching the subject, which is not surprising. However, the higher age of a teacher does not correlate to better teaching techniques—in fact, it is the other way around!

Roy (2001) studied the effectiveness of teacher education in the light of the goal of improving pupils' competencies in the Philippines. She argued that personal mathematical ideology not only influences the pedagogical style adopted by the teacher, but also affects pre-service teachers' reflective abilities.

Ribeiro and Carrillo (2011) opened a discussion on partial principles and foundations for optimal education. In this area, connecting university pre-service teacher education with school practice seemed to be an important direction. Neufeld (1993) presented a proposal for connecting theory and practice and replacing textbook-orientated methods by direct experience with pupils. Wollring (2003) showed that what is effective in the development of teacher competencies, and in linking theory and practice, is combining research on two levels—research in schools and research in the field of pre- and in-service primary teacher education.

In Albert's (2001) presentation, the idea was presented that a teacher should experience the teaching of mathematics in the way we want them to teach in their future career. Albert used the method of confronting teachers with challenges, and their mathematical knowledge and their self-confidence increased as a consequence. The challenge, or problem, should be such that the student gets involved easily and the difficulty increases gradually. The result of using this method was that pre-service teachers were not afraid of more challenging problems, and later used them with their pupils, and then shared their experience with other pre-service teachers. The presentation by Forgasz, Tobias, and Jones (1997) described how they conducted a university mathematics course for pre-service teachers in a constructivist way, in order to give the students hands-on experience with this way of teaching, thus making it easier for them to adopt this approach in their future practice. In their research, Taylor-Basil and Littler (1997) reported on what happened when students were given a specific scheme describing how to work in a rural school. The students found it very beneficial. They had the opportunity to take a closer look at the creation of the school education programme, which they would have to do in the future.

Kmetič (1999) showed in their presentation how teachers develop understanding of the concept-making process in children by monitoring teaching and analysing the learning process. By analysing pupils' errors while doing the operation of division, some aspects of pupils' difficulties and misconceptions were detected and were described. A sequence of effective activities for the removal of the error or misconception in the individual child was discussed, while this form of diagnosis was considered a special kind of formative assessment.

Ribeiro and Martins' (2009) presentation on in-service teachers confirmed that school practice can be influenced by appropriate in-service teacher education courses on teaching methods, lesson planning, and selecting appropriate tasks for pupils. It is generally true that in-service teacher education needs to be attended to consistently and continuously. An example of this form of in-service education is the *National Science Foundation, use of the North Carolina (NAEP)* project, focusing on support for in-service teachers in the areas of developing mathematical knowledge of content, understanding pupils' mathematical thinking, and understanding the importance of assessment for developing pupils' thinking; this was presented by Cifarelli et al. (2011).

Proposing programmes for pre-service primary teachers focusing on general didactical aspects

Some programmes implemented and researched in courses preparing pre-service teachers, or supporting in-service teachers, have already been mentioned above. Here we summarize other themes presented in nine presentations focusing on general didactical topics.

Barabash and Guberman-Glebov (2001) presented an interesting interdisciplinary project whose goal is to educate autonomous mathematics teachers for primary schools. Students in the programme were guided to develop their skills for lifelong learning by identifying their level of expertise. Teachers' skills include the ability to play roles in situations that are likely to occur in a regular classroom, according to the presentation by Lajoie and Pallascio (2001). They showed that teachers paid attention to the development of this ability in the course of their pre-service teacher education courses.

Ulovec Brunner et al. (2009) presented their finding that co-operation between a class teacher and a language specialist is of benefit to both parties. They enrich each other both in the field of content and in the field of pedagogical content knowledge. That is why a *Content and Language Integrated Learning* (CLIL) course is a beneficial course in pre-service teacher education. Moraová and Novotná (2017) in their presentation on CLIL lessons had developed future mathematics teachers' skills in lesson planning. They recommended that pre-service teachers should be guided towards specifying goals with regard to content, describing them accurately in their lesson plans, and implementing classroom activities that would lead to achieving the goals.

Sheets and Cifarelli (2011) presented a proposal for the implementation of activities that enrich pre-service teachers' experience by developing deep mathematical concepts through modelling activities in which students model their own teaching processes, and activities in which students become acquainted with progressively more complex sets of tasks. These experiences were said to provide 'scaffolding' for the preparation of mathematical concepts and situations in pre-service primary teacher training methods.

Tabach's (2011) presentation supported her research finding that the use of IT in mathematics lessons requires its inclusion in teacher education programmes, because prospective teachers never experienced the use of technology for learning mathematics when they were younger students. However, they could use technology in their life for social communication, for fun, and in their role as students in their current study programme. It was suggested that it is more effective to influence prospective teachers to integrate technology into their practice, rather than wait and attempt to change their practice as in-service teachers.

Saboya and Dufour-Janvier's (2013) presentation shares very positive results of a new activity—*exposés oraux (oral presentations)*—that they conducted with pre-service teachers. Thanks to it, students developed enriched vocabulary, solid knowledge of everyday language and the language of mathematics, and the habits of thinking and using gestures which will help them to overcome the difficulties faced by novice teachers in the first years of their careers.

Biton et al. (2017) reported a reaction to the dissatisfaction of teachers with normative assessment, and the support for alternative methods and creation of tools for assessment in teacher-training courses. Assessment plays an inseparable part in the teaching and learning of mathematics, and teachers are required, and want, to assess their students in an encouraging, valid, and reliable manner. However, they meet with various difficulties. For example, when teachers ponder over the question of the gap between the knowledge that a student displays in class and the knowledge that they reveal in the examination, this raises the issue of the lack of pedagogical knowledge and what are the professional tools that can effectively close these gaps. Similarly, when teachers in this research indicated that a numerical grading did not allow students to improve or correct themselves, because they cannot understand their mistakes from a grading alone, the need for alternative assessment tools was raised. Additionally, teachers claimed that other issues that needed immediate attention were class size, and variance in student understandings and performance in mathematics.

Scherer (2017) described two courses for pre-service teachers—*Substantial Learning Environments* (SLE) and *Diagnosis and Support*. In these courses, pre-service teachers consolidated knowledge of mathematical content, and developed the ability to react to pupils and the different strategies and ways of thinking they come with. Another objective of the courses is to develop pre-service teachers' abilities to work with special needs children in inclusive education.

Proposals of partial programmes for pre-service primary teachers focusing on arithmetic

In this section we present seven presentations that proposed programmes for development of pre-service primary teachers' knowledge and competencies in the area of arithmetic.

A presentation by Trch and Novotná (1991) described their arithmetic course for pre-service teachers, which focused on the development of the following abilities and skills: logical thinking, correct communication, generalization, and the ability to distinguish between situations where formal definitions of a mathematical term should be used and where only illustrations without formal generalization should be presented.

Koman (1995) pointed out an interesting phenomenon from a teacher education course. He analysed the heuristic strategies that pre-service teachers used in solving a non-standard word problem—and found that the students were more successful in solving word puzzles than in solving standard word problems. Where word puzzles are special kinds of word problems that require heuristic solving strategies, such as creating sequences of numbers and making use of the pattern, the students' success in solving these problems encouraged their intellectual self-confidence and consequently improved their attitude to mathematics.

Tržilová (1995) presented on the use of spreadsheets by pre-service primary teachers, which brought the possibility of solving some complex problems without the use of higher mathematics. For example, students worked with spreadsheets for finding the roots of quadratic equations by the formula, for creating a formula for finding the volume of some geometric solids, and for investigating how certain algebraic identities work. Quattro Pro software was used to get data and the students were able to use it without problems. The students utilizing the spreadsheet method made a lot of progress very quickly.

Way (2009), Chestnut (2011), and Capraro et al. (2011) pointed out a well-known problem in many countries, namely the lack of knowledge of pre-service teachers in the area of fractions and operations with them. They stress the need for pre-service teacher education programmes to work on in-depth knowledge in the area of teaching fractions supported by the use of effective technologies with appropriate representations.

Finally, Chapman's (2017) presentation on the topics of integers, including the understanding of the notation of numbers in the decimal system and knowledge of the method of teaching integers, provided insights into the fact that teachers' understanding of the numeration system is fundamental to providing appropriate and useful support for students to learn the base ten system and whole-number concepts in the elementary school curriculum. She suggested that prospective primary school teachers with an instrumental understanding of the numeration system can be helped to develop a conceptual understanding of it, through an inquiry task that enables them to experience conflict, puzzlement, and curiosity that lead to questioning and exploration to achieve this shift. She claimed that it was important to both challenge and build on what they know, so that they could see it in new ways, instead of replacing wrong understandings with correct understandings, as this could be demoralizing instead of empowering. The presentation reported that the pathway of learning is not linear, and cannot be predetermined. The presentation further suggested that the historical development of the base ten system helped teachers to gain insights of possible learning stages for students.

Proposals of partial programmes for pre-service primary teachers focusing on geometry

There have been a number of presentations describing the results of extensive research focusing on exploring the understanding of specific geometric concepts by pre-service teachers. These presented comparative research into the level of geometric knowledge and depth of understanding of certain concepts in two different countries. These will not be addressed here, but instead, four presentations that describe specific designs of a course on geometry for pre-service teachers are discussed.

Horálková and Jirotková (1991) proposed to adapt the content of a geometry course offering the highest theoretical views (axiomatic structure of geometry) to pre-service teachers with the aim of contributing to the development of their spatial imagination and ability to see connections. In their 1993 presentation, the idea was further supplemented by the idea of the development of a teacher's ability to grade tasks for their future pupils in terms of cognitive demands. The pre-service teachers learnt how to break a difficult 3D geometry task into a series of graded tasks. This meant that all tasks arise from the same initial problem, and are linked to one another, such that the preceding task prepares the solver for the solution of the following, more difficult problem. Such a series of graded tasks enables the teacher to diagnose students' cognitive obstacles, as well as giving the weakest students an opportunity to enjoy a feeling of success.

Littler (1995) worked with the idea of developing spatial imagination in pre-service teachers. Moreover, in his presentation he offered several topics in a logical conceptual sequence, that seem not only fun for teachers and pupils, but also didactically supported by the child's physiological development. Ribeiro, Carrillo, and Monteiro (2009) demonstrated in their example of a specific geometric chapter (examining the difference between squares and rectangles) that it is not enough to have subject knowledge when teaching young learners: instead, a deep pedagogical content knowledge is needed—in other words, knowledge of 'all' different ways of approaching the building of geometric concepts and doing it with pupils. We could also include here Swoboda (2001), which has, however, been discussed in an earlier section.

Proposals of methods suitable for development of pre-service and in-service teacher competences

The presentation by Selter (1991) illustrated with four lessons how it is possible to supplement university teaching—that is, lectures—with activities to activate students, to develop their communication and social interaction, so that teaching resembles as closely as possible the way the students should guide and teach their pupils in the future. Urbańska, Urbańska, and Rozek (1993) also focused on how to improve pre-service teacher communication skills. Their goal was to improve the teaching of mathematics at university through communication, by a shift from rules to understanding. Understanding is developed by the teaching methods for pre-service teachers proposed by Malicki and Sznajder (1995). They use functional methods, and analysis of mistakes and of children's difficulties. Bardelli (2011) also focuses on errors, their origin, and their interpretation, through analysis of pupils' solutions. He claims that it is essential to develop the teacher's ability to interpret an error, as it is one of the strong re-education tools. In his presentation, Novák (1995) proposed analysis of the course of didactic games, played by pre-service teachers. These analyses showed how pupils' motivation and calculation skills are developed, as well as their ability to solve tasks independently.

The presentation by Kmetič (2001) of a case study of university students, aged between 18 and 21, illustrated possible constraints for the understanding of mathematical concepts, and connections between them, as well as offering some potential reasons why students might be prevented from finding a proof for a given theorem.

In the Levenberg and Ophir (2001) presentation it was shown how work with a mathematical story helps to develop pre-service teachers' creativity, and how it supports their ability to formulate a mathematical story and a mathematical problem for a given group of numbers. Their study indicated a difficulty in phrasing a problem, and led teachers to adhere to textbooks, thus supporting the view that there was a need to prepare pre-service teachers to use a different approach for teaching the subject. A similar idea was presented by Bureš and Clanché (2009), who pointed out the importance of using problems, and their solving, as a tool in pre-service teacher education.

A presentation that contributed to the understanding of constructivist teaching was the in-depth analysis by Bureš, Havlíčková, and Novotná (2011) of communication between a teacher and third graders in didactical situations based on a learning environment supporting communication. Their presentation showed that, if university courses offer their students a large number of 'open' learning environments, they will have, in the future, a variety of pedagogical tools available to help them deal with open questions, support pupils' cognitive development, and teach in a constructivist way. In a contrasting presentation by Samková (2019), the Concept Cartoons method was suggested as providing an understanding of the different ways pupils grasp concepts and solve problems, and how these results can be interpreted. Samková also showed how this method supported formative assessment in school practice.

Conclusion

We hope that this chapter will inspire current teacher educators to think about what is essential in pre-service and in-service teacher education.

From SEMT presentations, we selected those focused on the education of prospective teachers and the development of their professional competencies. The presentations were divided into six groups according to common topics. However, in the presentations we can follow three key perspectives related to primary school teacher education.

The first perspective concerns content knowledge. The authors of some presentations addressed the question of *what?*—that is, what mathematics to teach, and what mathematics the teacher should have. These presenters expressed a strong need to change the mathematical content of courses for future and current teachers. Proposed changes mainly concerned efforts to adapt the mathematical difficulty of the course to the needs of the future teachers and their cognitive possibilities. The main goal of these efforts was to change the negative attitude of prospective teachers towards mathematics and mathematics teaching, which can be achieved by consistently linking theoretical knowledge with practical knowledge, which leads to student teachers learning what they will need in their future work with pupils.

The second perspective concerns didactical content knowledge. When considering the content of courses for primary teachers, an important question is *how*?—that is, how to present the curriculum to pupils, the didactics that are appropriate to lead to the development of mathematical knowledge in pupils' minds, and which theories support these didactics. It is important that student teachers experience such didactical procedures because they will then be able to apply them with understanding in their practice.

The third perspective concerns the cultivation of the teacher's personality, which should become a guide for their pupils' learning process. This means that the teacher should be able to ensure the conditions for the effective learning of their pupils. This perspective also includes the teacher's ability to create a safe classroom climate, the ability to communicate with the class, the ability to diagnose the cognitive level of pupils, and the ability to use formative assessment for their pupils' development.

All three of these perspectives must always be linked, and none can do without the others. For example, the trend of introducing formative and reflective assessment in schools does not mean that the teacher does not have to know the content of the primary mathematics curriculum and its overlaps with the lower secondary mathematics curriculum. Similarly, a teacher who has both knowledge of content and pedagogical content knowledge is better equipped to implement formative assessment effectively, as they are aware of where to guide the pupil to develop in both the cognitive and meta-cognitive spheres.

Acknowledgements

This paper was supported by the project MaTeK (Enhancement of Research Excellence in Mathematics Teacher Knowledge).

References

Albert, J. (2001). Working with elementary teachers on challenge problems in mathematics. In J. Novotná and M. Hejný (Eds) SEMT 41–43.

Bambico, T. A. (2001). Influence of teachers' personal variables to teachers' mathematics achievement and instructional skills: A study on the effectiveness of in-service training in the Philippines. In J. Novotná and M. Hejný (Eds) SEMT 43–47.

Barabash, M., and Guberman-Glebov, R. (2001). Inter-disciplinary seminar and graduate project in mathematics teaching for future primary school math teachers, based on self-instruction abilities. In J. Novotná and M. Hejný (Eds) SEMT 48–52.
Bardelli, M. (2011). The pedagogical content knowledge of primary mathematics teachers facing a student error. In J. Novotná and H. Moraová (Eds) SEMT 63–71.
Biton, Y., Hershkovitz, S., Hoch, M., Ben-David, B., and Fellus, O. (2017). Assessment issues that trouble mathematics teachers. In J. Novotná and H. Moraová (Eds) SEMT 63–71.
Bobik, E. (1993). How can history of mathematics help in teaching mathematics. In M. Hejný and J. Novotná (Eds) SEMT 66–67.
Bureš, J., and Clanché, P. (2009). Problem posing as a mathematical activity: The young teachers' perspective. In J. Novotná and H. Moraová (Eds) SEMT 81–87.
Bureš, J., Havlíčková, R., and Novotná, J. (2011). An analysis of a young teacher's communication with pupils in a mathematical environment fostering the combinatorial reasoning (a case study). In J. Novotná and H. Moraová (Eds) SEMT 72–78.
Capraro, M. M., Capraro, R. M., Rosli, R., Han, S., Harbaugh, A., and Moch, P. (2011). Teacher educator strategies for improving preservice teachers' knowledge of fractions. In J. Novotná and H. Moraová (Eds) SEMT 88–94.
Chapman, O. (2017). Prospective elementary school teachers' learning trajectory of the numeration system. In J. Novotná and H. Moraová (Eds) SEMT 138–148.
Chestnut, A. (2011). Using pedagogical content scaffolds evidenced in transcripts to measure teacher knowledge. In J. Novotná and H. Moraová (Eds) SEMT 95–102.
Chopin, M.-P., and Novotná, J. (2011). Contribution of the theory of didactical situations to mathematics education: Fundamentals and main concepts. In J. Novotná and H. Moraová (Eds) SEMT 361–362.
Cifarelli, V., Morge, S., Goodson-Espy, T., and Pugalee, D. (2011). Examining the math content and pedagogical knowledge of pre-service teachers: How Katelyn and Matthew made sense of students' problems. In J. Novotná and H. Moraová (Eds) SEMT 111–117.
Coufalová, J. (1991). The relation between theory and practice in the preparation of teachers. In J. Novotná (Ed.) SEMT 73–75.
Coufalová, J. (1993). Activity and creativity in the elementary teachers training. In M. Hejný and J. Novotná (Eds) SEMT 67–69.
Forgasz, H., Tobias, S., and Jones T. (1997). Students' perceptions of a pre-service teaching practicum: Preliminary findings from one Grade 6 classroom. In M. Hejný and J. Novotná (Eds) SEMT 83–87.
Fung, C. (2011). A practitioner's encounter with the mathematical knowledge needed for teaching in elementary school. In J. Novotná and H. Moraová (Eds) SEMT 134–141.
Gomes, H., Martins, F., and Ribeiro, C. M. (2011). Discussing early years' trainee teachers' MKT in geometry: A first approach concerning quadrilaterals. In J. Novotná and H. Moraová (Eds) SEMT 377–378.
Hodgson, B., Lajoie, C., and Poirier, L. (2011). The mathematical preparation of elementary school teachers: Facing the challenge from three complementary perspectives. In J. Novotná and H. Moraová (Eds) SEMT 365–366.
Horálková, J., and Jirotková, D. (1993). Experience with preliminary geometry for future teachers of lower grades. In M. Hejný and J. Novotná (Eds) SEMT 74–75.
Horálková, J., and Jirotková, D. (1991). Geometry in teachers´ education. In J. Novotná (Ed.) SEMT 78–79.
Hoskonen, K., and Pehkonen, E. (2005). Primary teacher students' views on their mathematics teachers in school. In J. Novotná (Ed.) SEMT 149–155.
Jirotková, D., and Zemanová, R. (2013). Student-teachers' empathy for pupils' thinking processes when solving. In J. Novotná and H. Moraová (Eds) SEMT 155–161.
Jirotková, D., Slezáková, J., and Vybíralová, T. (2022). Portfolio as a tool for forming student-teachers´ pedagogical belief. In J. Fejfar and M. Flégl (Eds) Proceedings of the 19th International Conference Efficiency and Responsibility in Education. Prague: Czech University of Life Sciences, Faculty of Economics and Management.
Kaslová, M. (2001). Competences to evaluate pupils. In J. Novotná and M. Hejný (Eds) SEMT 90–95.

Kmetič, S. (1999). From teacher observation to teacher reflection. In M. Hejný and J. Novotná (Eds) SEMT 151–154.

Kmetič, S. (2001). The confidence and competence of teachers concerning notion of proof. In J. Novotná and M. Hejný (Eds) SEMT 96–100.

Koman, M. (1995). Word Puzzles—future teachers—pupils relay in teacher education. In M. Hejný and J. Novotná (Eds) SEMT 96–100.

Kratochvílová, J. (2001). The analysis of one undergraduate student's project. In J. Novotná and M. Hejný (Eds) SEMT 101–104.

Lajoie, C., and Pallascio, R. (2001). Role-play by pre-service elementary teachers as a means to develop professional competencies in teaching mathematics. In J. Novotná and M. Hejný (Eds) SEMT 104–109.

Levenberg, I., and Ophir, S. (2001). Pre-service teachers facing literacy in mathematics. In J. Novotná and M. Hejný (Eds) SEMT 109–113.

Littler, G. H. (1995). Geometry for the teacher education student. In M. Hejný and J. Novotná (Eds) SEMT 18–23.

Malicki, T., and Sznajder, M. (1995). Realistic—functional method in mathematical training of primary school teaching students. In M. Hejný and J. Novotná (Eds) SEMT 138.

Marcinek, T. (2013). Using and assessing interpretation tasks in mathematics content courses for pre-service elementary teachers. In J. Novotná and H. Moraová (Eds) SEMT 196–203.

Moraová, H., and Novotná, J. (2017). Higher order thinking skills in CLIL lesson plans of pre-service teachers. In J. Novotná and H. Moraová (Eds) SEMT 336–345.

Neufeld, E. M. (1993). Young children enhance a teacher education course for elementary school mathematics. In M. Hejný and J. Novotná (Eds) SEMT 76–78.

Novák, B. (1995). Reflection of a didactic game in teacher training. In M. Hejný and J. Novotná (Eds) SEMT 138–141.

Perkkilä, P. (2001). Elementary school teachers' view of mathematics and their teaching practices. In J. Novotná and M. Hejný (Eds) SEMT 131–135.

Ribeiro, C. M., and Carrillo, J. (2011). Discussing Maria's MKT and beliefs in the task of teaching. In J. Novotná and H. Moraová (Eds) SEMT 290–296.

Ribeiro, C. M., and Martins, C. (2009). The impact of a professional development program in teacher's practice. In J. Novotná and H. Moraová (Eds) SEMT 212–219.

Ribeiro, C., Carrillo, J., and Monteiro, R. (2009). We teach what we know, but do we know what we teach? The practical 'distinction' between squares and rectangles in a primary school class. In J. Novotná and H. Moraová (Eds) SEMT 204–211.

Roy, A. (2001). Prospective elementary teachers reflexive competencies in mathematics using philosophical approach. In J. Novotná and M. Hejný (Eds) SEMT 140–144.

Saboya, M., and Dufour-Janvier, B. (2013). The complexity of training teachers in the practical aspects of education: The importance of 'exposés oraux'. In J. Novotná and H. Moraová (Eds) SEMT 281–289.

Samková, L. (2017). Planning and conducting inquiry based mathematics course for future primary school teachers. In J. Novotná and H. Moraová (Eds) SEMT 354–364.

Samková, L. (2019). Preparing future teachers for formative assessment: The case of Concept Cartoons. In J. Novotná and H. Moraová (Eds) SEMT 372–382.

Scherer, P. (2017). Preparing pre-service teachers for inclusive mathematics classrooms: Concepts for primary education. In J. Novotná and H. Moraová (Eds) SEMT 364–373.

Selter, C. (1991). Aktiv-entdeckendes und sozial Lernen im Mathematik-unterricht der Grundschule und in der Lehrerbildung. In J. Novotná (Ed.) SEMT 95–97.

Sheets, C., and Cifarelli, V. (2011). Developing instructional sequences in pre-service elementary mathematics teacher education. In J. Novotná and H. Moraová (Eds) SEMT 313–319.

Slezáková, J., Jirotková, D., and Kloboučká, J. (2017). Student portfolio as a tool for development of pre-service primary teacher's competencies in teaching mathematics. In J. Novotná and H. Moraová (Eds) SEMT 383–393.

Swoboda, E. (2001). How to prepare prospective teachers to teach geometry in primary education: Some remarks. In J. Novotná and M. Hejný (Eds) SEMT 150–154.

Tabach, M. (2011). Promoting technology integration among prospective mathematics teachers. In J. Novotná and H. Moraová (Eds) SEMT 328–335.

Tall, D. (1995). Cognitive growth in elementary and advanced mathematical thinking, In: Proceeding of 19th International Conference PME, Recife, Brazil.

Tall, D. (2001). What mathematics is needed by teachers of young children? In J. Novotná and M. Hejný (Eds) SEMT 26–35.

Taylor-Basil, V. B., and Littler G. H. (1997). Using final year teacher education students to evaluate the mathematical needs of small rural school and to help teachers to their wish to improve the mathematics within the school. In M. Hejný and J. Novotná (Eds) SEMT 92–96.

Tichá, M. (2011). The search for ways of enhancement of knowledge base for teaching. In J. Novotná and H. Moraová (Eds) SEMT 387–388.

Trch, M., and Novotná, J. (1991). The course of arithmetic for prospective primary school teachers. In J. Novotná (Ed.) SEMT 102–104.

Tržilová, D. (1995). Problem solving on spreadsheets. In M. Hejný and J. Novotná (Eds) SEMT 123–125.

Ulovec, A., Brunner, Ch., Brychta, A., Novotná, J., Hofmannová, M., Lotzfeldt, B., Jäpelt, A., Alvez, Y., Chesné, J.-F., Le Yaouanq, M.-H., Martucci, B., Favilli, F., and Peroni, R. (2009). Training courses for mathematics teachers' mobility. In J. Novotná and H. Moraová (Eds) SEMT 315.

Urbańska, A., Urbańska, E., and Rozek, B. (1993). Video records of student-teachers solving a problem as both a research and didactic means. Communication in the educational process. In M. Hejný and J. Novotná (Eds) SEMT 67–69.

Urbańska, E. (1991). About problems of the mathematical education of primary school teachers. In J. Novotná (Ed.) SEMT 105–108.

van Bommel, J. (2011). Preparing student teachers with MKT. In J. Novotná and H. Moraová (Eds) SEMT 336–344.

Vinner, S. (2011). What should we expect from somebody who teaches mathematics in elementary schools? In J. Novotná and H. Moraová (Eds) SEMT 31–41.

Way, J. (2009). Umm ... I think we're lost! Pre-service teachers' thoughts on developing children's understanding of fractions. In J. Novotná and H. Moraová (Eds) SEMT 248–254.

Wollring, B. (2003). Linking pre-service and in-service teacher training: Co-operative design and dissemination of working environments for primary mathematics. In J. Novotná and M. Hejný (Eds) SEMT 35–43.

SECTION 4
ASPECTS OF LEARNING MATHEMATICS

David Pugalee
University of North Carolina, Charlotte, USA

This section focuses on how learners learn mathematics and how this learning can be enhanced through the actions of teachers. The four chapters in this section are unique in that they address how this may be achieved through addressing the mathematical literacy of students, or enhanced through alternative approaches such as Japanese Lesson Study (LS), or through mathematical problem solving, extending to problem posing and enhancing creativity.

The first chapter, by Morge and Pugalee, focuses on mathematical literacy for students. Using a model recognizing the importance of processes and enablers that work together to build solid mathematical understanding in students, the chapter highlights the rich research spanning decades of SEMT, elucidating the roles of reasoning, problem solving, representing, and manipulating in students' mathematical learning. The discussion expands to provide perspectives from SEMT researchers that assist in building an understanding of the complex interplay between processes and enablers such as communication, technology, and values. The discussion capitalizes on the contribution of Chapman on the categories of learner-focused tasks to emphasize the importance mathematical tasks plays in this process. Reasoning and problem-solving research underscores how students build the types of thinking that engage them in the actual 'doing of mathematics'. Critical concepts of rich environments to promote reasoning and problem solving are described. The adeptness of students' mathematical representations in doing mathematics is approached from both the processes involved in using multiples representations and the role of curricula in supporting students' capacity for flexibility. SEMT research perspectives develop ideas on how students' understanding of algorithms and procedures supports the effective use of heuristics and tools in understanding concepts and approaching problems. The chapter continues with thinking around communication, technology, and values as important tools or enablers that facilitate mathematics learning. Through SEMT, researchers have presented work on how these tools can be used effectively as agents for substantive mathematical learning. Considering the SEMT research reinforces conceptions of mathematical literacy as developed in complex learning environments and the synergy between students' and teachers' actions.

The next two chapters have a focus on Japanese LS. Chapter 4.2, by Fujii, provides in-depth analysis of the traditional Japanese LS model as a framework for designing and adapting tasks. The model describes the Japanese problem-solving lesson within LS from the Project IMPULS (International Math-Teacher Professionalization Using Lesson Study). The work is based on the author's presentation at SEMT 2015. The details promote a clearer understanding of the collaborative work that goes into this process, illuminated through an exemplar of a live research lesson. Contrasting traditional professional development with LS accentuates how LS supports long-term growth for students' mathematical learning. The rich discussion around design underscores how problem-solving lessons are teacher-developed activities that are daily activities in Japanese schools and how effective LS supports growth as independent problem solvers and mathematical thinkers. This excellent discussion is complemented by the work of Doig and Groves in Chapter 4.3. This chapter presents insights from SEMT around Japanese LS. The authors summarize key SEMT papers with a basis on the structured problem-solving mathematics lesson within the Japanese LS framework. The presentations, a plenary address, and a workshop delivered at various SEMT symposia provided multiple professional learning opportunities through which participants were challenged to make a distinction between Japanese LS and structured problem-solving lessons. Another fruitful enrichment of participants' thinking was that effective Japanese LS is situated on teacher-led pedagogical development connecting teachers and researchers from diverse backgrounds. In-depth consideration of the principles of LS promoted debate around students' capacity to engage in long fertile end-of-lesson discussions. Both of these chapters draw attention to the superb quality of SEMT presentations regardless of the session format.

Finally, Chapter 4.4, by Cifarelli, M. Capraro, and R. Capraro, focuses on problem posing, pointing out how SEMT bridges theory-based approaches and practical application for teaching and learning. The work illustrates how young children can be supported to demonstrate mathematical actions that are quite sophisticated. Drawing from SEMT presentations, the authors conclude that SEMT presenters have consistently embraced the prominent roles that problem solving and problem posing play in children's mathematical learning. The discussion underscores that the SEMT presenters view young children as active learners who show self-determination in their actions and the SEMT problem solving and posing presentations stress critical thinking. The authors provide an interesting overview of the evolution of the problem-solving framework and the development of connections between problem solving and problem posing. The connections lead to productive discussions about SEMT presentations that actively investigated the effectiveness of specific instructional activities and pedagogies for problem solving or problem posing. The authors close in recognizing the high quality of SEMT presentations and how they have distinguished SEMT as a leading symposium for presenting research focused on the mathematics learning of young children.

4.1
Promoting mathematical literacy: Perspectives from SEMT

Shelby Morge
University of North Carolina Wilmington, USA
David Pugalee
University of North Carolina, Charlotte, USA

Introduction

What is mathematical literacy? Mathematical literacy is defined by the Programme for International Student Assessment (OECD 2018) as:

> an individual's capacity to formulate, employ and interpret mathematics in a variety of contexts. It includes reasoning mathematically and using mathematical concepts, procedures, facts and tools to describe, explain and predict phenomena. It assists individuals to recognise the role that mathematics plays in the world and to make the well-founded judgements and decisions needed by constructive, engaged and reflective citizens. (p. 67)

One way to organize our thinking around what it means to be mathematically literate is to consider a model. Pugalee (1999) proposed a model with processes on the outer circle and enablers on the inner circle (see Figure 4.1.1 below). The two circles are inter-related such that the enablers work together with the processes on the outer circle. We will use the processes and enablers of mathematical literacy in this model as an organizing component of the chapter.

The processes of mathematical literacy depend on students' engagement and reflective thinking that interact with enablers to support students' mathematical

Model of Mathematical Literacy

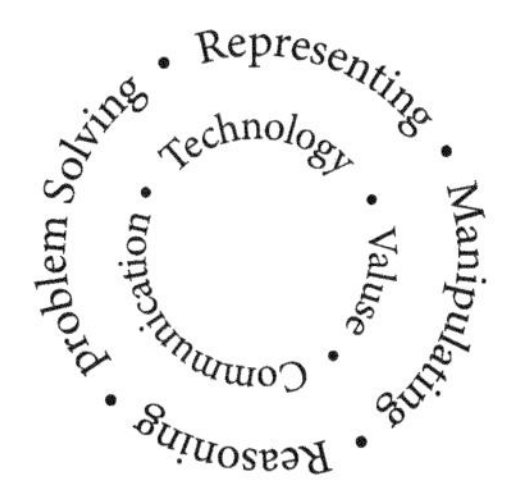

Figure 4.1.1 Model of Mathematical Literacy (Pugalee 1999, p. 20)

Shelby Morge and David Pugalee, *Promoting mathematical literacy: Perspectives from SEMT*. In: *Elementary Mathematics Teaching*. Edited by: Brian Doig, Jarmila Novotná, Berinderjeet Kaur and David K. Pugalee, Oxford University Press.
 DOI: 10.1093/oso/9780192869647.003.0013

Table 4.1.1 Categories and types of learner-focused mathematical tasks (Adapted from Chapman 2013, p. 16)

Reflection on real-world experience	Reflection on mathematical conceptions	Reflection on mathematical preconceptions	Reflection on mathematical thinking process	Reflection on mathematics learning experience
Identifying unspecified mathematics in experience Identifying a specified mathematical concept in experience Identifying use and/or meaning of mathematics in experience	Unpacking or applying prior mathematical knowledge	Revealing strong informal mathematical knowledge Revealing weak informal mathematical knowledge Predicting a mathematical outcome	Revealing a process of interpreting a mathematical concept Revealing a problem-solving process and/or strategy	Revealing affective problem-solving factors Revealing choice of tool to aid learning

development (values, communication, technology). This focus on processes also pushes us to recognize the work of Chapman (2013), who highlights the importance of tasks and the critical role of student reflection. Research on the nature of tasks and the way they are implemented provides significant direction in promoting students' mathematical learning. Table 4.1.1 summarizes categories and types of learner-focused tasks. Considering the nature of tasks provides connections to an engaging learning environment that can build students' mathematical understanding. As Chapman notes, these categories are intended to be illustrative. Likewise, our use of a model for mathematical literacy is intended to be a tool for promoting discourse around students' development of mathematical understanding and competence. Chapman's work underscores how mathematical processes, and the development of students' mathematical literacy, are co-dependent on relevant and authentic mathematical tasks that are implemented in ways that encourage active student engagement with mathematical concepts and problems.

Underlying these descriptions of mathematical reasoning is engaging in the process of doing mathematics. Mathematical reasoning involves thinking about tasks and making decisions about how to solve them, not following specified procedures. Kaur (2019) explained that providing students opportunities to do mathematics must engage them in significant time 'doing mathematics'. This includes engagement of students in reasoning and communicating mathematics. Tasks in textbooks are often unsuitable to engage students in the necessary reasoning. Kaur (2019) described 'what' strategies promote reasoning: what number makes sense, what's wrong, what questions can you answer, and what's the question if you know the answer. To emphasize the importance of this level of engagement, Nührenbörger et al. (2013) conducted a series of workshops 'to familiarize participants with examples and possibilities to discover mathematical relations' (p. 348). In three sessions, SEMT conference participants analysed tasks and manipulatives and reflected on video footage to consider how children use tasks and manipulatives to think about mathematics, thus facilitating the development of their mathematical reasoning. SEMT researchers have found that children of all ages can engage in some level of mathematical reasoning. Maj-Tatsis and Tatsis (2015) observed that two eight year old girls used different strategies and degrees of mathematical reasoning to solve tasks during an interview, but both were able to make conjectures and check their validity. One student was able to apply one of Polya's (1945) heuristics and both students applied prior experiences to problem solving.

Mathematical reasoning is not just important for students: teachers must also engage in mathematical reasoning to appreciate its value and facilitate effective learning for students. Cifarelli et al. (2011), interviewed pre-service elementary teachers (PSTs) and found that it was challenging for them to examine student work samples, but the PSTs seemed to recognize the value of asking students to explain how they solved problems. Certain teacher behaviours have been found to encourage mathematical reasoning in students. Cheeseman (2009) described five primary school teacher behaviours that challenge children to examine their mathematical thinking. The behaviours, identified from perspectives from the teacher, child, and researcher using classroom observations, videos of lessons, and interviews included: listening behaviours, behaviours that required sustained student thinking, questioning behaviours, mathematical pedagogical behaviours, and behaviours building social and emotional climate.

The process of problem solving

Problem solving involves the use of prior knowledge and skills in moving towards a resolution to a problem that lacks an apparent solution. Current views on problem solving see it as a dynamic process that allows the individual to move freely among various stages, which include, but are not limited to, problem posing, understanding the problem, making a plan, carrying out the plan, assessing progress considering the information in the problem, and evaluating whether the answer is logical (Wilson, Fernandez, and Hadaway 1993; Kulm 1991; Pugalee 1999).

Students at all levels can engage in problem solving, but it is not something that happens naturally. Teachers must provide tasks and classroom environments to encourage this process. An important element of establishing this classroom environment is context. Prusak and Hershkowitz (2019) described how problem solving is a context rather than a set of skills to be acquired. This context includes the iterated enactment of practices such as small group work, oral and written reporting, and teacher-led discussion. They claimed the emphasis on design and the culture of problem solving led Grade 3 students to engage in productive argumentation and develop meaningful understanding of the area concept. Martin et al. (2017) reported about fifth-grade students engaged in a Problem Based Learning (PBL) project that was created around a school fundraiser. They found the experience created an opportunity for students to be connected to authentic tasks, engage in multiple problem-posing and problem-solving strategies, share ideas with partners, develop number sense, and be reflective about their work. In addition, Savard and Polotskaia (2013) discussed an ethno-mathematical model emphasizing multiple contexts for problems: socio-cultural, citizenship, and mathematical—and language context in written texts. This ethno-mathematical model posited that the socio-cultural and citizenship contexts influence students' activity by providing access to the problem and connections to their mathematical knowledge and real-life experiences. The model also emphasized the importance of these contexts in moving students towards a mathematical analysis of the quantities and their mathematical relationship, promoting the development of new mathematical knowledge.

Researchers shared suggestions for establishing a problem-solving environment. For example, Doig and Groves (2017) described the critical elements of structured problem-solving lessons. They stressed the importance of understanding the entire process for delivering fully effective lessons. Polotskaia et al. (2019) conducted one-on-one interviews with remedial middle school students using word problems. The problems were first given with blank spaces in place of numbers to encourage students to explain what should be done to find an answer. Then students were given numbers to solve the problems in the traditional way. As a result of their findings, they encouraged teacher educators to emphasize the added value in paying careful attention to the combination of a relational approach to mathematics with a didactical *milieu*, and to work on meta-cognitive skills in order to understand what it is that precludes the use of appropriate strategies when solving problems.

The process of representing

Representing involves those skills that enable the individual to construct, and alternate between, various mathematical models, such as equations, matrices, graphs, and other symbolic or graphical forms. This process builds on the nature of mathematics as involving multiple representations of concepts: numerical, algebraic, graphical, and verbal (Mathematical Sciences Education Board 1996). Students should be adept at moving between various representations of mathematical situations and understand the information in whatever forms it is initially presented. Additionally, the ability to represent is vital for the individual to translate mathematical ideas into other representations so that they can communicate mathematical models and solve mathematical problems (Pugalee 1999).

The perspectives on representing include students' use of various representations to demonstrate thinking and understanding of mathematics. Lipovec and Podgoršek (2017) asked sixth-grade students to draw pictures of $\frac{2}{3}$ of 15 and 2^3. They found that approximately 60% of students represented each concept adequately. Procedurally orientated drawings prevailed for both concepts and were more likely to include mistakes. Sáenz-Ludlow (2017) focused on two fourth-grade students who participated in a year-long constructivist teaching experiment. At the beginning of the school year, the students were able to verbalize the memorized addition algorithm, unaware of the meaning of the position of digits in the numerals. Four months later, they constructed their own numerical diagrams (NDs) to mediate not only their decomposition of numbers into constituent units and the composition of units into abstract composite units but also their own ways of expressing their numerical strategies verbally. Their NDs supported their progressive conceptualization of number as a manifold of units as well as their reconstructed meaning of numerals (i.e. the place-value representation of numbers).

Research perspectives from SEMT also addressed how representations are presented in textbooks and used by teachers. For example, Macedonska (2003) reviewed the graphic representations used in textbooks and divided them into two categories: realistic pictures and graphic schemes. Both were considered essential codes in the process of transforming information connected with the learning of mathematics. The author found that if the textbook included more realistic pictures and schemes of abstract objects, it made it easier for students to develop knowledge and create a picture of mathematics. In contrast, classic geometrical pictures led to an appearance of knowledge in the form of a strict system of concepts and statements. Students' work with multiple representations builds both conceptual understanding and flexibility in approaching problems. In another example, Dabić and Milinković (2015) investigated how prospective teachers transmitted multiplication and the commutative property of multiplication through pictures. They found that the teachers knew a variety of representations of multiplication, but the most successful inferences were drawn from the combined area/array model. After analysing students' responses to interview questions, they noticed discrepancies between what teachers meant to convey through pictures and how children interpreted them. The array model for multiplication was the least comprehensible for students. The commutative property expressed in a number-line representation, area model, and set model were better visual aids than an equal group model. Dabić and Milinković concluded that more abstract models for the commutative property of multiplication were more effective in written visual communication between teachers and students. They suggested paying closer attention to representations that appear to be the most compelling to children instead of using representations that are easier for teachers.

The process of manipulating

Manipulating involves performing calculations and using algorithms and procedures successfully. Although the ultimate goal is the development of a deeper or conceptual understanding of mathematics, the ability to manipulate remains foundational in doing mathematics (Pugalee 1999).

Herman (2003) reminds us of the strong international research demonstrating the transition from arithmetic to algebra and the cognitive obstacles in algebraic thinking that promote conceptual blocks towards thinking algebraically. One of the problems identified is the limited work in generalizing arithmetical structures. Experiences manipulating with symbols, algorithms, and procedures contribute to mathematical sense making and generalization (Pugalee 1999). Herman (2003) notes that these experiences contribute to insights into problem structures at the arithmetic level and provide a transition to the use of algebraic elements.

The use of heuristics is a major component of manipulating. Heuristics are viewed as strategies and frameworks to promote students' mathematical processes. This includes Polya's problem-solving framework that strategically guides students in their thinking (Polya 1945): understanding the problem, devising a plan, carrying out the plan, and looking back. Students' use of heuristics often involves shortcuts or procedures that are designed to be efficient. Applying heuristics involves flexibility and critical thinking. Tambunan (2018) found that heuristic strategies positively influence students' understanding of concepts, creativity, mathematical communication, problem solving, and reasoning ability.

Students learn to manipulate mathematics according to the characteristics of a given problem or situation (Břehovský et al. 2013), and these strategies must be developed as part of students' repertoires. These SEMT researchers concluded that students might limit the use of heuristics such as graphical representation (GR) in some situations because they have had no opportunities to learn these strategies. They also found that the GR strategy resulted in higher success, leading the researchers to posit that this strategy results in developing visual imagination. This work underscores how manipulation, through the application of framework, heuristics, and strategies, contributes to students' success in mathematics. Even short-term work with heuristic strategies such as analogy, guess–check–revise, systematic experimentation, problem reformulation, solution drawing, way back, and use of graphs of functions, can have positive results on student's problem solving (Novotná et al. 2014).

Manipulatives become important tools for developing conceptual understanding because of the 'minds-on' nature of students' explorations when appropriately supported by teachers. Procházková (2015) describes the use of paper folding, sticks, tiles, geoboards, and square grids as effective in identifying differing levels of understanding of geometrical concepts according to Hejny's (1993) theory, to area tasks. Similarly, Larsson (2015) highlights how students' manipulation with numbers in multiplication connects to an understanding of multiplication as an operation of repeated addition with commutativity, as is the case with addition. So, manipulation becomes an active process through which students develop mathematical understanding of concepts as well as competence with procedures.

Enablers: Tools to facilitate mathematics learning

Enablers facilitate the 'doing of mathematics' (Pugalee 1999). They interact with processes in ways that are directly related to students' learning of mathematics. Enablers include communication, technology, and values. This is by no means a comprehensive list, and each of these is complex, from the perspective of students' learning as well as teachers' practices.

The enabler of communication

Communication is a tool that assists students in the development of mathematical understanding (Pugalee 1999). The importance of communication in mathematics is emphasized in the National Council of Teachers of Mathematics (NCTM) process standards (2000) and is a key element of effective mathematics instruction. Students' communication about mathematics includes organizing their thinking, communicating it to others, analysing and evaluating the thinking of others, and using precise mathematical language.

A focus on communication has been noted as essential to learning in curriculum materials. Levenson, Barkai, and Larsson (2013) reviewed curriculum documents for their respective countries in a search for explanation-related terminology. The authors identified six different functions of explanations within the Swedish and Israeli curriculum documents: description of one's thinking process, responses to 'why' questions, interpretations, steps in directing new explorations leading to generalizations, justifying solutions or processes, and as a means of communication. They proposed that teachers use the findings as an organizational tool, helping them to incorporate explanations in their lessons to promote various goals. In addition, Poirier and Bisaillon (2015) described the role of communication in learning mathematical language and reasoning, as articulated by Québec's mathematics curriculum. They used classroom activities to demonstrate how teachers implement rich and significant mathematics that stimulates discussions, debate, and negotiation of meaning between students and how these social experiences influence development of mathematical language and reasoning.

Teachers who put specific structures in place to support communication have been successful. Pugalee and Seifert (2013) found that sixth-grade students' abilities to write in mathematics are influenced positively by the teacher's emphasis on written communication as part of regular classroom instruction. In their research, the teacher introduced a rubric on how written work would be evaluated and students responded to a problem or prompt that required them to describe their mathematical thinking almost daily. Students' responses provided a source of information for teachers to assess how their students learn and think about mathematics. In another example, three fourth-grade teachers provided their students with packets including multi-step word problems that asked for description, explanation, and evaluation (Martin et al. 2015). Teachers modelled the activity and students completed the packets independently. The authors analysed students' responses to the writing prompts and found that students engaged in organizational strategies that appeared to help them think through their understanding. Students' writing revealed their levels of mathematical understanding and areas of weakness.

The enabler of technology

The mathematically literate individual must be able to use technological tools in investigating mathematical ideas and in finding solutions to mathematical problems (Pugalee 1999). Technology is one of the six mathematics principles in the

Principles and Standards documents and is considered essential in teaching and learning mathematics (NCTM 2000, p. 24).

Technology should be an integral part of teachers' instructional tools for deepening students' understanding. Grandgenett, Harris, and Hofer (2010–2011) proposed seven 'genres' of mathematics activities in which teachers can combine the use of technology with effective pedagogy to support students in learning important mathematics content. These genres use active words to emphasize the dynamical and student-centred environment: consider, practise, interpret, produce, apply, evaluate, and create. Various technology tools can be used to implement the mathematics activities, but they must be strategic and meaningful in order to deepen student understanding. Krauthausen (2013) provided an overview of digital tools and their uses and made recommendations for their use with a conceptual focus. Krauthausen suggested the primacy of didactics over media by focusing on what is conceptually desirable and then selecting the appropriate digital tool. In addition, it was suggested that teacher professionalization should focus on a conceptually grounded integration into classroom practice, include subject didactics, and focus and limit the selection of digital tools to high-quality products. In 2011, Hershkovitz and Tabach encouraged this focus on meaningful technology activities during workshops in which teachers engaged in the lesson study cycle to reflect on the possibilities for professional development in their own countries. This included a focus on examining the affordances and constraints of technology tools, reviewing videos of teachers implementing a technology tool while paying attention to teachers' choices in terms of the educational environment and the student's learning potential, and discussing the educational potential of future use of technology tools.

SEMT presenters have described various digital tools available for mathematics instruction that focus on specific mathematical content. What is notable throughout is the integration of digital tools with effective instructional practices. With the structure and support of the teacher, students are able to use the tools to think with and explore mathematical ideas. Huhmann, Höveler, and Eilerts (2019) shared results from the development and use of a digitally supported learning environment for pentominoes with children in the third year of primary school. They focused on similarities and differences between processing pentomino tasks with an analogue versus a digital learning environment. The results showed that problem reduction could be successfully counteracted by the developed digital aid. The authors suggested further development of the learning environment to counteract representational volatility. In another content example, Favilli (2017) shared information about a micro-world created in a mathematics classroom where technology supported the introduction and development of the concept of greatest common divisor of two non-zero natural numbers. In the unit, different algorithms were presented using dynamic software. Favilli concluded that students' interaction with technology helped create a learning environment where communication (student–student, student–teacher, student–technology) played a key role. The discovery of the software features and underlying algorithms encouraged students' conjecturing and investigating mathematical ideas. Also, different semiotic representations of mathematical concepts were promoted. Ruthven (2007) reviewed results of a 'calculator-aware' number (CAN) curriculum that was created in collaboration with primary school teachers in England

and Wales. The goal of the curriculum was to develop a new approach to teaching 'numbers' using calculators in a practical and investigational format. The author examined various pedagogical strategies and other wider uses of calculators. The CAN curriculum impacted students' proficiency with and attitudes towards mental calculations. Analysis of students' calculator use when tackling a realistic number problem showed how effective use of calculators requires mastery of operating procedures, a grasp of underlying mathematical ideas, and the development of distinctive calculator methods. The author suggests that a calculator-aware approach cannot simply be improvised around a conventional curriculum.

NCTM (2014) cautions against teachers sharing procedures for using tools or technology to solve problems without giving students opportunities to problem solve or make connections using mathematical reasoning. For this reason, it is important for teachers to become familiar with technology tools. Tabach (2011) shared reflections of pre-service teachers enrolled in a course designed to enhance their pedagogical geometry content knowledge. The course involved a series of activities in which prospective teachers engaged as learners and then as teachers in a pencil-and-paper environment, and then in a dynamic geometry environment. Findings revealed reported changes in the prospective teachers' knowledge and attitudes; reported changes in the knowledge and attitudes of the students the prospective teachers worked with; and, finally, holistic reactions to the use of technology. Also, Bakos and Sinclair (2019) investigated six primary teachers' interactions with an iPad touchscreen application. The teachers engaged with tasks and participated in discussions related to the tasks and the use of TouchTimes (TT). The authors noted shifts in thinking related to the order of multiplier and multiplicand that were provoked by a TT artefact sign and involved a new way of conceptualizing multiplication. There was also a change in thinking related to the equal sign, though the focus was more on the importance of helping students develop a better understanding of it. They also noted a change in thinking about the appropriate age for teaching multiplication provoked by observing kindergarten children learn from and come to know how to use TT, particularly in terms of their unexpected use of certain mathematical discourse.

The enabler of values

The idea of values is expanded to include one's emotions, beliefs, and attitudes towards doing mathematics and towards the nature of mathematics itself (Pugalee 1999). The learner must come to appreciate the role of mathematics in today's society as well as the historical, social, cultural, and scientific contexts of its development (NCTM 1989). Sullivan, Clarke, and Clarke (2013, p. 8) refers to this as productive disposition- habitual inclination to see mathematics as sensible, useful, and worthwhile, coupled with a belief in diligence and one's own efficacy. Littler (1999) argues that a more positive attitude increases students' understanding of mathematics as well as their questioning of methods, and their learning to work cooperatively and communicate about their work. These lead to increased enjoyment of mathematics. Values are also noted as central to raising awareness of gender and cultural issues around opportunities for learning mathematics (Erickson and Marcinek 2019; Moraová and Novotná 2015).

In one study, researchers found that students' feelings of insecurity affect their performance, highlighting the need for awareness of the establishment of norms and feedback tools (Symeonidis, Tatsis, and Kaldrymidou 2015). They also point out that there is a continuous re-negotiation and re-establishment of norms so that there is an effort to improve students' attitudes, perceptions, and emotions to positively impact their mathematical performance. Clarke (2001), in discussing the Early Numeracy Research Project, identified key growth-orientated outcomes, which included notable changes in children's positive attitudes towards mathematics, characterized by enthusiasm, stronger confidence, and risk taking. Some key takeaways from the project were that insights from one-to-one interviews with students are a powerful basis for understanding what students know and can do, having high expectations for all children and encouraging a climate that all can 'do mathematics' results in positive effects, and encouraging children to share thinking is valuable. This research makes clear that the values that students develop towards mathematics can be largely influenced by the mathematical opportunities they have in classrooms.

Levenson and Barkai (2017) employ complexity theory to consider the multi-faceted system of the classroom with individuals who are highly connected and where decisions and actions affect all. They demonstrate that shared knowledge and insights promote the development of creativity for diverse learners. In these environments, dynamic interactions among participants show that creativity can be for all, but that diversity can actually contribute to the promotion of mathematical creativity in heterogeneous classrooms. Bertazzoni and Marchini (2005) conducted an experiment with primary students over an entire year to investigate whether using non-standard problems would improve students' attitudes towards mathematics but also their performance. The data showed better results in routine problem solving, with improvements in arithmetic, geometry, proportional thinking, probability, set theory, etc. The research team concluded that the working environment with non-standard problems resulted in pupils' improved attitudes towards mathematics. Values cannot be underestimated as a critical variable in students' mathematics learning.

Conclusion

Mathematical literacy continues to be a continuing thread throughout the SEMT research. Our review has highlighted some of the interrelationships within the processes and enablers as well as between them. We observed that most perspectives shared in this chapter include more than one process or enabler, demonstrating the connectedness within each circle of the model. In addition, we observed the two circles are inter-related such that the enablers work together with the processes in the outer circle. This demonstrates the importance of connections in the development of mathematical literacy. Processes and enablers are integral components to effective learning and teaching of mathematics.

Considering complexity theory as advocated by Levenson and Barkai (2013) as a framework for viewing classrooms as complex systems, the perspectives shared in this chapter provide synergy for considering learning environments and students' and teachers' interactions and actions within those environments that contribute to mathematical literacy. As such, the summaries of perspectives and research presented in this chapter speak to a robust programme for considering state of the art contributions to students' mathematics learning.

References

Bakos, S., and Sinclair, N. (2019). Exploring the semiotic potential of TouchTimes with primary teachers. In J. Novotná and H. Moraová (Eds) SEMT 53–62.

Bertazzoni, B., and Marchini, C. (2005). Improving classroom environment by problem solving. In SEMT 2005: Proceedings of Symposium on Elementary Maths Teaching, 78–85. Prague: Charles University.

Břehovský, J., Eisenmann, P., Ondrušová, J., Přibyl, J., and Novotná, J. (2013). Heuristic strategies in problem solving of 11-12-year-old pupils. *Journal on Efficiency and Responsibility in Education and Science*, 7(1).

Chapman, O. (2013). Engaging children in learner-focused mathematical tasks. In J. Novotná and H. Moraová (Eds) SEMT 9–20.

Cheeseman, J. (2009). Challenging children to think: Teacher behaviours that stimulate children to examine their mathematical thinking. In J. Novotná and H. Moraová (Eds) SEMT 11–23.

Cifarelli, V., Morge, S., Goodson-Espy, T., and Pugalee, D. (2011). Examining the math content and pedagogical knowledge of pre-service teachers: How Katelyn and Matthew made sense of students' problems. In J. Novotná and H. Moraová (Eds) SEMT 111–117.

Clarke, D. (2001). Understanding, assessing and developing young children's mathematical thinking: The experiences of the Early Numeracy Research Project. In J. Novotná and M. Hejný (Eds) SEMT 7–14.

Dabić, M., and Milinković, J. (2015). Teachers' representations of multiplication: Do children understand them? In J. Novotná and H. Moraová (Eds) SEMT 99–106.

Doig, B., and Groves, S. (2017). Mathematical tasks stimulating student problem solving strategies. In J. Novotná and H. Moraová (Eds) SEMT 479.

Erickson, D., and Marcinek, T. (2019). Raising pre-service elementary school teachers' awareness of the images of mathematics and mathematicians as they are depicted in popular culture. In J. Novotná and H. Moraová (Eds) SEMT 163–173.

Favilli, F. (2017). Mirror curves and basic arithmetics. In J. Novotná and H. Moraová (Eds) SEMT 168–176.

Grandgenett, N., Harris, J., and Hofer, M. (2010–2011). An activity-based approach to technology integration in the mathematics classroom. *NCSM Journal of Mathematics Education Leadership*, 13(1): 19–28.

Herman, J. (2003). Analysis of pathways from Arithmetic to Algebra at the pre-algebraic level. In J. Novotná (Ed.) SEMT 87–90.

Hershkovitz, S., and Tabach, M. (2011). Instruction and technology. In J. Novotná and H. Moraová (Eds) SEMT 363–364.

Huhmann, T., Höveler, K., and Eilerts, K. (2019). Counter-acting representational volatility in geometry class through use of a digital aid: Results from a qualitative comparative study on the use of pentomino learning environments in primary school. In J. Novotná and H. Moraová (Eds) SEMT 211–220.

Hejný, M. (1995). The understanding of geometrical concepts. In Proceedings of the 3rd Bratislava International Symposium on Mathematical Education, BISME-3. pp. 52–64.

Kaur, B. (2019). Opportunity to learn: Reason and communicate mathematically in the elementary class. In J. Novotná and H. Moraová (Eds) SEMT 9–19.

Krauthausen, G. (2013). Digital tools in elementary mathematics education. In J. Novotná and H. Moraová (Eds) SEMT 33–44.

Kulm, G. (1991). *Assessing Higher Order Thinking in Mathematics.* Washington, DC: American Association for the Advancement of Science.

Larsson, K. (2015). Connections for learning multiplication. In J. Novotná and H. Moraová (Eds) SEMT 202–211.

Levenson, E., and Barkai, R. (2013). Mathematical creativity in elementary school. In J. Novotná and H. Moraová (Eds) SEMT 346–347.

Levenson, E., Barkai, R., and Larsson, K. (2013). Functions of explanations: Israeli and Swedish elementary school curriculum documents. In J. Novotná and H. Moraová (Eds) SEMT 188–195.

Levenson, E., and Barkai, R. (2017). Promoting mathematical creativity in heterogeneous classes. In J. Novotná and H. Moraová (Eds) SEMT 42–52.

Lipovec, A., and Podgorošek, M. (2017). Students' visual representation of fractions and exponentiation. In J. Novotná and H. Moraová (Eds) SEMT 319–327.

Littler, G. H. (1999). What mathematics is needed for or gained from reality. In M. Hejný and J. Novotná (Eds) SEMT 184–185.

Macedonska, L. (2003). Graphic representations and forming of mathematical understanding at the pre-defining stage. In J. Novotná (Ed.) SEMT 184.

Maj-Tatsis, B., and Tatsis, K. (2015). Reasoning processes of two girls during problem solving. In J. Novotná and H. Moraová (Eds) SEMT 222–231.

Martin, C., Lambert, R., Wang, C., Polly, D., and Pugalee, D. K. (2015). Supporting learning mathematics through writing: Formative assessment and instructional tool. In J. Novotná and H. Moraová (Eds) SEMT 240–249.

Martin, C., Lambert, R., Wang, C., and Polly, D. (2017). Supporting mathematics learning through project based learning: A fifth grade case study. In J. Novotná and H. Moraová (Eds) SEMT 328–336.

Mathematical Sciences Education Board. (1996). *The Preparation of Teachers of Mathematics: Considerations and Challenges*. Washington, DC: National Research Council.

Moraová, H., and Novotná, J. (2015). Teaching maths in English at primary school level: Utopia, nightmare or reality. In J. Novotná and H. Moraová (Eds) SEMT 249–257.

National Council of Teachers of Mathematics (NCTM). (1989). Curriculum and Evaluation Standards for School Mathematics. Reston, VA: NCTM.

National Council of Teachers of Mathematics (NCTM). (2000). Principles and Standards for School Mathematics. Reston, VA: NCTM.

National Council of Teachers of Mathematics (NCTM). (2014). Principles to Actions: Ensuring Mathematical Success for All. Reston, VA: NCTM.

Novotná, J., Eisenmann, P., Přibyl, J., Ondrušová, J., and Břehovský, J. (2014). Problem solving in school mathematics based on heuristic strategies. *Journal on Efficiency and Responsibility in Education and Science*, 7(1): 1–6.

Nührenbörger, M., Söbbeke, E., Häsel-Weide, U., Tubach, D., and Wissing, E. (2013). Discovering and reasoning mathematical relations in manipulatives and tasks. In J. Novotná and H. Moraová (Eds) SEMT 348–349.

Organization for Economic Co-operation and Development (OECD). (2018). Preparing our youth for an inclusive and sustainable world. The OECD PISA global competence framework. Paris. Retrieved 24/04/2024, from: ww.oecd.org/pisa/Handbook-PISA-2018-Global-Competence.pdf

Polotskaia, M., Savard, A., Cavalcante, A., and Fellus, O. (2019) Middle school students' difficulties in problem solving and their roots in elementary education: What went wrong for them? In J. Novotná and H. Moraová (Eds) SEMT 315–324.

Polya, G. (1945). *How to Solve It.* Princeton, NJ: Princeton University Press.

Poirier, L., and Bisaillon, N. (2015). The role of language and communication in the learning of maths in primary school: Some examples. In J. Novotná and H. Moraová (Eds) SEMT 384–385.

Procházková, I. (2015). Developing the concept of area with the Hejný method. In J. Novotná and H. Moraová (Eds) SEMT 41–51.

Prusak, N., and Hershkowitz, R. (2019). Task designed for promoting the concept of area in a problem-solving environment: The story of one activity. In J. Novotná and H. Moraová (Eds) SEMT 325–333.

Pugalee, D., and Seifert, A. (2013). Including writing to expand cognitive level for mathematical tasks. In J. Novotná and H. Moraová (Eds) SEMT 257–264.

Pugalee, D. (1999) Constructing a model of mathematical literacy. *The Clearing House*, 73(1): 19–22, doi.org/10.1080/00098659909599632

Ruthven, K. (2007). Towards a calculator-aware number curriculum. In J. Novotná and H. Moraová (Eds) SEMT 9–21.

Sáenz-Ludlow, A. (2017). Children's progressive construction of number and numerals by means of numerical diagrams. In J. Novotná and H. Moraová (Eds) SEMT 345–353.

Savard, A., and Polotskaia, E. (2013). Word problem solving task management and students' access to mathematics: Cases in elementary education. In J. Novotná and H. Moraová (Eds) SEMT 290–298.

Sullivan, P., Clarke, D., and Clarke, B. (Eds) (2013). *Teaching with Tasks for Effective Mathematics Learning*. New York: Springer.

Symeonidis, N., Tatsis, K., and Kaldrymidou, M. (2015). Norms in decision-making and in shaping attitudes, emotions, and performance. In J. Novotná and H. Moraová (Eds) SEMT 322–330.

Tabach, M. (2011). Promoting technology integration among prospective mathematics teachers. In J. Novotná and H. Moraová (Eds) SEMT 328–335.

Tambunan, H. (2018). Impact of heuristic strategy on students' mathematics ability in high order thinking. *International Electronic Journal of Mathematics Education*, 13(3): 321–328. doi.org/10.12973/iejme/3928

Wilson, J., Fernandez, M., and Hadaway, N. (1993). Mathematical problem solving. In P. S. Wilson (Ed.) Research Ideas for the Classroom: High School Mathematics, 57–58. Reston, VA: NCTM.

4.2

Designing and adapting tasks in the Japanese problem-solving lesson within Lesson Study

Toshiakira Fujii
Tokyo Gakugei University, Japan

Introduction

The history of Lesson Study in Japan spans more than a century (Makinae 2010). For Japanese educators, Lesson Study is like the air that they breathe, part of everyday school life. Lesson Study came to the attention of educators outside of Japan primarily through the publication of *The Teaching Gap*, which described findings from the Third International Mathematics and Science Study (TIMSS) video study that focused on the eighth-grade mathematics lessons in the USA, Germany, and Japan (Stigler and Hiebert 1999). Chapter 7 in particular, entitled 'Japan's approach to the improvement of classroom teaching', provoked enormous interest, not only in Lesson Study as collaborative work among teachers, but also in the typical structure of Japanese mathematics lessons. Since then, many mathematics teachers, and teacher educators, around the world have been involved in Lesson Study and also have tried to adapt or adopt the Japanese model of problem-solving lessons. This paper tries to clarify an embedded key factor of the Japanese model of teaching mathematics in the context of Lesson Study, particularly focusing on activities involved in the process of planning a lesson or, more precisely, on designing the task for a lesson.

Lesson Study components and process

Lesson Study is an approach to teacher professional development (PD) that differs sharply from the PD practices common in other countries. Liptak (cited in Lewis, 2002, p. 12) contrasted Lesson Study with the traditional PD as practised in the USA, as in Table 4.2.1 below.

Lesson Study begins with a question, not with an answer prepared by someone else. Identifying this question, which becomes the research theme for Lesson Study, is the first step in the process that can be illustrated as in Figure 4.2.1. The research theme is developed through consideration of the reality of students' current state vis-à-vis long-term goals for their learning and development.

The Lesson Study process is illustrated in Figure 4.2.1 as having five steps. Borrowing from Lewis's (2002) and Lewis and Hurd's (2011) descriptions, each step can be summarized as follows:

Toshiakira Fujii, *Designing and adapting tasks in the Japanese problem-solving lesson within Lesson Study*. In: *Elementary Mathematics Teaching*. Edited by: Brian Doig, Jarmila Novotná, Berinderjeet Kaur and David K. Pugalee, Oxford University Press. © Oxford University Press (2024). DOI: 10.1093/oso/9780192869647.003.0014

Table 4.2.1 Contrasting views of professional development (Liptak, cited in Lewis 2002, p. 12)

Traditional	Lesson study
Begins with answer	Begins with question
Driven by outside expert	Driven by participants
Communication flow: trainer—teachers	Communication flow: among teachers
Hierarchical relations between trainer and teachers	Reciprocal relations among learners
Research informs practice	Practice is research

Figure 4.2.1 The process of Lesson Study (Fujii, 2014, p.113)

1. *Goal setting*: Teachers consider long-term goals for student learning and development. They identify gaps between these long-term goals and current reality and decide the research theme.
2. *Lesson planning*: Teachers collaboratively plan a research lesson designed to address the goals. They prepare a lesson plan, or more accurately a 'lesson proposal', a document that describes the research theme, content goals, connections between the current content and related content from previous and later grades, rationale for the chosen approach, a detailed design for the research lesson, and anticipated student thinking, etc.
3. *Research lesson*: One team member teaches the research lesson, while the other members of the planning team, staff members from across the school, and, usually, an external knowledgeable other observe and collect data. In the case of district-wide and nationwide Lesson Study, observers would come over from a broader range of areas.
4. *Post-lesson discussion*: Observers share data from the lesson in a formal lesson colloquium, highlighting successes and issues concerning student learning, disciplinary content, lesson and unit design, and broader aspects of teaching and learning.
5. *Reflection*: Not only the instructor, but also other members in the school consolidate and document what was learnt from the Lesson Study cycle. They write a report or bulletin that includes the original research lesson plan, student data

from the research lesson, and reflections on what was learnt so that this knowledge can be shared with a larger number of people, as well as formulating new questions for the next cycle.

This differs from other, similar schemes in other publications that only have four steps (e.g. Lewis 2002; Lewis and Hurd 2011). The inclusion of a fifth step more accurately portrays how Japanese teachers progress through the Lesson Study cycle.

On the other hand, Takahashi and McDougal (2016) have proposed a six-step model of Lesson Study by adding the *kyozai kenkyuu* between the first and second steps. *Kyozai kenkyuu* is literally translated by saying it is 'as study or research on teaching materials that helps teachers gain knowledge and insight into mathematics and student thinking'. One of the important qualities of *kyozai kenkyuu* located in the context of Lesson Study is the collaboration among teachers. Actually, Takahashi and McDougal have coined a new term, collaborative lesson research (CLR), instead of Lesson Study to better explain the essence of Lesson Study for people outside of Japan. The term 'collaborative' implies that at the heart of Lesson Study is collaboration among teachers. Doig and Groves (2011) also emphasized the collaborative nature of Lesson Study using the term 'community of inquiry', as 'it (Lesson Study) offers teachers the opportunity to develop professional communities of inquiry' (p. 90).

Outside of Japan, the third and fourth steps in the Lesson Study cycle, the research lesson and post-lesson discussion, being the most visible parts, attract people's attention. This leads to a typical misconception of Lesson Study: that the research lesson and post-lesson discussion are Lesson Study. Indeed, Lesson Study is sometimes introduced, as an open lesson, by a veteran teacher 'jumping in' to another teacher's classroom (Takahashi 2013, p. 84). But each step in the Lesson Study cycle is closely related to the others, and the third and fourth steps in particular are based on the first and second steps.

One of the misconceptions about Lesson Study that I often observe is that a mathematics research lesson, the third step in the Lesson Study cycle, has to be a structured problem-solving lesson. When you try to incorporate Lesson Study into another country, research lessons can be any style of lessons. For example, the TIMSS video study (1995) shows that the United States, Germany, and Japan each have a unique lesson pattern. Although each lesson in a country seems to be different, if you look at 20 or 30 lessons in that country, you will notice that there is certainly a common pattern. Lessons are a cultural and historical activity, and there are lesson patterns for every country. Thus, when you conduct a Lesson Study in your country, it may not be necessary to follow another country's lesson style, such as structured problem solving. The reason you find most research lessons in Japan being aimed at developing structured problem-solving lessons, and you do not see many exercise-orientated lessons, is that structured problem-solving lessons require careful designing and skilful teaching.

When introducing Lesson Study in other countries, you may aim for Japanese problem-solving lessons resulting in research lessons because you have concluded that this style of mathematics lessons is appropriate and desirable for your country's classrooms. Still, it is possible to develop a new and innovative approach, other than structured problem solving, to improve teaching and learning that is more suitable for your own country.

People often argue that seeing is believing; thus, they may believe that watching an exemplary lesson on video may be enough to teach it in your classroom. However, teachers who do not have a clear concept of structured problem-solving lessons may not be able to identify critical elements of structured problem solving by just watching a video of the lessons. That often yields misunderstandings. Only if you know what Japanese problem-solving lessons are, can you clearly see the components and structure of Japanese problem-solving lessons, and see where they were inadequate or which parts were excellent. Therefore, watching exemplary classroom videos may be a good idea for teachers to understand the approach, but it is not reasonable to expect them to teach using it in their classrooms. Instead of watching a video, participating in workshops on Lesson Study, such as the SEMT 2009 and 2015 workshops, and reading about the structured problem-solving lesson would seem to be more meaningful and more appropriate (Doig, Groves, and Macháčková 2009; Doig and Groves 2015).

The Japanese problem-solving lesson: Teaching mathematics through problem-solving

The Japanese problem-solving lesson seems to be unique or distinctive to foreign people's eyes in many ways. It is a lesson in which students work on their own to solve a given task for which they have not been shown solution strategies, and critically analyse and reflect on their own and other students' solutions. It is based on a 'one task for one lesson' policy that is distinguished from other problem-solving lessons in Asia such as Hong Kong and Singapore. The Japanese approach to mathematics instruction, as described by Stigler and Hiebert (1999), was a 'structured problem-solving lesson'. I prefer to describe this teaching method as teaching mathematics *through* a problem-solving lesson. In order to avoid misinterpretations, let me describe in detail each phase of teaching mathematics through a problem-solving lesson in Japan.

First of all, teaching mathematics through problem solving consists of four components, and takes up time in the lesson, indicated in parentheses, as follows:

1. Presenting problem for the day (understanding the problem from the children's point of view) (5 to 10 minutes)
2. Problem solving by the students (10 to 20 minutes)
3. Comparing and discussing (*neriage*) (10 to 20 minutes)
4. Summing up by the teacher (*matome*) (5 minutes).

Presenting the problem for the day

Some people may misinterpret 'presenting the problem' as providing a detailed explanation or demonstration of the procedures involved in solving the given problem. However, this is not actually the case. In *The Teaching Gap*, Stigler and Hiebert (1999) write, 'the (Japanese) teacher presents a problem to the students without first demonstrating how to solve the problem', and, 'we saw that a feature we hardly noticed before is perhaps one of the most important features of U.S. lessons—that the teacher almost always demonstrates a procedure for solving problems before assigning them to students' (p. 77).

In Japan, 'present the problem' means to make students understand the context of a given task and the corresponding mathematical conditions that would be used to solve that task. Japanese educators distinguish between 'teaching how to solve the task' and 'problem-solving-based mathematics instruction' in which students learn how to solve the problem by solving it themselves. Therefore, it is important to choose the task carefully. If chosen well, the task allows for the important new mathematical ideas to emerge in classroom discussion.

Students work to solve the problem

While students work to solve the task on their own, the teacher in the classroom is extremely busy. The teacher must move among the students, practising what we call in Japanese *kikan-jyunshi*, which means purposeful scanning or monitoring. *Kikan-jyunshi* is not trying to consult with every student. So, what do Japanese teachers consciously do during *kikan-jyunshi*? In Japan, the teacher:

- Identifies how each student is solving the problem according to the anticipated solutions, including typical incorrect solutions. These are already listed in the lesson plan.
- Takes notes on the general atmosphere and how most students are solving the problem.
- If necessary, gives hints to slower learners and asks faster learners to come up with another way of solving the problem, etc.
- Plans what the teacher should do during the compare and discuss (neriage) phase, such as, which strategies and students are presented first, how to compare or contrast each strategy, and so on.

One of the most difficult aspects of *kikan-jyunshi* is the need to anticipate potential student solutions during lesson planning. This is particularly difficult for novice teachers. However, I think this practice of trying to prepare all possible solutions improves Japanese teachers' teaching abilities to see the mathematics through the eyes of the student and consider how best to facilitate their understanding of the underlying concepts or skills. Furthermore, it highlights how lesson planning, step 2 of the Lesson Study cycle in Figure 4.2.1, is a critical factor in determining the quality of mathematics teaching. Anticipating solutions enriches a teacher's knowledge on the topic and deepens their understanding of their students' cognitive nature and abilities. During *kikan-jyunshi*, if a teacher discovers an unexpected solution, the teacher may ask the student about the thought process and record the solution. The post-lesson discussion follows after the lesson as step 4 of the Lesson Study cycle, and the quality of the task will be discussed and evaluated. Teachers discuss not only how likely students were to attempt the anticipated solutions listed in the lesson plan but, more importantly, how students actually move towards the goal of the lesson by solving the task and comparing and contrasting solutions.

Kikan-jyunshi is important in creating a rich and profound comparison and discussion stage for the problem-solving lessons. The teacher needs to look beyond individual solutions and grasp how the class, as a whole, is approaching the task. Moreover, while engaging in *kikan-jyunshi*, the teacher has to decide which student will present first and think about how to direct the class discussion. This means that the teacher must configure the next phase of the scenario: the comparison and discussion phase. In the post-class discussion, teachers discuss the quality of *kikan-jyunshi* based on evidence and data collected during the research lesson.

Compare and discuss (*neriage*)

The third phase of a lesson, called *neriage* in Japanese, assumes that students will arrive at different solution methods, and therefore focuses on a comparison and discussion of those different solution methods. During the *neriage* phase, students' solutions are typically presented from naïve, or simple, to sophisticated, on the blackboard. Therefore, the teacher should be careful, as certain students may realize why they are always the first to present. Teachers must also carefully plan how everything will be written on the blackboard in order to best compare and contrast students' solutions. The blackboard is used to organize both thought processes and results. Japanese teachers used to be reluctant to use overhead projectors, in contrast with US teachers, because they are small and therefore unsuitable for showing complete thought processes.

The *neriage* phase is difficult for novice teachers. They can usually give positive comments to each student, but they have a difficult time pulling out points and clarifying points that the students don't understand. Sometimes, students will not be able to distinguish between two similar solutions. In such cases, the teacher should suggest how these two solutions differ significantly from each other. Or sometimes it's the opposite issue: two solutions appear to be different but are actually the same. *Neriage* provides an opportunity for teachers to foster mathematical ways of thinking in their students. In other words, it is a chance to focus on process. Mathematical ways of thinking cannot be demonstrated by simply reviewing the definition of concepts; they can only be truly experienced through problem-solving activities. Therefore, Japanese teachers teach mathematics through problem solving. This kind of lesson encourages students to think mathematically and helps them become independent learners. These are the ultimate goals of school education.

Sometimes the *neriage* phase may appear to be implemented well, but is actually not. These cases typically are a result of a one-to-many correspondence between teacher and students describing their solutions during discussion. Takahashi (2008) described this phenomenon as 'show-and-tell'. The heart of *neriage* is reflective discourse (Doig, Groves, and Fujii 2011). Although rather a classic citation, *neriage* should be an actualization of Vygotsky's Zone of Proximal Development (Ohtani 2014). Therefore, during *neriage*, teachers must see the students' potential as clearly as possible and help them move towards their potential from their starting point on their own—that is, ascertain what they can do without help.

Summarizing the lesson (*matome*)

Fujii et al. (1998) identified that US teachers tend not to summarize lessons at the ends of classes. Without this phase, the lesson ends with students only being able to take satisfaction in their individual work. By including a summary at the end of the lesson, every student can also feel satisfaction regarding the mathematical concepts explored.

Matome, or 'summarizing the lesson', is brief, but critically important. It reflects the teacher's value system and helps impart this to students. Students should listen carefully to the teacher's final words. If the teacher concludes a lesson with simply a review of mathematical procedure, students may feel that procedural knowledge is what matters most. However, if teachers sum up the lesson by reviewing how students constructed their ideas, it sends the message to students that how they think is also important. In other words, during *matome*, the teacher should review not only content but also process. *Matome* should include comments by the teacher concerning the mathematical and educational values of the task and lesson. An embedded key factor of the Japanese model of teaching mathematics through problem solving may be hidden here.

A closer look at lesson planning: Designing and adapting tasks

In the second step in the Lesson Study process—that is, 'lesson planning'—the Japanese teachers spend a lot of energy and time crafting a *gakushushido-an*. *Gakushusido-an* is usually translated as 'lesson plan', but since 'plan' misleadingly implies only one or two pages of fixed script, the term 'proposal' is better. *Gakushusido-an* should have been translated into *lesson proposal*, but it seems to be too late. So, I use the term 'lesson plan' in this paper, although its format is more like a lesson proposal.

Typical lesson plan components and structure

The contents of the typical lesson plan give clues about the task design process. Although the details vary from school to school, and even from teacher to teacher, Lewis (2002) notes that a typical lesson plan, or proposal, for a research lesson in Japan, consists of the following:

- name of the unit
- unit objectives
- research theme
- current characteristics of students
- learning plan for the unit, which includes the sequence of lessons in the unit and the tasks for each lesson
- plan for the research lesson, which includes:
    - aims of the lesson

  - teacher activities
  - anticipated student thinking and activities
  - points to notice and evaluate
  - materials
  - strategies
  - major points to be evaluated
  - copies of lesson materials (e.g. blackboard plan, student handouts, visual aids)
- background information and data collection forms for observers (e.g. a seating chart).

The explicit inclusion in the lesson plan of the tasks for each lesson in the unit indicates how important the tasks are believed to be, and that the authors of the lesson plan think carefully about their sequence within the unit.

Teachers starting *kyozai kenkyuu* with textbooks

As a typical activity for designing tasks, Japanese teachers, in the second step of the Lesson Study cycle, start *kyozai kenkyuu* with textbooks. There is a reason why textbooks play a critical role in designing tasks. Since Japan has a National Course of Study, textbooks are authorized by the Ministry of Education, Culture, Sports, Science and Technology. There are six publishing companies that publish the mathematics textbooks. It is in a way natural for teachers to compare each textbook's treatment of the same content. Investigating the textbooks leads to studying of the teacher's manuals. The teachers' manual that comes with the textbooks does not just include answers to problems in the textbook: it also includes approaches to teaching the content, explanations of the solutions, and typical misconceptions of students regarding lesson content.

Besides textbooks and teachers' manuals, there are many resources for *kyozai kenkyuu* readily available for teachers, such as books and journals, and even lesson plans written by other teachers. Because Japanese teachers write the final reports at the fifth step of the Lesson Study cycle and the final reports includes not only reflections on what was learnt but also the lesson plans for the research lessons and student data, sometimes a detailed transcription of the research lesson is included. The final reports are usually printed and distributed in the district where the school is located for free, or sometimes they are published commercially. Therefore, teachers in Japan have easy access to published lesson plans and detailed information concerning the research lessons. That is a benefit for Japanese teachers, as Corey et al. (2020) have pointed out.

Tasks in the Lesson Study process

The lesson that is the focus here was titled 'Who is the fastest?' for the introduction of a formula for speed. It was held at a small school with only one class at each grade level.

Four lesson planning meetings were held: 21 May (30 minutes), 6 June (90 minutes), 11 June (20 minutes with a final commentator), 18 June (90 minutes) and the research lesson held on 3 July 2013. The lesson plan came to 15 pages (an English version of the lesson plan is available through the IMPLUS website—see Note 3 below).

The first meeting included five regular members: two classroom teachers for Grades 5 and 6, the music teacher, the art teacher, and the teacher for mathematics. The reason the music teacher and the art teacher were included was that they were teaching the Grade 5 and 6 students, and therefore the regular members were the upper-year-level team. The Grade 6 teacher was the leader of the school research steering committee and taught the research lesson.

The lesson planning meetings followed the framework of teaching mathematics through problem-solving lessons. Therefore, the discussions could be aligned with the four phases of teaching mathematics through problem-solving lessons. For example, at the second meeting, a discussion on how students might grasp the given task (15 minutes) was related to phase 1, Presenting the problem for the day; discussion about likely student responses (14 minutes) was related to phase 2, Problem solving by the students; discussion about how to organize the comparison and discussion period (15 minutes) was obviously related to phase 3, Comparing and discussing; and discussion about how to conclude the lesson (5 minutes) was related to phase 4, Summing up by the teacher.

Tasks in designing and planning for the lesson: Appropriateness of the task

If we want to teach the concept of speed and how to calculate speed through a problem-solving lesson, what is the lesson like? The typical textbook task in Japan is below in an English language version (Figure 4.2.2).

Using the English translated version of the task, the lesson would begin by considering who is faster, Takumi or Akira, since the distances they have run are the same. Students may conclude that Takumi is faster than Akira. Then, students consider who is faster, Hitoshi or Akira, since the times they ran are the same. Students may conclude Hitoshi is faster than Akira. So eventually the task of the lesson is who is faster, Takumi or Hitoshi, since both the distances and the times are different.

Teachers would engage in detailed discussions about the task itself, including which numbers to use and why. However, the selection of numbers is not always from a purely mathematical point of view. For example, in this school, teachers thought

Distance and time run

	Distance (m)	Time (sec)
Takumi	80	18
Hitoshi	100	20
Akira	80	20

Figure 4.2.2 An English version of a Japanese task (Translated by the author)

about numbers both in terms of their students' reality and from a procedural or calculation point of view. The teacher who would teach the research lesson said:

> The child A in the problem can run 40 metres in only 6 seconds. In my class there is no such fast runner. However, I decided to use these numbers, because these numbers are easier for children to calculate.

The planning team decided to use the task, including data from the three students, as the most important task that students would think of in the lesson. However, beside the main task they planned to use three supplementary tasks: S1, S2, and S3. S1 (Figure 4.2.3) is to use for students to notice that only the time (seconds) data is not enough to decide who is faster. S2 (Figure 4.2.5) is to consider which method is more efficient for expressing speed with numbers. S3 (Figure 4.2.6) is for students to recognize that people in daily life use a distance *per* second method rather than a time *per* distance method. The sequence to show these tasks is S1, Main task (Figure 4.2.4), S2, and then S3, as shown below (each task is from the English version of the lesson plan).

When the planning team was considering using data from six students in S2, time and distance data for the first three people in the problem (A, B, C) were not changed, but data for two people (E, F) were changed from E (42 metres in 6.7 seconds), F (28 metres in 4.9 seconds) to E (45 metres in 6.5 seconds), F (50 metres in 8 seconds),

	Time (Second)
A	6
B	6
C	5

Figure 4.2.3 Supplementary task S1 (Photograph by author)

	Distance (m)	Time (Second)
A	40	6
B	30	6
C	30	5

Figure 4.2.4 Main task for the lesson (Photograph by author)

	Distance (m)	Time (Second)	Place
A	40	6	
B	30	6	
C	30	5	
D	35	5.5	
E	45	6.5	
F	50	8	

Figure 4.2.5 Supplementary task S2 (Photograph by author) 300km per hour 320km per hour

300km per hour 320km per hour

Figure 4.2.6 Supplementary task S3

in order to provide some faster speeds. Original numbers for D, E, and F were considered hard for students to calculate. Teachers also worried about having decimal fractions as the result of calculations. However, they decided to keep the numbers and let students use calculators if they wanted. Teachers did a lot of calculation to decide what numbers are appropriate for students.

Anticipated student solutions

Student solutions to the task are closely related to the curriculum, because curriculum-related content is expected to be a resource for students to solve tasks. However, the location of the teaching of speed has not been stable within the upper grade curriculum. At present, the formula for speed is introduced in the fifth grade. When the research lesson was held, speed was taught in the sixth grade. In the second meeting there was the following exchange:

TEACHER A: Students learned how to arrange to get the same numbers for time or distance, didn't they?

TEACHER B: Yes, I suppose. However, the idea of a common multiple was learned a long time ago from the students' point of view.

TEACHER C: Probably they forgot the procedure to find the common multiple (5th grade).

TEACHER B: When they learned division of decimal numbers, they learned the idea of per-unit (5th grade). It's the same thing here. However, the idea of per-unit was not learned in the context of comparing things.

PRINCIPAL: The idea of per-unit quantity was applicable for comparing crowdedness (5th grade). That is a mathematical way of thinking that could be applicable for Speed (6th grade).

Teachers spent a lot of time discussing the students' response to the main task of the research lesson. These discussions usually began by considering what was most likely from the class as a whole. They then went on to consider likely responses from students who were rather slow learners and those who were fast learners.

At the third meeting, teachers pretended to be students in order to solve the main task of the lesson from the students' point of view. Through this mock lesson activity, teachers confirmed the plausibility of the four anticipated solutions already written in the lesson plan:

1. Finding a common multiple of distance to compare.
2. Finding a common multiple of time to compare.
3. Finding the amount of time *per* metre to compare.
4. Finding the distance *per* second to compare.

Concerning how to deal with slow learners, the teacher had already decided to provide hints to students who wanted them during the individual problem-solving period. The planning team discussed specifically what should be on these hint cards. While a hint card suggesting using common multiples was reasonable from the teachers' initial point of view, they no longer thought this might be the case when they imagined, or visualized, the lesson. They thought this strategy would eventually be rejected in favour of a better strategy: finding the distance *per* second.

One teacher said, 'Students might ask the teacher, "Why did you not give me the best hint, if you knew?"' The other teachers agreed that was likely to happen. So, they discussed how to let students notice the '*per* second' strategy. Finally, teachers thought of using 30 metres and 5 seconds as the data. 'It divides beautifully'. If the teacher asks a question such as, 'Five seconds to go (30m), so if it were one second how far could you go?', then students may be able to notice the idea of *per* second. 'It will work', one teacher said, and another said, 'It looks fine'. Eventually, the teacher decided to suggest using the '*per* second method' to solve the task using the data of 30 metres and 5 seconds. The teacher clearly anticipated that some students would calculate 30÷6 and 30÷5 to get the distance *per* second. Teachers also considered how to deal with fast learners in the lesson.

Besides 'beautifully divided' numbers, teachers deliberately chose awkward numbers for the additional speed data of persons D, E, F for the task S2. The teacher explained, 'I want students to say that it is awkward to calculate common multiples among them.' She deliberately chose numbers that would push students to calculate distance divided by time.

The task in action: The task is implemented in a research lesson

The lesson started by showing task S1 (Figure 4.2.7)—the three students' records of time in a sprint race (insufficient information)—and the teacher asked students, 'Who is the fastest?' As the planning team teachers had expected, some students responded quickly saying, 'The smallest value is the fastest.' But at the same time, other students reacted soon by saying, 'We can't tell who is the fastest because we don't know if they ran the same distance.' The teacher picked up such student voices, and brought up the idea of how to compare the running speed of these children—that is, the distance (travel distance) needs to be set the same. Then the main task was given, providing information of the distance those children ran and helping them to recognize that the

Figure 4.2.7 The task S1 (Photograph by author)

speed of those children can be compared when either the distance or the time they run is made the same.

The main task (Figure 4.2.8) that the students are dealing with in this lesson tells them that two out of the three children run different distances and times, so it is not easy to compare the speed of the three children. At this instance, students did come up with a question: 'When both the time and distance of running are different, how can we compare them?' In this lesson, the question that arose from the students was used to establish a main task that did help the students to engage in autonomous learning through problem solving.

The main task was finally clarified as 'How can we compare between A and C when both the time and distance of running are different?' Up to this point, the teacher spend 8 minutes and 11 seconds. These steps can be seen in Figures 4.2.9 and 4.2.10 above. Then students started to solve the task individually. While the students worked on the task, the teacher did *kikan-jyunshi*: purposeful scanning or monitoring, in

Figure 4.2.8 The main task (Photograph by the author)

Figure 4.2.9 We can compare between A and B, and between B and C (Photograph by the author)

Figure 4.2.10 How can we compare between A and C? (Photograph by the author)

order to identify anticipated solutions, or even an unexpected solution with deep thought behind it. The duration of individual problem-solving activity was about 12 minutes, during which the teacher did *kikan-jyunshi*. During the *kikan-jyunshi* the teacher gave a whiteboard to selected students to write down their solutions. These whiteboards would be used for the whole-class discussion.

Then the comparing and discussing, which is the third component of the structured problem-solving or teaching mathematics through problem-solving lesson, began. Eventually, four strategies—which were exactly the same as the anticipated solutions in the lesson plan—were displayed: see Figure 4.2.11 below.

The teacher did not present these four solutions at once, but step by step. First, the teacher put the whiteboard that had anticipated solution (1) on the extreme left-hand

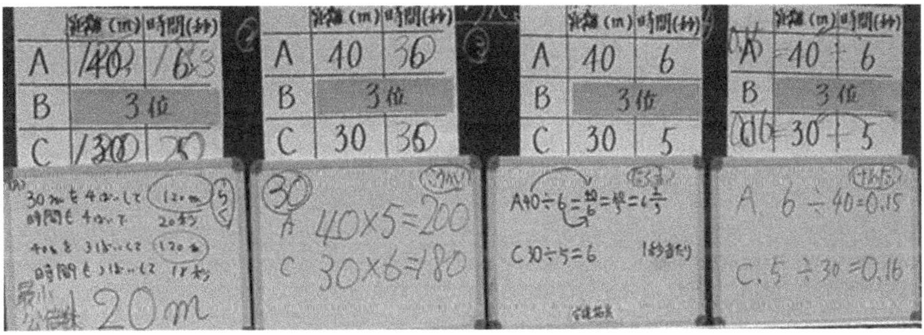

Figure 4.2.11 Four solutions were presented at the blackboard (Photograph by the author)

side of the blackboard. And the teacher asked some volunteers to explain this solution. One student came to the front and interpreted this solution (see Figure 4.2.12).

In other words, the student who wrote the solution did not explain how and why he did it. Classmates understood the idea of the solution was to make the distance the same. Then the teacher said, 'I know someone did a similar thing, could someone

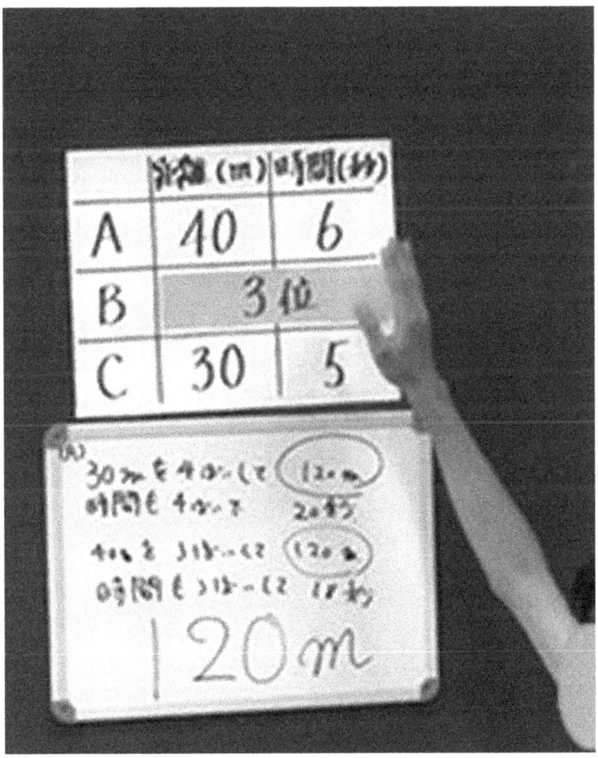

Figure 4.2.12 I can explain my friend's solution (Photograph by the author)

imagine that?' Then, another student responded, saying that instead of the distances being the same, we could use a common multiple of time.

Then a second solution was presented and explained by somebody who did not write the solution. After that, two solutions were presented simultaneously, and one of them was chosen by the teacher, who said, 'Again, I know someone did a similar thing, I mean, it is to consider the time, right? So, can anyone guess which one was focused on the time?'

The third solution displayed was explained by a student who did not write that solution (see Figure 4.2.13). The fourth, and final, solution was presented in the context of considering another *per* quantity—this was *per* metre. The manner of explanation by the student was the same.

When the four solutions were arrayed on the blackboard, the teacher asked students if there were common features among the four solutions. Students came up with the answer that the four solutions were all trying to get same quantity. The teacher also asked for common features between (1) (4) and (2) (3). Students responded to this by saying (1) (4) are focused on distance, and (3) (4) are focused on time. The teacher finally asked for common features between (1) (2) and (3) (4). Students responded to this by saying (3) (4) used a *per* unit strategy, and (1) (2) use a *per* common multiple strategy. At this point, the teacher had spent about 24 minutes.

Then, once 46:57 minutes had passed from the beginning of the lesson, the supra-mental task S2 was given. Students were surprised, saying, 'Eh?!' However, students were not asked to calculate all the data given to determine who was the fastest. Instead, students were asked which strategy they preferred to use. Students came to the front for about three minutes and put their name cards under their favourite strategy. You can see name cards under each strategy, except strategy (1), in Figure 4.2.11. At 50:20, for two minutes, students explained which solution they preferred and why. The reason why they avoided choosing the (1) strategy was that it would be too awkward

Figure 4.2.13 She did focus on the time (Photograph by the author)

to deal with the common multiples for numbers, as they would become very large. Through the S2 task, students understood that as the number of data increases, a common multiple strategy is awkward, so the *per* unit strategy is better. However, both *per* time and *per* distance are equally appropriate so far. Therefore, finally, S3 was given for students to notice that people use the *per* time strategy to express speed, since, if we express speed by using the *per* time strategy, we get larger numbers for faster speeds.

The goal of the discussion in this lesson was for the students to search for more effective ideas. By helping students to have viewpoints such as 'concise', 'clear', 'accurate', and 'general', the teacher guided students to search for 'more effective ideas', and then helped them understand how we express the speed that is consistent with our feelings. That is, speed is expressed using the distance travelled in a unit amount of time, because, by doing so, we can express more quickly with larger numbers, so it's easy to imagine how fast it is. This idea seemed to be appreciated by the class, and met with their agreement. In fact, in the final stage of the lesson, one student said, 'Expressing bigger number is faster', and another student said, 'To make the same condition is important', and finally one student said, 'I can see the same idea!!' What she meant was that the idea of making the same condition is the same as making a common denominator in order to add two fractions that have different denominators.

The task in evaluation: The task is evaluated at the post-lesson discussion

After comparing three students' data, A, B, and C, in the main task, another three pairs of data, D, E, and F, were added for students to compare. In the post-lesson discussion teachers argued about whether these additional data were useful or not. The arguments were based on how students actually responded to the task in the lesson.

The arguments progressed from evaluating the task to modifying the task. In fact, the final commentator, the 'knowledgeable other', suggested more direct ways to manipulate numbers to identify a faster speed without calculating six pairs of numbers. He gave the example, shown in Table 4.2.2, of two pairs of numbers in the context of population density.

The final commentator suggested using these numbers instead of the six pairs of numbers that were used in the research lesson, because some students struggled to carry out the calculations in the time available, and then missed the educational value

Table 4.2.2 An example of two pairs of numbers used in the context of population density, recommended by the knowledgeable other

Pool	Area (m^2)	Number of people
A	200	15
B	400	45

of the task and the whole-class discussion. The post-lesson discussion provided a context for revising the task used at the research lesson, since points missed in planning meetings were revealed in the post-lesson discussion. This shows how the planning meetings of the Lesson Study cycle are closely related to the research lesson itself, and also to the post-lesson discussion.

The post-lesson discussion provided a context for revising the task used at the research lesson. However, this does not imply that re-teaching is necessarily part of Japanese Lesson Study. Based on their experience, Japanese teachers know that if students are different, then their reactions will be different. They understand that a lesson is itself an organic system, and is not like a machine. A non-organic system, such as a car, is composed of parts that may be easily replaced. However, in organic systems, like a lesson, each part supports the whole ecology. In the case of the school focused on here, important ideas missed in planning meetings were revealed in the post-lesson discussion. Teachers then regretted that their *kyozai kenkyuu* (see, e.g., Watanabe, Takahashi, and Yoshida 2006) was not profound enough or broad enough to cover the idea, but they did not move to re-teach the lesson with a revised lesson plan. In other words, Japanese teachers' attitude towards research lessons and lesson plans is that their best lesson plan should be implemented at a research lesson, and that a research lesson is the proving ground for teachers (Lewis and Tsuchida 1998).

Discussion

Numbers in the task are quasi-variables

Task design in lesson planning always seeks appropriate numbers to be used for the task. However, close attention to the specific numbers does not mean that teachers are sticking to a concrete level of thinking or encouraging students to think concretely. On the contrary, teachers consider the general aspect of the numbers—their quasi-variable aspects. A quasi-variable is a number deliberately used in a general way, so that it serves as a representative of many numbers, just as a variable would (Fujii and Stephens 2001, 2008; Fujii 2008, 2010). Numbers are often chosen based on their quasi-variable power, or how well they can demonstrate a general truth—a general truth that is brought out during the whole-class discussion.

In the discussion among the teachers in *The Teaching Gap*, the teachers agreed that the choice of numbers would influence which strategies the students would use when solving the problem. The task in this instance was subtraction with regrouping for first-graders. There are a total of 36 subtraction tasks of one-digit numbers from two-digit numbers (less than 20) using regrouping: 18–9, 17–9, 17–8, 16–9, 16–8, 16–7, . . . , 11–3, and 11–2. Out of these 36 tasks, all the six mathematics textbook companies in Japan chose the same task to start with, which is 13–9 or 12–9. This may look strange for non-Japanese mathematics educators, but for Japanese mathematics educators this is actually a result of accumulated *kyozai kenkyuu* findings that started with 13–9 or 12–9 as the best tasks to begin with. The numbers in the task are quasi-variables. The task may lead to the subtraction–addition strategy, and eventually for students to understand using the base ten system for calculation is reasonable because numbers are expressed in the base ten system.

For the speed lesson, three pair of numbers were used for the main task; those numbers were chosen specifically so that students might notice that we might be able to compare them if we made one of them the same. In other words, when teachers are designing the task focusing on numbers, they are trying to specify the general aspects of concrete numbers. The general nature needs to be connected to the educational values or the big ideas. The subtraction task may lead to the big idea—that is, that the efficiency of calculation is reflected by the number representation system, or the efficient calculation is always using how numbers are represented. The speed lesson task implies the big idea that making the same base is the principle for comparing something. To identify the educational values, the teachers need to do *kyozai kenkyuu.* Through a visible feature of the designing tasks activity, teachers could understand the *kyozai kenkyuu* process more profoundly. The designing tasks activity may lead, surely, to the strengthening of teachers' content knowledge and deepen their understanding of the nature of Japanese Lesson Study itself. Such teachers' attitudes might be a tacit factor of the Japanese model of the problem-solving lesson and Lesson Study.

Designing tasks in the context of Lesson Study: Planning goes with evaluation

In the context of Lesson Study, the task is not judged based on some abstract determination of whether it is good for teaching a certain skill or concept, but based on concrete evidence from the research lesson of how the students responded to it. The task is later evaluated during the post-lesson discussion. For the subtraction lesson, the teacher who implemented the research lesson confessed that she was very disappointed because she could not get a variety of student solutions on the board; in particular, she had hoped to see the subtraction–addition method presented (Fernandez and Yoshida 2004). These comments at the post-lesson discussion show that a task cannot be evaluated solely on its mathematical merits, but should be judged based on its actual effects on student thinking and learning. This is characteristic of task design and task evaluation in the context of Lesson Study in Japan. For the 'who is faster?' lesson, three pairs of data points were added for students to compare, but at the post-lesson discussion teachers argued over whether these additional data were useful or not. The arguments were based on how students actually responded to the task in the lesson. Designing and adapting tasks, in lesson planning, goes with this form of evaluation. This is one of the distinguishing aspects of Lesson Study.

Conclusion

Problem-solving lessons in mathematics are not special event lessons, but daily activities in Japanese schools. Therefore, the design of one task for one lesson and how to use it is the focus of the mathematics teaching, and that is also teachers' daily regular PD activity. In other words, task design is not in the hands of specialist designers, but rather it is the central focus for teachers' planning and development (Watson and Ohtani 2015). Considering tasks, in the teaching mathematics through problem-solving lesson, situated in Lesson Study, we can see some crafting features of task design by teachers.

A common misconception about problem-solving lessons is that the solving the task is the main point. Such misconception leads to a focus on goals such as 'students can do X' or 'students understand X'. But a 'Japanese problem-solving lesson' is not just about finding the solution to a problem. It is well and good that students can do X, but X should contain some value, and the value should be appreciated by students. The values are not only mathematical values but also more general values for human beings. Therefore, a Japanese problem-solving lesson is beyond simply a sense-making activity. We teach mathematics *through* problem solving, because we want our students to become independent problem solvers and, more importantly, to become independent thinkers.

The author hopes that by making these aspects of designing tasks visible, this chapter will contribute to helping educators outside Japan appreciate the full richness of teaching mathematics through problem-solving lessons situated in Lesson Study, and better understand how it can improve teaching and learning.

Note 1: This paper is based on my presentation at SEMT, Charles University, Prague, the Czech Republic, 15 August 2015.

Note 2: IMPULS (International Math-Teacher Professionalization Using Lesson Study) is the active international organization of Lesson Study in mathematics, established in 2011 at Tokyo Gakugei University, Tokyo, initially funded by the Ministry of Education, Culture, Sports, Science and Technology of Japan, as a six-year project: see httw.impuls-tgu.org/libnoterp://wwary/measurement/index.html.

The mission of IMPULS is two-fold. First, IMPULS and its network of laboratory professionals learn about authentic Japanese Lesson Study, and thereby prepare them to create Lesson Study systems in their own countries for long-term, independent, educational improvement in mathematics teaching. Second, IMPULS conducts research projects examining the mechanics of Japanese Lesson Study in order to maximize its impact on schools in Japan.

Note 3: Also, I use the term *kikan-jyunshi* instead of *kikan-shido* intentionally. With *kikan-shido*, there are (Japanese) teachers who interfere with children's problem-solving activities, talk to them, and provide extra help. The term *shido* means 'instruct'; therefore, they think they must teach something during *kikan-shido*. To avoid this misinterpretation, I use the term *kikan-jyunshi* (*jyunshi* means to inspect/patrol), although it is a classical term in Japanese education.

Acknowledgements

This work was supported by JSPS KAKENHI Grant Number JP12345678.

References

Corey, D., Williams, S., Monroe, E., and Wagner, M. (2020) Teachers' knowledge of student mathematical thinking in written instructional products. *Journal of Mathematics Teacher Education*, 24(3): 613–639.

Doig, B., Groves, S., and Macháčková, J. (2009). Lesson Study: Could it work for you? In J. Novotná and H. Moraová (Eds) SEMT 269–270.

Doig, B., and Groves, S. (2011). Japanese Lesson Study: Teacher professional development through communities of inquiry. *Mathematics Teacher Education and Development*, 13(1): 77–93.

Doig, B., Groves, S., and Fujii, T. (2011). The critical role of task development in Lesson Study. In L. C. Hart, A. S. Alston, and A. Murata (Eds) *Lesson Study Research and Practice in Mathematics Education*, 181–199. Dordrecht, The Netherlands: Springer.

Doig, B., and Groves, S. (2015). Reaching for the heavenly bead of teaching through structured problem solving. In J. Novotná and H. Moraová (Eds) SEMT 117–127.

Fernandez, C., and Yoshida, M. (2004). *Lesson Study: A Case of a Japanese Approach to Improving Instruction through School-Based Teacher Development*. Mahwah, NJ: Lawrence Erlbaum.

Fujii, T., Kumagai, K., Shimizu, S., and Sugiyama, S. (1998). A cross-cultural study of classroom practices based on a common topic. *Tsukuba Journal of Educational Study in Mathematics*, 17: 185–194.

Fujii, T., and Stephens, M. (2001). Fostering an understanding of algebraic generalisation through numerical expressions: The role of quasi-variables. In H. Chick, K. Stacey, J. Vincent, and J. Vincent (Eds) Proceedings of the 12th ICMI Study Conference. The Future of the Teaching and Learning of Algebra, 258–264. Melbourne: University of Melbourne.

Fujii, T., and Stephens, M. (2008). Using number sentences to introduce the idea of variable. *Algebra and Algebraic Thinking in School Mathematics. National Council of Teachers of Mathematics*, Seventeenth Yearbook, 127–140.

Fujii, T. (2008). Knowledge for teaching mathematics. Plenary Talk at the 11th International Congress on Mathematical Education, Monterrey, Mexico, 6–13 July.

Fujii, T. (2010). Designing tasks in the Japanese Lesson Study: Focusing on the role of the quasi-variable. Proceedings of the 5th East Asia Regional Conference on Mathematics Education, 86–93. Tokyo, Japan.

Fujii, T. (2014). Theorizing Lesson Study in mathematics education as an emerging research area (2): Identifying components and its structure of lesson study (in Japanese). Proceedings of Second Annual Spring conference of Japan Society of Mathematical Education, (pp. 111–118).

Lewis, C., and Tsuchida, I. (1998). A lesson is like a swiftly flowing river. *American Educator*, 22(4): 12–17; 50–52.

Lewis, C. (2002). *Lesson Study: A Handbook of Teacher-Led Instructional Change*. Philadelphia, PA: Research for Better Schools, Inc.

Lewis, C., and Hurd, J. (2011). *Lesson Study Step by Step: How Teacher Learning Communities Improve Instruction*. Portsmouth, NH: Heinemann.

Makinae, N. (2010). The origin of Lesson Study in Japan. In Y. Shimizu, Y. Sekiguchi, and K. Hino (Eds) The Proceedings of the 5th East Asia Regional Conference on Mathematics Education: In Search of Excellence of Mathematics Education, vol. 2, 140–147. Tokyo: Japan Society of Mathematics Education (JSME).

Ohtani, M.(2014). Construction zone for the understanding of simultaneous equations: An analysis of one Japanese teacher's strategy of reflection on a task in a lesson sequence. In Frederick K.S. Leung, K. Park, D. Holton, and D. Clarke (Eds) *Algebra Teaching around the World*, 113–128. Dordrecht, The Netherlands: Sense Publishers.

Stigler, J., and Hiebert, J. (1999). *The Teaching Gap: Best Ideas from the World's Teachers for Improving Education in the Classroom*. New York: The Free Press.

Takahashi, A. (2008, July). Beyond show and tell: Neriage for teaching through problem-solving–Ideas from Japanese problem-solving approaches for teaching mathematics. In *11th international congress on mathematics education in Mexico (Section TSG 19: Research and development in problem solving in mathematics education)*, Monterrey, Mexico.

Takahashi, A. (2013). Investigation of the mechanism of Lesson Study as the core of the mathematics teacher professional development (in Japanese). Proceedings of First Annual Spring Conference of Japan Society of Mathematical Education, 83–87.

Takahashi, A., and McDougal, T. (2016). Collaborative lesson research: maximizing the impact of lesson study. *ZDM—Mathematics Education*, 48: 513–526.

Watanabe, T., Takahashi, A., and Yoshida, M. (2006). Kyozai Kenkyuuu: A critical step for conducting effective Lesson Study and beyond. In F. Arbaugh and P. Taylor (Eds) *Inquiry into Mathematics Teacher Education*. Association of Mathematics Teacher Educators (AMTE) Monograph Series, vol. 5, 131–142.

Watson, A., and Ohtani, M. (2015). *Task Design in Mathematics Education*. Cham, Switzerland: Springer. doi.org/10.1007/978-3-319-09629-2

4.3
Japanese Lesson Study: Insights from SEMT

Brian Doig
Deakin University, Australia
Susie Groves
Deakin University, Australia

Introduction

Stevenson and Stigler's (1992) book, The *learning gap: Why our schools are failing and what can we learn from Japanese and Chinese education*, caused a sensation because it revealed the large differences between the teaching of mathematics in the United States of America, and that in China and Japan. This was followed by Stigler and Hiebert's (1999) book, *The teaching gap: what educators can learn from the world's best teachers*, based on results from the Third International Mathematics and Science Study (TIMSS) video study. In this study the intention was to videotape 100 lessons at 8th grade, in the United States of America, Germany, and Japan, and analyse differences in pedagogical approaches. The researchers, and Stigler and Hiebert, found that many procedures were vastly different in the three countries. Also, it became evident, that in Japan teachers, everywhere applied the same pedagogical practices. In the end only 80 German classes and 50 Japanese classes were recorded.

However, at this time, Makoto Yoshida's doctoral dissertation (Yoshida, 1999; Fernandez & Yoshida, 2004) became available. Thus, Stigler and Hiebert's (1999) accounts of Lesson Study were timely and effective in raising mathematics educators' interest in this 'new' pedagogical practice. Other publications revealing Japanese pedagogy became available – see, for example, Lewis and Tsuchida (1998) and especially Lewis (2002) who had been an exchange student in Japan herself and wrote from an 'insider's perspective'.

From these small beginnings, more and more papers and books have become available to an international audience of mathematics educators, and at the 1998 Stellenbosch meeting of the International Group for the Psychology of Mathematics Education (PME) the authors of this chapter met Professor Toshiakira Fujii, who invited us to visit him in Tokyo to see for ourselves what went on in Japanese primary schools.

Meeting Professor Fujii, going to Japan, and observing lessons, encouraged us to pursue learning about, and using, Japanese Lesson Study on our return to Australia.

Brian Doig and Susie Groves, *Japanese Lesson Study: Insights from SEMT*. In: *Elementary Mathematics Teaching*. Edited by: Brian Doig, Jarmila Novotná, Berinderjeet Kaur and David K. Pugalee, Oxford University Press. © Oxford University Press (2024). DOI: 10.1093/oso/9780192869647.003.0015

What is Japanese Lesson Study?

Essentially Japanese Lesson Study is a form of continuing, rigorous, teacher professional development, practised since the end of World War two, and universal in Japan. Darling-Hammond, Wei, Andree, Richardson, and Orphanos (2009) suggested that, 'relatively few U.S. teachers engage in intensive professional collaboration around curriculum planning' (p. 22) This situation was not restricted to USA teachers, and thus it is not surprising, that international attention has turned to the less familiar, but apparently more successful, professional development practices, such as Japanese Lesson Study (JLS). For example, Bussi and Ramploud (this volume) give some details of their experience with lesson study in Italy.

Unlike many Western professional development initiatives, richly funded and sometimes mandated, Lesson Study in Japan is neither funded nor mandatory. Essentially school-based, and organised by teachers themselves, it pervades primary school education, and to a lesser extent lower secondary school education, across the country, with teachers researching their own practice in school-based communities of inquiry (see, for example, Groves, Doig, & Splitter, 2000; Doig & Groves, 2011).

Lewis (2002) contrasted current teacher professional development in Western nations and the Japanese Lesson Study approach. Table 4.3.1 shows the comparison that highlights the differences quite starkly. The first difference, which we in the West rarely consider, is the notion that teacher research might be a sensible approach to teacher professional development. The idea of starting with a question along the lines, for example, of 'How can we better introduce students to a two-digit subtraction idea (algorithm)?' may never arise because our curricula have set standard approaches, of the 'Show an example, and now practice' variety, which is quite ironic, given that we, in the West, attribute rote learning to Asian nations, but never to ourselves.

The second difference is the use, in the West, of outside experts, as opposed to teachers within the school, working together. Of course, some of the teachers need to be more experienced than others to avoid the trap of the 'blind leading the blind'. A middle course would be to have a more experienced teacher from another school to help. The 'traditional' approach, judging by Lewis' table, suggests that this model is less than satisfactory, whereas the Lesson Study model, being teacher driven, addresses teacher needs and develops the teachers involved.

Table 4.3.1 Contrasting Views of Professional Development by Liptak, cited from Lewis (2002)

Traditional Professional Development	Lesson Study
Begins with an answer	Begins with a question
Driven by outside 'expert'	Driven by participants
Communication flow: trainer to teachers	Communication flow: among teachers
Hierarchical relations between trainer and teachers	Reciprocal relations among learners
Research informs practice	Practice is research

The inclusion of direction of information flow, the third table entry, once stated, is obvious, but it is not usually considered as an aspect of teacher professional development that might not be optimal, or one that should be changed.

The difference between being taught, even by an expert, and learning from one's own practice, would appear to be a fundamental difference between Western and Asian teacher professional development practices. Perhaps our approach is an adult version of teaching to children, using the well-known 'sage on the stage' version of teaching.

The last entry in the table clearly reinforces the view of Fujii, that Japanese teachers research their practice, and thus are able, as a group, to improve teacher performance and thus student achievement.

Fujii used this table very effectively in his SEMT presentation, to encourage the SEMT audience to question their notions of 'what is good practice' in their own situations.

Lewis (2002) described the Japanese Lesson Study Cycle as having four phases:

(1) goal setting and planning, including the development of the Lesson Plan;
(2) teaching the 'research lesson';
(3) enabling the lesson observation; and
(4) the post-lesson discussion.

On the other hand, Fujii, in his presentation offered a slightly different structure, as shown in Figure 4.3.1.

The fourth point has many far-reaching consequences (see, for example, Lewis & Tsuchida, 1998). It can be said that research lessons make participants and observers think, quite profoundly, about specific and general aspects of teaching. In Fujii's phrase, it means that the lesson is observed by a '100 eyes' (Fujii, personal communication, 2005) rather than simply a small number of observers, as is the case in some adaptations.

In the Fujii version there is a review (the fifth point) by the Lesson Study planning group, who write a review of their experience and findings, and may include points

Fig. 4.3.1 The process of Lesson Study (Fujii, 2015, p. 11)

addressed by the 'knowledgeable other' who spoke at the post lesson discussion. The review may be for school use, or published and shared more widely with other teachers and schools. This, of course, provides a rich resource for teachers all around the country, as well as overseas.

At the Japanese School of Melbourne, Japanese children are offered a regular Japanese education, so that when they return to Japan they can fit back into school with ease. There the authors watched a lesson that introduced the concept of the circle to Year 3 children and later came across the identical lesson in Japan!

In Japan, Lesson Study occurs across many curriculum areas, mainly at the primary school level, and to a lesser extent junior secondary. Lesson Study can be a single school based, district wide, or even at a national level. Whichever it is, a planning team from within the school, considers the aim of teaching, say, mathematics, and plans a lesson that will facilitate the learning or understanding of the school aim. The goal may be general, as in 'improve student thinking', or very specific as in 'to 'understand the (mathematical) meaning of a circle' (as a locus).

Lesson Study in mathematics

In mathematics, the research lesson usually follows a lesson pattern called a 'structured problem-solving lesson'. According to Shimizu (1999), major characteristics of such lessons include:

> *hatsumon* – a single thought-provoking question or problem, that students engage with, and that is the key to students' mathematical development and making of mathematical connections;
>
> *kikan-shido* – sometimes referred to as the 'purposeful scanning' that takes place while students are working individually, or in groups, which allows teachers not only to monitor students' strategies, but also to orchestrate the summary of students' solutions in the next phase of the lesson;
>
> *neriage* – the 'kneading' stage of a lesson, that allows students to compare, polish, and refine solutions through the teacher's orchestration of discussion and probing of student solutions; and
>
> *matome* — the summing up and careful review of the students' discussion in order to guide them to higher levels of mathematical sophistication.

While this may seem a trite description of the lesson model for mathematics, the lesson aim, the selected, single, problem posed for solution, and the four phases combine to make for a powerful pedagogical experiment, created by teachers, taught and observed by teachers, and evaluated by teachers.

Furthermore, although Japanese Lesson Study has been widely emulated internationally, most of those who have adapted the practice, have failed to take account of the critical role played by the single mathematical task set by the teacher for the students to solve. This is discussed further below.

Japanese Lesson study at SEMT 2009

The opportunity to offer a workshop at SEMT in 2009 convinced the authors to provide participants with an experience of the Japanese Lesson Study cycle, albeit in a severely truncated form, and to reflect on the process and look at the possibilities it might provide for professional development in their own countries. The workshop participants were to be split into groups of about eight people, so that there would be more than one possible lesson plan developed, and that some discussion of variations would be enabled.

The intended structure for the three workshop sessions was as follows:

> Session 1: Introduction to Lesson Study: its different formats; the research lesson, the post-lesson discussion, and view a Japanese video example of a structured problem-solving lesson. Further, discussion was to help consider Japanese Lesson Study as professional development, and its merits.
>
> Session 2: Participants work in groups to develop a lesson plan for a 45-minute structured problem-solving research lesson. This involves identifying goals for the lesson, establishing the *hatsumon*, considering anticipated student responses, and planning for the *neriage* phase of the lesson. Discussion of lesson plans developed by different groups. Synthesis of the lesson plans into a single lesson plan to be taught by a Czech colleague, Jana Macháčková, in Session 3.
>
> Session 3 Jana Macháčková teaches the lesson (45 minutes), which was to be video recorded if possible. Post-lesson discussion and workshop debriefing (30 minutes) also video recorded if possible. Reflection: Where to from here? Could it work for you? (15 minutes).

There was an organizational difficulty encountered because SEMT was held during the August School holidays, so that getting a full class of, say 20 students, was not possible. In the event, Jana Macháčková, a local primary school teacher, was able to persuade five of her class to give up an afternoon of their holidays to attend a mathematics class! Another issue that was not anticipated, nor resolved, was that the video-recordings were produced with software that was borrowed, and became unavailable to the authors and participants after the workshop.

Session one was conducted with the content as described above: that is, an introduction to the definition of Japanese Lesson Study and the structured problem-solving lesson in mathematics. Participants in the workshop viewed a video-recorded lesson from Japan, which generated much discussion and formed a powerful basis for the following sessions. The authors explained to the participants the meaning of the various Japanese words, in particular the *hatsumon*, or task, as this was to form a large part of Session two.

Session two started with participants forming groups, with members from many different backgrounds and nationalities. As the students were around ten years of age, this meant examining upper primary school curricula for a suitable focus for the planned lesson. That is, their task was to explore mathematics primary curricula across as many countries as could be accessed *via* the Internet, looking for suitable content for the lesson. This examination of curricula, and accompanying pedagogy, and materials, was expected to make participants aware of, and practise, *kyozaikenkyu*, the Japanese term for a deep exploration of the mathematical content of the lesson, in terms of placement within the curriculum, and pedagogical aspects of the content, as well as possible relevant teaching materials. Japanese educators place a strong emphasis on task selection, "[but] this effort is largely ignored by non-Japanese adapters of Lesson Study, possibly because the effort involved may be almost invisible, in the way that 90% of an iceberg is invisible, with all of our attention going to its visible tip" (Doig, Groves, & Fujii, 2011, p. 182).

This deep exploration would, in theory, lead to the selection, or creation, of a *hatsumon*: that is, the single task for the lesson that would address the requirements set out above. The focus for the proposed lesson was to be on division, as this fitted the year level of Jana's students and was appropriate for the Czech curriculum for these students.

An interesting aspect, of the mathematics curricula internationally, discovered by the participants, was that curricula spent time on simple division with whole number quotients, but over-looked division with fractional quotients. These fractional quotients were at secondary school level, and in a sense, arrived unannounced: they simply appeared without comment. An example of this is the division of 3 by 4, giving a quotient of three-quarters. This apparently had escaped the notice of curriculum writers as well as mathematics educators. Thus, a lesson introducing this possibility was suggested by Jana, who, as the teacher of the lesson, had the final say on the lesson topic. In Japan, the group of teachers planning the research lesson would make this decision. The participants worked in their groups, proposing ways to introduce the task, anticipating students' solutions and devising their group's lesson plans. Many ideas and suggestions were made and discussed during this session, but time, and the impending third session, meant that a decision had to be made, so Jana was left with the task of producing the final lesson plan.

In the Third session of the workshop, Jana taught the lesson, in Czech for the benefit of the students, and translated it, on the run, into English for the benefit of the workshop participants, which was a very demanding experience.

The *hatsumon*, written on the blackboard, as in Japan, was the following:

> Three pizzas are to be shared between four people.
> How much pizza does each person receive?

The students were given, A3 sheets of paper with three large circles drawn on them, representing the three pizzas, and black marker pens with which to write or draw their thinking and solution. However, the students were also asked to see how many different ways they could find, or express, their solutions to the question. The educators had, as in a true Japanese Lesson Study fashion, considered how many different ways students might find a solution, but came up with only three possible solutions, and some participants expressed the opinion that the students would not find more than one.

However, as it turned out, the five students came up with some 25 different approaches for a solution to the *hatsumon*, although not all of these were correct.

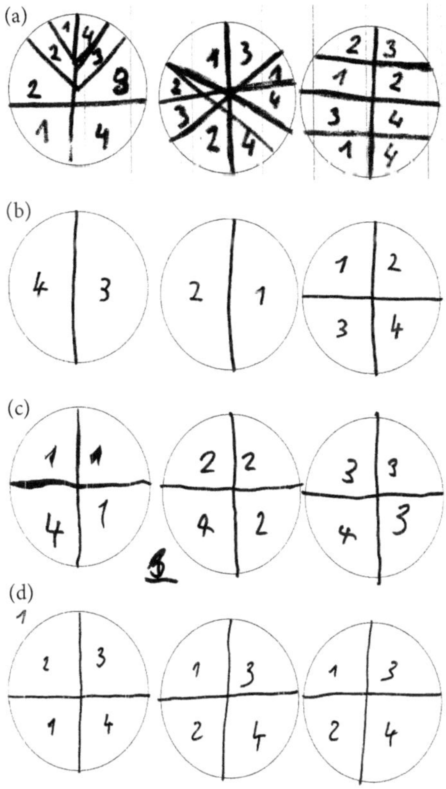

Fig. 4.3.2 Student responses to sharing the pizzas

Some of these are shown in Figure 4.3.2. In the first response (a) the student has been clearly distracted by the word 'pizza' and has cut the pizzas as you might with a knife, and forgotten the implied 'equal shares' involved with division.

For the next response, (b), the student has divided two pizzas into halves, thus giving a half to each of the four people, and then divided the remaining pizza into quarters. This gives a total of three-quarters to each person.

The next response, (c), has all three pizzas divided into quarters, and these quarters numbered to show the allocation to people. The numbering suggests that the student knew the result, as the pizza parts are numbered as three-quarters to person 1, then person 2, and then person three, leaving exactly three-quarters for person 4.

However, response (d) shows that the student has shared the pizzas, one by one, into quarters, and then allocated them by numbering the quarters. The hint, of course, was that there were four people to share the pizzas, and thus a division of the pizzas into quarters was a sensible approach.

In preparation for the lesson discussion (*neriage*) phase, of structured problem-solving lessons, it is usual for teachers to ask selected students to display their solutions on the board and explain their reasoning. Example responses are often selected to start with the least sophisticated on the far left, and end with an example of the most sophisticated response on the right. This phase of a structured problem-solving lesson aims to provide students with the opportunity to view a range of strategies or solutions to the *hatsumon*, and to clarify their own ideas by explaining to the other students and allow students to learn from one another. In this lesson, all 25 solutions were displayed, but the discussion focussed on a selection made by the teacher.

In the final phase, the *matome*, the teacher, in this case Jana, reviews the solutions, emphasising relationships and differences, using this to summarise the lesson and draw out the mathematical ideas underpinning the lesson.

After this, Jana asked the students what they thought about the lesson. Even though they had given up part of their Summer holidays to attend the lesson, most agreed that it was fun, and interesting, and different to their normal mathematics lessons.

The lesson now finished, and the students left the room, and the SEMT participants were left to have a discussion about the experience. A common comment was on the amazing performance from Jana, teaching in one language (Czech), and translating and commenting on the procedure as she went, in a second language (English).

Other comments from participants focused on the fact how a single task, the *hatsumon*, could generate so many solution strategies, and provide the stimulus for a long, student driven, discussion. This mainly reflected the participants' experience in their own countries, where a great deal of 'busy work' was needed to occupy the students for the length of the mathematics period. It was noted that although Japanese Lesson Study has been widely emulated internationally, most of those who had adapted the practice, had failed to take account of the critical role played by this single mathematical task, the *hatsumon*, which is set for the students to solve (see, for example, Doig *et al.*, 2011). However, Barkai and Levenson (2015) correctly point out that "as students explain their thinking, they practice reasoning mathematically and deepen their understanding of mathematical concepts" (p. 54), which is a direct consequence of the quality of the *hatsumon*, and its intended function within the lesson. Further, they claim that "in addition to offering their own explanations, students [should be] encouraged to listen and evaluate others' explanations" (p. 54) as stated in the Israeli, and many other, curriculum materials.

It was agreed that the lesson was a success mathematically, and that the structured problem-solving format was both interesting and effective. Participants were still surprised by what they had discovered about various curricula that they had examined as part of their *kyozaikenkyu* experience, and that this had given them pause to consider their own mathematics curriculum in a new light.

Japanese Lesson Study at SEMT 2015

There was more than one presentation at SEMT 15 addressing aspects of Japanese Lesson Study. These included a presentation by Professor Fuji, of Tokyo Gakugei University, and a presentation by the authors.

Presentation by Professor Fujii

Despite the title of his address (*Designing and adapting tasks in the Japanese Lesson Study: Focusing on the role of the quasi-variable*) he highlighted three aspects of Japanese Lesson Study that were new to many in the SEMT audience. In introducing his topic, he outlined the lesson study cycle of teachers' planning and the development of a mathematics structured problem-solving lesson. The address had three major foci, which are described below.

In the first section of his presentation Fujii focused, in part, on *koyzaikenkyu*, and gave examples of the effort to which Japanese teachers went in their investigation to develop the *hatsumon*. This effort had surprised the teachers in our earlier 2009 workshop.

Fujii cited, from a book chapter of which he was a co-author, which held that:

> Although kyozaikenkyu is recognized as a critical part of Lesson Study by Japanese educators, teachers outside Japan often neglect it. Japanese educators place a strong emphasis on task selection, [but] this effort is largely ignored by non-Japanese adapters of Lesson Study, possibly because the effort involved may be almost invisible, in the way that 90% of an iceberg is invisible, with all of our attention going to its visible tip (Doig *et al.*, 2011, p. 182).

Further, Fujii gave a detailed example of the depth to which *kyozaikenkyu* was undertaken by Japanese teachers. The example concerned the choice of the very first two-digit subtraction problem, involving borrowing, to set for students, where the subtrahends are selected from the numbers from 11 to 18, and the minuends are single digit numbers from 2 and above. This gives 36 possible combinations, however, the two most used questions are 13–9 and 12–9.

He explained that:

> The reason why 13 – 9 or 12 – 9 is the first task is that the minuend 9 is close to ten, and it is easier for the student to separate 13 into 10 and 3 and subtract 9 from 10 and then add the difference to 3. $13 - 9 = (10 + 3) - 9 = (10 - 9) + 3$. We call this the subtracting-adding strategy. (Fujii, 2015, p. 16)

The use of 9 clearly encourages students to consider breaking 13, or 12, into a 10 and a remainder. This leads naturally to two-digit subtrahends and a vertical subtraction algorithm. He had earlier observed, ruefully, that 'from non-Japanese eyes it seems to be a bit too detailed to discuss the exact numbers to be used' (p. 14).

His second focus, however, stemmed from this precision of discussion and thought, as it leads to considering that although 'the number is obviously a concrete one, even so it could imply a generality ... and the general aspects of a concrete number is described by the term "quasi-variable" (Fujii, 2003). The example, from Carpenter and Levi (1999), was "[the] mathematical expression such as:

$$78 - 49 + 49 = 78.$$

Students were asked whether they thought this was a true number sentence, and all but one answered that it was. Thus, this type of "number sentence, or group of number sentences, that indicate an underlying mathematical relationship which remains true with whatever the numbers used are" (p. 15) represents a concrete number acting almost as a variable, and therefore the term Fujii used was *quasi-variable*.

A third focus of Fujii's was on the selection of the single task, *hatsumon*, for a structured problem-solving lesson. According to him, this task is critical to the outcomes of the lesson and therefore:

> The task should be satisfied with [sic] at least the following conditions:
>
> - It is appropriate and mathematically valuable in terms of the aims of the lesson.
> - It is to attract the interest of the students.
> - It is the appropriate level of difficulty.
> - It has a potential of having several ways to solve the task.
> - It has a potential of applicability to other mathematical problems or real life problems. (p. 13)

While the first three conditions would be satisfied internationally, the last two conditions are not so commonly raised outside of Japan. The fourth condition is rarely mentioned, as most teachers, outside Japan, usually have a solution strategy in mind when using a particular task. With respect to the last, application to real life may be considered in most countries, the idea that the task should have potential for other mathematical learning seldom is. The examples, given above with respect to 13–9 and 12–9, are examples of this type of task. The task itself has a clear aim for the current lesson, but it also has a direct possibility for introducing a two-digit subtraction algorithm.

A similar task, known to Fujii, but not mentioned in his presentation, was presented by Doig and Groves, also at SEMT 15, in a paper entitled *Reaching for the Heavenly Bead of teaching through structured problem solving*. While this will be described later in this chapter, it suffices to say now, that the task to be presented is a good example of one that has important consequences for later mathematics.

Fujii presented the aforementioned Table 4.3.1, as part of his explanation of why Lesson Study is a different form of teacher professional development than that practised in the West. In Lesson Study, teachers ask themselves a question about what needs to be done to improve their own teaching and their students' learning, which contrasts with the international view that it is about fixing a problem, often determined by those outside the classroom.

The second point made in Table 4.3.1 underscores this 'fixing' approach in the traditional form, while in Lesson Study it is clear that the process is teacher led and maintained. A further difference, shown in row three of the table, explores the transmission of new ideas or techniques. In Lesson Study there is embedded a collective determination of what needs to be done, how it should be done, and what the outcomes are.

Finally, the attitude that Lesson Study takes is that teacher practice is a place for research into pedagogy, not that some independent research should be the only basis for teacher development. Thus, with both external research and Lesson Study, teachers have the most to gain.

Fujii's use of this table allowed SEMT 15 attendees not only to understand what Lesson Study is, and how it is different, but also to consider whether the traditional approach is a better way or not. Discussion among participants, later during SEMT, seemed to suggest that there was a place for both approaches.

In closing his presentation, Fujii made the point that, outside Japan, a common misconception about problem-solving lessons was that they sought merely to have the students find the correct answer, thus a 'focus on goals such as 'students can do X' or 'students understand X' [but a] *problem solving lesson* is not just about finding the solution to a problem' (p. 17, our italics). In other words, there needs to be a larger educational value in the lesson, such as developing problem-solving capability, in addition to the more immediate mathematical benefit.

Structured Problem Solving: A Research Report

As a follow-up to the successful Japanese Lesson Study workshop, at SEMT in 2009, a presentation was offered by the authors in 2015. However, this time the focus was particularly on the *hatsumon*, the critical task for a structured problem-solving lesson in mathematics, as this has been an issue with adaptations of Japanese Lesson Study (see, Fujii, 2015), where he stated that the *hatsumon* should address the five criteria listed in the previous section.

In the presentation, entitled *Reaching for the Heavenly bead of teaching through structured problem solving*, the analogy made was that in the Japanese *soroban* (abacus) the beads are divided into four Earth beads with a value of 1, and above a horizontal dividing bar, one Heaven bead worth five. This arrangement is identical to the Roman abacus, (Hogben, 1968). Thus, the Earth beads represent fingers, while the Heaven bead has the value of a hand.

So, what are "the Earth beads – the essential elements – of the structured problem-solving lesson?" (Doig & Groves, 2015, p. 120). The authors claimed that there were a total of five beads that make up a structured problem-solving lesson, and, like the *soroban* beads, they were closely connected to one another. The Earth beads are: the lesson planning; the task selection; blackboard display; whole-class discussion, which lead to the teacher review and *denouement*, the Heavenly Bead. Not only are the 'beads' highly related, but also like the *soroban*, they form an ecological whole, wherein each part is necessary for the 'life' of the system.

Of particular interest is the *neriage*, or whole class discussion, phase of the lesson – the 'heart' of the lesson – which begins after students struggle with the problem posed and come up with their solutions. "During the process, a teacher highlights important mathematical ideas and concepts for students to reach the goals of the lesson" (Takahashi, 2008, p. 5). Typically, this stage takes upwards of half the time of a lesson, which is very different to the usual end of class discussion conducted in most Western schools.

The whole-class discussion is created by the students, and the display and organization of student solutions on the blackboard, for the whole class to discuss, helps the teacher to orchestrate a discussion that draws the students along a narrative path to new learning: this is the phase of the lesson where learning is expected to occur, and be visible to the teacher and to other observers (see, Figure 4.3.3, below). Shimizu (1999) describes this as "the process of polishing students' ideas and of developing an integrated mathematical idea through the whole-class discussion ... The teacher's role is not to point out the best solution but to guide the discussion toward an integrated idea" (p. 110). This goal is achieved through the teacher reviewing and carefully commenting on the solutions displayed on the blackboard. These displayed solutions are used to show the students how each solution attempt addressed the solving of the task, but also how they differ, usually in terms of their mathematical sophistication.

Figure 4.3.3 below, shows a mathematics structured problem-solving lesson where aspects of the lesson plan are shown on the blackboard arrangement, with explanatory notes added.

Clearly, the details in this plan show a great deal of the integration of the parts, not only on the blackboard, but also in the comments about the processes in the lesson. Observers at the lesson often take photographs of the blackboard at the end of the lesson, whence the lesson can be 'read' quite readily.

As Doig and Groves suggested in their SEMT presentation (2015):

> It is clear that the lesson is based on a symbiotic relationship between the goals of the lesson, the selected task, the anticipated student responses, and the actual student responses. Each part of this relationship is necessary for the overall plan to succeed, and deliver effective student learning (p. 125).

That is to say, in order to achieve the Heavenly bead of an effective mathematics structured problem-solving lesson, it is important to pay attention to the details of all of the Earth beads that, together, contribute to the effective achievement of this goal. In the case of Japanese Lesson Study, not undertaking the work of understanding the complete process leads to a less than fully effective adaptation (see, for example, Felner, 2007). Thus, in the SEMT 15 paper, we tried, not only to give participants an over-arching view of Japanese Lesson Study, but also to provide enough detail that would allow interested researchers and teachers to try, for themselves, an adoption that would have high fidelity with the original model. As Isoda and Olfos (2021) point out, "the Japanese approach ... does not imply just a method of teaching like the scaffolding used to construct a building ... [but that e]very component is explained by the theories and used for designing the classroom" (p. 10).

Implementing structured problem-solving mathematics lessons through lesson study

Encouraged by the positive reception of the 2009 Japanese Lesson Study SEMT workshop, the authors, and two of their university colleagues, implemented a longer professional development progamme for primary teachers in Melbourne, Victoria.

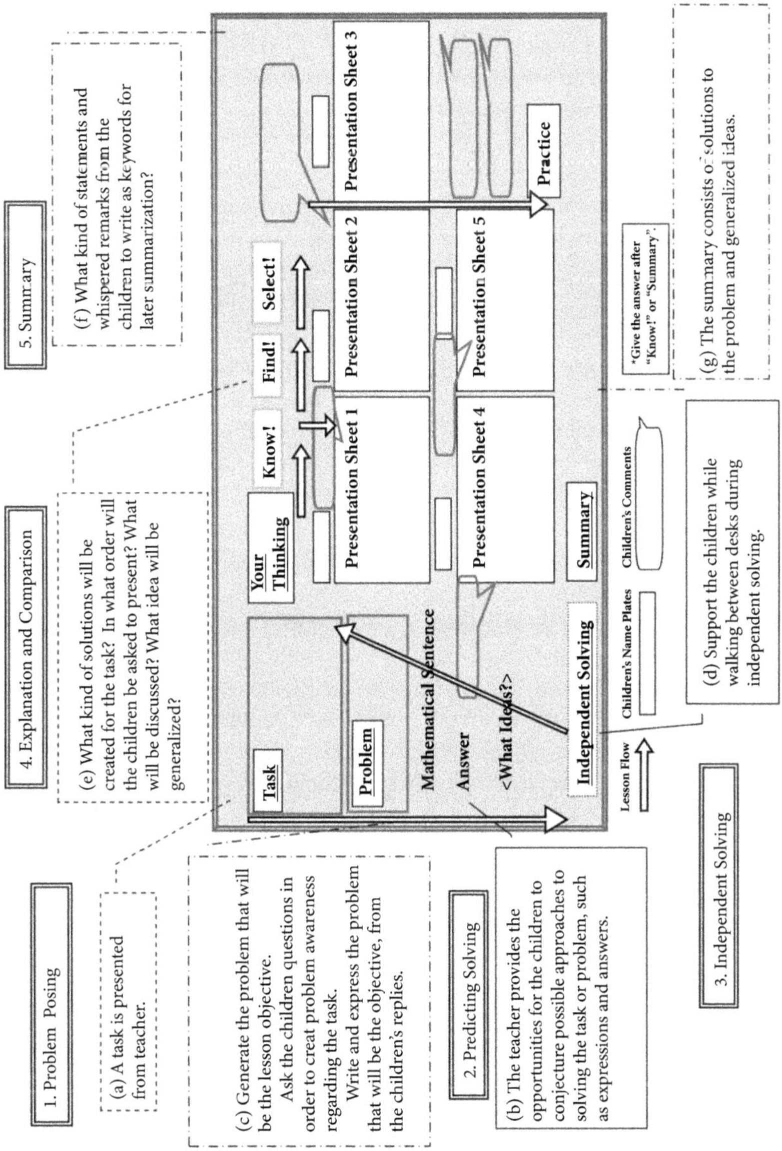

Fig. 4.3.3 A lesson plan formatted by using the blackboard in the problem-solving approach (Isoda & Olfos, 2021, p.11)

The aim of the project was to work with a group of teachers, in a school network, and provide sufficient professional development and support to implement structured problem-solving mathematics lessons, with the longer-term aim of their continuing beyond the life of the project. This research project, *Implementing structured problem-solving mathematics lessons through lesson study*, was conducted in 2012, and was based around three schools from a Melbourne school network. A cross-school model was chosen to enable a group of teachers to explore ways in which key elements of Japanese Lesson Study could be embedded into mathematics teaching and professional learning at more than one school.

In all, ten participants – teachers and numeracy coaches – were involved in the project. Numeracy coaches are teachers, who have been nominated by their school principal, to undertake continuing professional learning for other school staff through mathematics curriculum leadership and coaching. This professional learning model involved professional interactions with other coaches in the network of schools.

All participants in the project took part in an initial professional development day, which was an orientation to Japanese Lesson Study, after which they were divided into two cross-school groups: these were self-named the *Bobbies* and the *Matomes*. Each group consisted of three Year 3 or Year 4 teachers – one from each school – and two numeracy coaches. Over the life of the project, each group completed two lesson study cycles, one in each of Terms 3 and 4 (the last half of the school year, July to December), planning their own research lesson, based on the problems that were provided by the research team (see Notes below). The two groups planned each of their research lessons during four two-hour planning sessions.

One member of each group was selected to teach the research lesson on behalf of the planning group. All members of both planning teams, key staff at each school, together with all interested teachers who could be released from their classes, observed the teaching of the planned lessons and took part in the post-lesson discussions. In all between 20 and 30 people, including members of the leadership teams from other schools, observed each research lesson and took part in the post-lesson discussions. Staff from the regional office, mathematics educators, and an outside expert, the 'knowledgeable other' were also present. The knowledgeable other presented a final commentary on the lesson at the end of the post-lesson discussion, in order to give an over-arching expert view of the planning and execution of the lesson, and pointers to future possibilities (see, Acknowledgements, below).

The participants thought, initially, that nearly eight hours to plan a lesson was too long, but because of the depth of discussions around the *hatsumon*, the *kyozaikenkyu*, many were amazed at what they learned from other teachers and the numeracy coaches. There was a lot of time spent searching relevant curriculum documents and textbooks, but ultimately, the groups decided to use *hatsumons* from Japanese, or other sources, provided by the researchers.

A further discussion point, showing the difference between the Victorian and Japanese teachers, was the time devoted to the whole-class discussion. Many teachers did not think that the students would be able to discuss for such a lengthy time (about half the lesson duration). One sceptical teacher, whose class was not to be used for the research lesson, tried her students with the *hatsumon*, in essence, running a structured problem-solving lesson, and reported to her group that she had been completely wrong: her students had found many different solution strategies and indeed were able to discuss their work at length. She became one of our strongest supporters!

The outcomes from the project were very encouraging in terms of teacher attitude to an alternative pedagogy, and their respective school principals offered to continue the process for other staff members at the end of the project. For those teachers who taught the structured problem-solving lessons on behalf of their group, all were subsequently promoted.

Eliciting student explanations in interactive whole class discussions

At SEMT 17 the authors conducted a workshop that focused squarely on the 'whole-class discussion' phase of a structured problem-solving lesson. The aim was to offer an extended professional development experience focused on *neriage* and use these experiences to "explore anticipated teacher-student interactions and ways to elicit rich student explanations" (Doig, & Groves, 2017, p. 479). It is in this phase that students compare, and refine the displayed and discussed solutions, which leads to the *matome* (or climax) of the lesson. It is then that the mathematical goal of the lesson becomes clear. The whole-class discussion is an area where adopters of Lesson Study outside Japan have often experienced difficulties in moving beyond 'show and tell' (Takahashi, 2008).

The first session, focused on the key features of a structured problem-solving mathematics lesson, included participants taking part in a 'mock' lesson, followed by an analysis of the discussion phase of the lesson, in order to see how one can avoid the 'show and tell' typical of many classroom end of lesson discussions, and instead promote "a community of inquiry, where learners drive their own learning" (Doig & Groves, 2017, p.480).

The second session turned to examine the types of interaction that promote mathematical reasoning and the development of mathematical ideas (see, for example, Groves & Fujii, 2008; Okazaki, Kimura and Watanabe, 2016). Participants viewed short video extracts from Australian primary school classrooms, and analysed the ways in which the teachers elicited students' explanations during *neriage*. In the final session, a group of participants acted as 'teachers' and planned, and conducted, a structured problem-solving lesson. Other participants played the role of the students for this lesson.

Our aim was to raise the issues, and look at solutions, with *neriage* that had been revealed in our earlier project: that is, teachers did not believe that their students could provide sufficient content for a sustained discussion, nor that the students could cope with such an extended discussion between themselves and the teacher.

Conclusion

Despite many obstacles, the initial 2009 SEMT workshop had a degree of success. This was partly due to the professional ability of the Czech teacher, as she demonstrated that, what many participants had thought to be impossible, was in fact achievable.

An aspect that caused much discussion was the lack of curriculum support for the introduction of rational quotients in division across many countries. However, the Czech students in the workshop class did not seem to be fazed by the 'new' result of division but took it on board with ease. This may well have been because of the graphic presentation of the *hatsumon*, as the student solutions, while showing a range of ideas, were mostly represented graphically rather than numerically. This is typical of many other Japanese *hatsumon* too.

One aspect of the 2009 SEMT workshop, and the Victorian project, was that the distinction between Japanese Lesson Study and structured problem-solving lessons appeared to be unclear to many participants. This led to the 2015 SEMT presentation that aimed to clarify the distinction. While Japanese Lesson Study occurs in most subject areas, the structured problem-solving lesson is found only in mathematics. This is to say, that one can have Japanese Lesson Study without a structured problem-solving lesson, and one can teach mathematics using a structured problem-solving lesson outside of Lesson Study. Thus, there is a distinction between teacher professional development and a particular, mathematical, pedagogy.

Overall, our sense is that Japanese Lesson Study, with its focus on teacher-led pedagogical development, is well received by teachers and researchers from disparate backgrounds.

The 2017 SEMT workshop appeared to convince some, but not all participants, of their students' capacity to participate in long and educationally fruitful end of lesson discussions. However, our experience with Victorian teachers has shown that trying *neriage*, within the structured problem-solving lesson approach, does convince teachers that their students have this capacity. Using a 'mock lesson' may not be as powerful as experimenting with your own students, but at the very least it provides an existence proof experience. We are hopeful that there will be further opportunities to offer presentations of this nature at future symposia.

Notes

The *hatsumon* for the lessons were printed in the upper left-hand corner of an A3 sheet of paper, orientated in landscape mode. Each student had their own copy to work on, and all ideas and thinking were to be recorded on the sheet.

The Bobbies' *hatsumon* for the first cycle structured problem-solving lesson was:

Here is a picture made up of matchsticks.
How many matchsticks are there altogether?
How can you show your thinking using the diagram?

The Bobbies' *hatsumon* for the second cycle structured problem-solving lesson was:

Here is a diagram of some dots.

Can you work out how many dots are in the diagram without counting them one-by-one?

Please make sure that you show your thinking in the space provided.

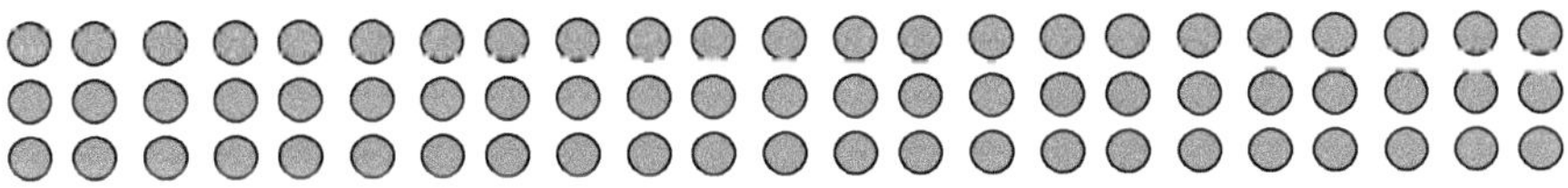

Acknowledgement

The *Implementing structured problem-solving mathematics lessons through lesson study* project was funded by the Centre for Research in Educational Futures and Innovation, Deakin University. We gratefully acknowledge the contributions of participating teachers, numeracy coaches, students, schools and the '*knowledgeable others*', Dr Max Stephens, University of Melbourne, and Professor Toshiakira Fujii, Tokyo Gakugei University, Japan.

References

Barkai, R., & Levenson, E. (2015). Roles of explanations in Primary school mathematics classrooms: prospective teachers' perceptions. In J. Novotná & H. Moraová (Eds.) SEMT 53–63.

Carpenter, T., & Levi, L. (1999). Developing conceptions of algebraic reasoning in the Primary grades. Paper presented at the Annual Meeting of the American Educational Research Association, Montreal, Canada.

Darling-Hammond, L., Wei, R., Andree, A., Richardson, N., & Orphanos, S. (2009). Professional learning in the learning profession: A status report on teacher development in the United States and abroad www.nsdc.org/news/NSDCstudy2009.pdf Retrieved 12 May 2021

Doig, B., S. Groves, S., & J. Macháčková, J. (2009). Lesson study – could it work for you? In J. Novotná & H. Moraová (Eds.) SEMT 269–270.

Doig, B., & Groves, S. (2011). Japanese Lesson Study: Teacher Professional Development through Communities of Inquiry. Mathematics Teacher Education and Development, 13(1), 77–93.

Doig, B., & Groves, S. (2011). Reaching for the Heavenly Bead of teaching through structured problem solving. In J. Novotná & H. Moraová (Eds.) SEMT 117–127.

Doig, B., & Groves, S. (2017). Eliciting student explanations in interactive whole class discussions. In J. Novotná & H. Moraová (Eds.) SEMT 479–480.

Doig, B., Groves, S., & Fujii, T. (2011). The critical role of task development in Lesson Study. In L. Hart, A. Alston & A. Murata (Eds.), Lesson Study Research and Practice in Mathematics Education: Learning Together (pp. 181–199). Dordrecht: Springer.

Felner, R. D., Seitsinger, A., Brand, S., Burns, A., & Bolton, N. (2007). Creating Small Learning Communities: Lessons From the Project on High-Performing Learning Communities About "What Works" in Creating Productive, Developmentally Enhancing, Learning Contexts. Educational Psychologist, 42(4), 209–221.

Fernandez, C., & Yoshida, M. (2004). Lesson study: A case of a Japanese approach to improving instruction through school-based teacher development. Mahwah, NJ: Lawrence Erlbaum.

Fujii, T. (2003). Probing students' understanding of variables through cognitive conflict problems: Is the concept of a variable so difficult for students to understand? In Proceedings of the 2003 Joint Meeting of PME and PMENA (Vol. 1, pp. 49–65).

Fujii, T. (2005). [Personal Communication '100 eyes'].

Fujii, T. (2015). Designing and adapting tasks in the Japanese Lesson Study: Focusing on the role of the quasi-variable. In J. Novotná & H. Moraová (Eds.) SEMT 9–18.
Groves, S., Doig, B., & Splitter, L. (2000). Mathematics classrooms functioning as communities of inquiry: Possibilities and constraints for changing practice. Paper presented at the Proceedings of the Twenty-fourth Conference of the International Group for the Psychology of Mathematics Education, Hiroshima.
Groves, S., & Fujii, T. (2008). Progressive discourse, Neriage, and some underlying assumptions. Paper presented at the 41st Annual Conference of Japan Society of Mathematical Education, Tokyo.
Hogben, L. (1968). The wonderful world of mathematics. London: Macdonald and Company.
Isoda M., Olfos R. (2021) Introduction: Japanese Theories and Overview of the Chapters in This Book. In: Isoda M., Olfos R. (eds) Teaching Multiplication with Lesson Study. Springer, Cham. https://doi.org/10.1007/978-3-030-28561-6_1
Lewis, C. (2000). Lesson study: The core of Japanese professional development. Paper presented at the Annual Meeting of the American Educational Research Association, New Orleans, LA.
Lewis, C. (2002). Lesson Study: A Handbook of Teacher-Led Instructional Change. Philadelphia: Research for Better Schools (RBS).
Lewis, C., & Tsuchida, I. (1998). A Lesson is Like a Swiftly Flowing River: Research lessons and the improvement of Japanese education. American Educator, 14–17 & 50–52.
Okazaki, M., Kimura, K., & Watanabe, K. (2016). Types of interaction that promote or hinder the narrative coherence of a mathematics lesson. Paper presented at the 40th Annual Meeting of the International Group for the Psychology of Mathematics Education, Szeged, Hungary.
Shimizu, Y. (1999). Aspects of mathematics teacher education in Japan: Focusing on *Teachers' Roles*. Journal of Mathematics Teacher Education 2, 107–116.
Stevenson, H. W., & Stigler, J. (1992). The learning gap: Why our schools are failing and what can we learn from Japanese and Chinese education. New York: Summit Books.
Stigler, J. W., & Hiebert, J. (1999). The teaching gap: What educators can learn from the world's best teachers. New York: Free Press.
Widjaja, W., Vale, C., Groves, S., & Doig, B. (2017). Teachers' professional growth through engagement with lesson study. Journal of Mathematics Teacher Education, 20(4), 357–383. doi: 10.1007/s10857-015-9341-8
Yoshida, M. (1999). Lesson study: A case study of a Japanese approach to improving instruction through school-based teacher development. Doctoral dissertation, University of Chicago.

4.4
Thirty years of international progress in young children's problem solving and posing

Victor Cifarelli
University of North Carolina, Charlotte, USA
Mary Capraro
Texas A&M University and Aggie STEM, USA
Robert Capraro
Texas A&M University and Aggie STEM, USA

Introduction

In preparing to write this chapter, we reviewed all of the International Symposium on Elementary Mathematics Teaching (SEMT) proceedings, starting with the initial symposium in 1991 and concluding with the most recent in 2019. In offering our review, we wish to provide a few caveats that describe the guidelines we followed in choosing presentations upon which to comment. First, while acknowledging that all SEMT presentations have consistently valued the processes of problem solving (PS) and problem posing (PP) as playing important roles in mathematics learning, we focused on presentations that trace the role these important learning processes play in the mathematical PS of young children. So, for example, although we observed that almost all of the SEMT workshops and poster sessions espouse a prominent role for problem-based learning, we restricted our search to presentations of investigative analyses that explicitly address either PS or PP (or both). Second, we did not distinguish among the many different ways in which presenters consider and define PS and PP. For example, constructivist and embodied cognition perspectives define PS and PP in substantially different terms, which have important implications for the kinds of analyses that are conducted: constructivist researchers consider problems in terms of perturbations to existing action schemes, whereas researchers following an embodied cognition perspective view problems as situated within a learner's interaction with objective tasks. By making distinctions about the nature of the problem, these different perspectives may not agree on the important roles played by PP and PS in the learning of mathematics. Not surprisingly, these different emphases often lead to different analyses. Trying to sort out and explain these kinds of differences among the various research paradigms is well beyond the scope of this chapter.

Victor Cifarelli, Mary Capraro, and Robert Capraro, *Thirty years of international progress in young children's problem solving and posing*. In: *Elementary Mathematics Teaching*. Edited by: Brian Doig, Jarmila Novotná, Berinderjeet Kaur and David K. Pugalee, Oxford University Press.
DOI: 10.1093/oso/9780192869647.003.0016

SEMT as a leader in advocating for studies of sophisticated PS and PP processes of young children

In this section we review studies from the early years of SEMT (1991–1993) to make the point that, from its inception, SEMT has been a leader in advocating for analyses that explain sophisticated PS processes of young children. In so doing, we look to demonstrate how this emphasis has been sustained throughout the years 1995–2019 of the symposia and thus helped lay the foundation for the sophisticated analyses presented during the more recent years of the symposium.

In our review of the PS presentations at SEMT, we make some general observations of the body of work over the years 1991–2019. First, we were impressed that, from the very beginning, in 1991, SEMT presenters have been consistent in their assumptions about young children as active learners and the mathematics that young children demonstrate in their learning. Specifically, SEMT presenters have always viewed young children as active learners who demonstrate self-determination in their actions. Hence, the proceedings from SEMT have always provided substantial accounts of learning that seemed to improve with the advent of each successive symposium. Second, although we found some examples of the more traditional heuristics analyses of PS, made popular in the 1970s and 1980s, we found that most SEMT presentations of PS and PP focused on PS as involving critical thinking, thus making possible a more comprehensive analysis of PS than was presented in the aforementioned heuristics-based presentations. In the following paragraphs, we present some highlights of how SEMT presenters have advocated for studies of sophisticated PS and PP processes, starting from the first symposium in 1991 and proceeding to the most recent in 2019.

The early years of SEMT 1991–1993

In the first two SEMT symposia, in 1991 and 1993, we identified four presentations of PS that we found to be both interesting and contemporary for the times (Hošpesová 1991; Stehlí ková 1993; Szetela 1991, 1993). These presentations were early examples of our view that SEMT presenters have been strong advocates of research that recognize the PS of young children and at the same time formed a nice contrast to traditional quantitative approaches to studying PS. For example, Szetela (1991) examined how students in Grades 4 and 5 verbalized their PS and thus provided an early comprehensive analysis of PS. The analysis focused on the students' sense making while solving a set of non-traditional tasks, and thus highlighted the importance of exploration in the context of PS. This was a very sophisticated view of PS for the times. In particular, in addition to having students complete PS tasks, the students were also required to assess the hypothetical solutions of peers, a technique that we now employ routinely in PS studies (Driscoll and Bryant 1998; Brown and Clark 2006). In 1993, Szetela elaborated this framework to include a focus on the role of student engagement to motivate critical thinking, a topic of interest in mathematics education that continues to the present day.

The studies by Hošpesová (1991) and Stehlí ková (1993) provided important analyses of how a focus on the mathematical misconceptions of young children yields important information for teachers to utilize as they design instructional activities.

Although earlier studies of the misconceptions of students have been criticized as being based on deficit models of learning (Ay 2017), the studies by Hošpesová (1991) and Stehlíková (1993) utilized examples of student errors as a starting point for designing a framework within which teachers could make positive interventions to help students develop more productive conceptions of their PS situations. These early studies emphasized the importance of students building strong understandings of the problems they face.

SEMT 1995–2019

New directions in PS were ushered in at SEMT 1995. In considering the remaining years of SEMT, we identified the SEMT 1995 symposium as particularly significant in that it served as an important theoretical foundation of mathematics learning for the symposia that followed. Specifically, the SEMT 1995 symposium included an impressive set of nine PS presentations. We view these studies as prominent because they were conducted at a time when there appeared to be a consensus forming in mathematics education that we had learnt all that we needed to know about PS (Schoenfeld 1992), a view that coincided with Lester's (1994) commentary, in the *Journal of Research in Mathematics Education*'s Anniversary Issue, which argued that research in PS had become less of an area of interest among mathematics education researchers. In light of these assessments about the lessening importance of PS research, the SEMT 1995 presentations offered a strong rebuttal to this view, and thus represented a renaissance for PS research, reminding us that there was still much to be learnt about PS. These presentations thus provided good cause to rethink Lester's view that interest in research on PS was waning.

Among the highlights from the SEMT 1995 were presentations that advocated for more comprehensive analyses of PS, which included both analyses of microscopic processes and macroscopic analyses of how a student's emerging ideas guide their actions throughout PS (Lemut and Dettori 1995; Novotná 1995); presentations that advocated for the use of challenging problem tasks, to shed light on how problem solvers formulate, develop, and test hypotheses, while engaged in PS (Hošpesová 1995; Semadeni 1995); and studies that focused on PS as the evolution of informal action to resolve problem situations (Irwin 1995; Kaslová 1995). Not surprisingly, these fresh perspectives on PS inquiry were not acknowledged as areas of need in the Lester or Schoenfeld commentaries. Perhaps the birth of the modern perspectives of PS occurred in 1995 at SEMT in the, then, new Czech Republic.

The advent of these perspectives was accompanied by more nuanced analytic methods and strategies. We want to mention the implications of a number of these studies as early influences on the more sophisticated analyses we have seen in recent years. For example, in calling for comprehensive analyses of PS, Novotná (1995) highlighted the importance of tracing how the solver's evolving ideas organize and structure their PS and how these evolving idea structures ultimately contribute to students developing powerful mathematical knowledge. The implications of these foci are found in the sophisticated analyses of PS presented at more recent SEMT symposia (see Brehovsky et al. 2015; Prusak and Hershkowitz 2019; Maj-Tatsis and Tatsis 2015). In particular, Brehovsky et al. (2015) and Maj-Tatsis and Tatsis (2015) examined the mathematical thinking processes of 11- and 8-year-old students, respectively, and represent two of the earliest analyses to focus on advanced mathematical thinking (such as conjectures, generalizations, justifications, and refutations) in young children. In addition, the Prusak and Hershkowitz (2019) research demonstrated how implementing rich PS tasks within classroom instruction may promote inquiry-based learning.

This foundational work led to a more resolute and thematic approach to unpacking PS. Several presentations called for analyses that focus on how problem solvers interpret and resolve the problems that they face during PS. Among these, Irwin (1995) reminds us that PS is often much more complex than the mere completion of tasks, and that a problem's solution may be more a resolution of a cognitive conflict, than a step-wise application of mathematical rules and skills.

Consistent with the idea that PS is more elaborate than first thought, Kaslová (1995) advocated for studies that shed light on the informal actions of solvers in problem situations, arguing that these idiosyncratic actions often precede formal actions, an idea found in later presentations of the PS and PP of third- and fourth-graders conducted by Cifarelli and his colleagues (i.e. Cifarelli 2009; Cifarelli et al. 2017; Cifarelli, Stephan, and Pugalee 2015). Presenters (Lemut and Dettori 1995) also argued for the value of maintaining a multiple-perspective approach to study PS (e.g. perspectives such as semiotics and Realistic Mathematics Education). Additionally, Fonová (1995) presented a single-subject qualitative analysis of a seven-year-old's PS within a fairy tale micro-world, an analytical approach that was quite progressive for the times and figured prominently in later SEMT symposia (Mamede 2003; Stehlí ková 1999). Finally, Koman's (1995) presentation on word puzzles explained the importance of considering self-reflective action in our PS analyses, a theme also found in Semadeni (1995), who cited the importance of considering how solvers self-generate and test hypotheses about potential activity in the course of their continuing PS. Semadeni's work is particularly noteworthy because it represented the first presentation at SEMT where the importance of PP during PS is mentioned, a topic that has been featured prominently in later SEMT presentations (Bevan, Williams, and Capraro 2019; Bureš and Clanché 2007; Calabrese, Williams, and Capraro 2019; Cifarelli 2009; Cifarelli et al. 2015). Hence, the presentations from SEMT 1995 served as important theoretical foundations for the sophisticated analyses we now commonly see at SEMT.

SEMT as a leader in explaining how PS contributes to mathematics learning

A long-term challenge for researchers of PS is to document its connection to mathematical learning. One would think that, with all of the research we have conducted, using a variety of theoretical perspectives to examine PS, since Polya's (1945) seminal work on the subject, we would have constructed a compelling account of how PS contributes to mathematics learning (Lester 2013; Schoenfeld 1992). There are several reasons why no such account has yet appeared. We mentioned earlier in this chapter the challenge of reaching a consensus about issues involved in PS and PP among various perspectives. In addition, there has always been a debate over how PS should be viewed as a goal of mathematics teaching. For example, Lester (2013) identified a common quandary about the teaching of PS that all mathematics educators have pondered: should we consider PS as a set of processes we teach students to nurture and embolden their own mathematical autonomy, or should we view PS as a means through which mathematics is learnt? Lester had more to say on the matter:

> Furthermore, there is the issue of whether problem solving is intended as the end result of instruction, or the means through which mathematical concepts, processes, and procedures are learned. Or, to put it another way, should teachers adopt 'teaching for problem solving,'—an ends approach—or 'teaching *via* problem solving'—a means approach? In my view, the answer to this question is that both approaches have merit; problem solving should be both an end result of learning mathematics and the means through which mathematics is learned. (p. 246)

We must admit that SEMT has certainly not settled the debate about the role of PS in mathematics learning. However, we will point to some SEMT presentations that argue each side of the issue (PS as a goal of instruction versus PS as a means to learn mathematics). In addition, we will review research that we believe offers an important link between PS and learning, namely PP.

In our review we found several SEMT presentations that take the view that PS can be a means through which to learn mathematics. Stehlíková (1999) and Back, Piggott, and Pumfrey (2003) provided examples where PS is embraced for its potential to influence students' learning. In their work with children aged 10 to 12, they found that the use of rich mathematical tasks provided authentic conceptual challenges for students, the negotiation of which had the potential to structure and re-structure students' mathematical knowledge. Noteworthy in these presentations was the view that structuring and re-structuring can occur in the course of PS, a view of mathematics learning that was quite dynamic for the times.

In contrast, we found SEMT presentations that took the view that teaching PS as an end goal is useful because it can empower students to become autonomous in their learning. Following this theme, several SEMT presentations examined the use of advanced mathematical processes as part of the students' repertoire of PS tools. Maj-Tatsis and Tatsis (2015) hypothesized that young children employ reasoning processes every bit as mathematically sophisticated as those of older children. Their presentation focused on the PS of a pair of eight-year-olds solving non-routine mathematics problems. Their observations included examples of the students' development of conjectures, generalizations, justifications, and validation of justifications.

Strategies for solving problems were also explored in SEMT presentations. Gökcek and Azak (2015) provided a substantial analysis of the PS of sixth-graders that extended previous research on 'Guess and Check' and 'Working Backwards', two popular PS strategies employed by young children. In addition, Mamede (2003) provided a rich analysis of two first-grade students using calculators to validate their emerging hypotheses about potential problem solutions. Taking the two results together, we can see that there is a clear evolution in thinking about both strategies for unpacking PS as well as unpacking how students process the PS experience and learn to draw conclusions and select solution pathways.

Many of the SEMT presenters proffered an ambitious and fast-paced evolution of the PS theoretical framing. The ambitious theoretical scope of PS processes as described in these presentations could be found as integrated veins in other studies we reviewed. For example, Brehovsky et al. (2015) addressed the 'importance of [the] experimental heuristic strategies' (p. 72) of 11-year-old students solving non-traditional problems. Specifically, they examined systematic experimentation strategies in students' PS that included systematic 'Guess and Test', formulation and revision of conjectures, and introduction of auxiliary information to solve problems. Although the findings contained few examples of these strategies in students' PS (with most students using a Working Backwards PS strategy), the presenters provided an impressive conceptual framework with which to examine how PS has an impact on students' autonomous actions. This framework may be the earliest evidence that underlies some of the most contemporary work on PS, that of creativity.

Translational evolution of the PS framework

Although PP was not a new idea at the time of the first SEMT symposium, it was often relegated to a secondary place in the research and education community. We draw attention to SEMT presentations that examined the role that PP can play in PS, with particular focus on possible connections between these processes, and how they might contribute to students' learning. The controversy lies primarily in the idea that, for PP to be prominent, students must already have strong language skills, understand both mathematical relationships and mathematical vocabulary, and possess the syntactic ability to join these ideas in meaningful ways (Capraro, Capraro, and Rupley 2010, 2011). Because these ideas were often considered the purview of teachers of reading, PP was often only used in upper grades, where students have fully developed the necessary semantic and syntactic skills.

When we think of PP, we are reminded of the challenges that come with trying to establish important connections between PS and PP. According to Silver (1994), although we may intuitively consider connections between PP and PS as different, but related, aspects of mathematical activity, establishing the meaningful links between the two (if they exist) has proved most difficult. With the exception of English's (1997) research on the PP of fifth-graders, and the work of Silver and Cai (1996), there has been little research on the PP of young children.

The SEMT presentations by Bureš and Clanché (2009) and Cifarelli (2009) considered PP to be naturally related to PS. In particular, Cifarelli (2009) found that the process of solving a problem presents to the solver opportunities for new questions to emerge. This finding is consistent with Brown and Walter's (1993) claim that:

> we need not wait until after we have solved a problem to generate new questions; rather, we are logically obligated to generate a new question or pose a new problem in order to solve a problem in the first place. (p. 114)

As the theory continued to evolve, there was an interest in students self-scaffolding their learning. That is, students building on what they learnt to make leaps into new learning. Cifarelli (2009) observed that a pair of fourth-graders engaged in PS tasks proceeded to pose new questions and problems to solve; PP was seen as a way for solvers to do the same. This finding showed that young students were able to build their own new PP/PS knowledge based on what they had just learnt. Thus, PP and PS may be viewed as naturally related in the sense that, in order to solve the original problem, the solver generates additional questions or problems that must be addressed.

Links between mathematical proficiency and PP and PS have been slow to emerge. More recently, the work of Bevan, Williams, and Capraro (2019) and Calabrese, Williams, and Capraro (2019) represent important studies of the next iteration of PP theory. In fact, PP plays an essentially important role in early elementary mathematics acquisition. The Calabrese, Williams, and Capraro (2019) presentation demonstrated, with great detail, how PP strategies can be used by second- and fourth-grade students to improve their rate of mathematical learning, and the speed at which they acquire fairly complex mathematical semantics and syntax. The Bevan, Williams, and Capraro (2019) presentation established a clear relationship between students' improved mathematical attitudes and their mathematical achievement when they learnt to pose problems. They supposed that students may have achieved some level of mathematical power and, perhaps, even enjoyed learning about PP. They also hypothesized that the improvement in students' mathematical attitudes was also a cause of their improved mathematical achievement. These two presentations, when taken together, show that PP supports PS and facilitates mathematics learning. Finally, the presentations by (Calabrese, Williams, and Capraro (2019) and Bevan, Williams, and Capraro (2019) included substantial analyses with relatively large numbers of students (N = 137 and N = 53, respectively) as compared to prior, primarily theory-driven work.

The effectiveness of specific instructional activities on PS or PP

SEMT presenters have actively investigated certain instructional activities and pedagogies. It is important to note that, outside the United States of America, the idea of didactics is often used synonymously with pedagogy: therefore, in this work we too honour that practice. Despite finding many such SEMT presentations from the symposia, we chose two to discuss, because they strongly connect with most of the important themes we have summarized in this chapter.

The idea of mathematical culture comes up as an important theme in SEMT presentations. This is a broad construct that has been disaggregated into sub-components (Baroody, Baroody, and Coslick 1998; Rowan and Robles 1998), and mathematical identity (Esmonde 2009; Ma and Singer-Gabella 2011). The presentation of Prusak and Hershkowitz (2019) described an ambitious study of the effectiveness of using instructional design research to implement a PS culture within a third-grade mathematics classroom. The study on which the presentation was based came from a year-long project on the mathematical learning of third-graders. Their presentation included a detailed discussion of the theoretical framework that drove the design of the following instructional activities: creating collaborative situations, socio-cognitive conflicts, providing tools for checking hypotheses, and inviting students to reflect on their solutions. These constructs are closely aligned with the development of mathematical power and mathematical identity. Prusak and Hershkowitz (2019) presented a detailed case study in which small groups of third-grade students explored, and solved, problems involving the geometric fact that the enlargement of a square by a linear scale factor of n will increase its area by a factor of n^2. The related instructional activities were designed to trigger productive peer argumentation that has impacts on students' conceptual learning. But what makes this noteworthy is that it made use of educational software.

The activity of these presenters encouraged students to collaborate as well as engage in using a digital tool in their learning. The students were asked, specifically, to work in pairs or triads, and to reach an agreement about their solutions. These collaborations led students to compare solutions, and because students were requested to justify their solutions, these justifications naturally created socio-cognitive conflicts. As the presenters document, in the illustrative examples of student PS, the students used their constructed diagrams as a heuristic to help them justify their solutions. The use of the free digital geometry software GeoGebra served as a tool for checking their hypotheses. The analysis of the data showed the positive impact of using this tool. Hence, the presentation showed the importance of the design and of the culture of PS that led students to engage in productive argumentation and, as a result, gain a meaningful understanding of the concept of area. The presenters argued persuasively about the acquisition of mathematical power and identity, the effectiveness of instilling a PS culture, and for promoting conceptual learning. The primary outcome, from this year-long study of third-graders, was that the theoretical model promoted students' understanding of the mathematical concepts and assisted students in becoming more capable and competent as mathematical problem solvers.

The link between PS and PP is a more recent advance by SEMT presenters

The presentation by Bevan, Williams, and Capraro (2019), discussed previously, presented a persuasive analysis of how the implementation of PP within second- and fourth-grade classrooms led to increased motivation and a greater interest in mathematics as well as an improvement in mathematical achievement. Through their presentation, a convincing connection between the students' PP and an increase in their academic achievement was documented. The design of their study made use of two important factors. The first was the targeted instruction, by a trained instructor, to use PP to stimulate mathematics interest. The second was the careful design, and scaffolding, of mathematical tasks to build competence and mathematical power. As a result, this is one of very few examples that explain how PP has impacts on PS performance. The presenters' goal was to investigate how PP can increase students' achievement in, and positive attitudes towards, mathematics, as well as the relationship between PP and PS. Importantly, they examined the important role that PP can play in classroom learning, arguing that PP activities present creative-thinking opportunities for students because they allow for student ownership in learning, leading to meaningful changes in students' mathematical attitudes and achievement. Utilizing PP activities in the mathematics classroom allows teachers to apply strategies that foster creative thinking, thus improving students' attitudes and achievement.

Teachers can encourage creativity in the mathematics classroom and test students' knowledge and understanding of mathematical concepts through PP and giving students the opportunity to write their own word problems. In addition, students can examine and reflect upon the problems they have created to determine if their problems are realistic and solvable (Bevan, Williams, and Capraro 2019, p. 83).

The design for the Bevan, Williams, and Capraro (2019) research was straightforward: elementary school students (N = 53) in the second and fourth grades were assigned to either a PP or a PS intervention. A quasi-experimental design was used in order to determine the relationship between improved mathematical attitude in PP activities and mathematical achievement. One group was taught using PP strategies, and the other group was taught using only PS strategies. Students were administered an attitude survey and achievement assessment at the beginning and end of the intervention period of ten weeks. The achievement assessment included both PS and PP items. Results showed a positive, statistically significant relationship ($p < 0.001$) between increased mathematical attitude and mathematical achievement in students who were involved in the PP activities. Results suggest that allowing for creative-thinking activities leads to both positive attitudes towards mathematics and increased mathematical achievement among students as compared with those having only experiences with PS.

Conclusion

In this chapter we addressed how presentations at SEMT, over the 30 years of symposia, have embraced the view that PS and PP play prominent roles in the mathematical learning of young children. Although we provided only a small fraction of the high-quality presentations at SEMT symposia, we believe they provide an effective snapshot of how SEMT has distinguished itself as a leader in showcasing research on the mathematical learning of young children. We expect that as SEMT looks to its next 30 years, the symposia will continue to provide a venue for contemporary analyses of the mathematical learning of young children.

References

Ay, Y. (2017). A review of research on the misconceptions in mathematics education. In Shelley, M., and Pehlivan, M. (Eds) *Education Research Highlights in Mathematics, Science and Technology 2017*, 21–31. Ames, IA: ISRES Publishing.

Back, J., Piggott, J., and Pumfrey, J. (2003). Using non-standard problems to challenge preconceptions: Can they extend knowledge? In J. Novotná (Ed.) SEMT 43–45.

Baroody, A., Baroody, A. J., and Coslick, R. (1998). *Fostering Children's Mathematical Power: An Investigative Approach to K-8 Mathematics Instruction*. Abingdon: Routledge.

Bevan, D., Williams, A., and Capraro, M. (2019). Strike a pose: The impact of problem posing on elementary students' mathematical attitude and achievement. In J. Novotná and H. Moraová (Eds) SEMT 80–88.

Brehovsky, J., Eisenmann, P., Novotná, J., and Pribyl, J. (2015). Solving problems using experimental strategies. In J. Novotná and H. Moraová (Eds) SEMT 72–81.

Brown, C., and Clark, L. (Eds) (2006). *Learning from NAEP: Professional Development Materials for Teachers of Mathematics*. Reston, VA: National Council of Teachers of Mathematics.

Brown, S. I., and Walter, M. I. (Eds) (1993). *Problem Posing: Reflections and Applications.* New York, NY: Psychology Press.
Bureš, J., and Clanché, P. (2009). Problem posing as a mathematical activity: The young teachers' perspective. In J. Novotná and H. Moraová (Eds) SEMT 88–95.
Calabrese, J., Williams, A., and Capraro, M. (2019). Problem-posing strategies: Showcasing elementary student responses. In J. Novotná and H. Moraová (Eds) SEMT 105–113.
Capraro, R., Capraro, M., and Rupley, W. (2010). Semantics and syntax: A theoretical model for how students may build mathematical mis-understandings. *Journal of Mathematics Education,* 3(2): 58–66.
Capraro, R., Capraro, M., and Rupley, W. (2011). Reading-enhanced word problem solving (REPS): A theoretical model. *European Journal of Psychology of Education*, 27(1): 91–114.
Cifarelli, V. V. (2009). Examining the problem posing of 4th graders: How Corrine and Austin made sense of their problems. In J. Novotná and H. Moraová (Eds) SEMT 111–117.
Cifarelli, V., Stephan, M., and Pugalee, D. (2015). Exploring the problem posing of third graders. In J. Novotná and H. Moraová (Eds) SEMT 91–98.
Cifarelli, V., Stephan, M., Pugalee, D., and Wang, C. (2017). Implementing an instructional sequence for solving net worth problems. In J. Novotná and H. Moraová (Eds) SEMT 148–157.
Driscoll, M., and Bryant, D. S. (1998). *Learning about Assessment, Learning through Assessment.* Washington, DC: National Academies Press.
English, L. D. (1997). Development of fifth-grade children's problem-posing abilities. *Educational Studies in Mathematics*, 34: 183–217.
Esmonde, I. (2009). Ideas and identities: Supporting equity in cooperative mathematics learning. *Review of Educational Research*, 79(2): 1008–1043.
Fonová, A. (1995). Experimenter-pupil verbal interaction (Analysis of one interesting case). In M. Hejný and J. Novotná (Eds) SEMT 77–79.
Gökcek, T., and Azak, S. (2015). Sixth grade students' strategies for solving mathematical problems. In J. Novotná and H. Moraová (Eds) SEMT 127–136.
Hošpesová, A. (1991). Solving word problems by children aged six to ten. In J. Novotná (Ed.) SEMT 21–26.
Hošpesová, A. (1995). The analysis of children's solutions of word problems. In M. Hejný and J. Novotná (Eds) SEMT 85–86.
Irwin, K. (1995). Using conflict in word problems to understand decimal fractions. In M. Hejný and J. Novotná (Eds) SEMT 87–90.
Kaslová, M. (1995). Problem solving strategies. In M. Hejný and J. Novotná (Eds) SEMT 93–96.
Koman, M. (1995). Word puzzles – future teachers – pupils relay in teacher education. In M. Hejný and J. Novotná (Eds) SEMT 96–100.
Lemut, E., and Dettori, G. (1995). Representations in problem solving activities. In M. Hejný and J. Novotná (Eds) SEMT 90–93.
Lester, F. (1994). Musings about mathematical problem-solving research: 1970–1994. *Journal of Research in Mathematics Education*, 25(6): 660–675.
Lester, F. (2013). Thoughts about research on mathematical problem-solving instruction. *The Mathematics Enthusiast*, 10(1): 245–278.
Ma, J., and Singer-Gabella, M. (2011). Learning to teach in the figured world of reform mathematics: Negotiating new models of identity. *Journal of Teacher Education*, 62(1): 8–22.
Maj-Tatsis, B., and Tatsis, K. (2015). Reasoning processes of two girls during problem solving. In J. Novotná and H. Moraová (Eds) SEMT 222–231.
Mamede, E. (2003). Children's beliefs about the use of calculator in exploratory problems solving processes. In J. Novotná (Ed.) SEMT 108–112.
Novotná, J., (1995). Analysis of text comprehension in pupils' solutions to word problems via atomic analysis. In M. Hejný and J. Novotná (Eds) SEMT 114–117.
Polya, G. (1945). *How to Solve It.* Princeton, NJ: Princeton University Press.
Prusak, N., and Hershkowitz, R. (2019). Task designed for promoting the concept of area in a problem solving environment – the story of one activity. In J. Novotná and H. Moraová (Eds) SEMT 325–333.

Rowan, T. E., and Robles, J. (1998). Using questions to help children build mathematical power. *Teaching Children Mathematics*, 4(9): 504–510.

Schoenfeld, A. (1992). Learning to think mathematically: Problem solving, metacognition, and sense making in mathematics. In D. A. Grouws (Ed.) *Handbook of Research on Mathematics Teaching and Learning*, 334–370. Reston, VA: National Council of Teachers of Mathematics.

Semadeni, Z. (1995). Developing children's understanding of verbal arithmetical problems. In M. Hejný and J. Novotná (Eds) SEMT 27–32.

Silver, E. (1994). On mathematical problem posing. *For the Learning of Mathematics*, 14(1): 19–28.

Silver, E., and Cai, J. (1996). An analysis of arithmetic problem posing by middle school students. *Journal for Research in Mathematics Education*, 27(5): 521–539.

Stehlíková, N. (1993). Word problems pupils' results in the Czech Republic and England. In M. Hejný and J. Novotná (Eds) SEMT 52–54.

Stehlíková, N. (1999). Some observed phenomena of pupils' abilities to structure and restructure mathematical knowledge during specific mathematical tasks. In M. Hejný and J. Novotná (Eds) SEMT 167–170.

Szetela, W. (1991). A study of verbalization in problem solving in grades 4 and 5. In J. Novotná (Ed.) SEMT 58–61.

Szetela, W. (1993). Communication of critical thinking in problem solving in grade 4. In M. Hejný and J. Novotná (Eds) SEMT 99–102.

Afterword

Brian Doig
Deakin University, Australia
Jarmila Novotná
Charles University, Faculty of Education, Prague, Czech Republic
Chercheur titulaire, CeDS, Université Bordeaux, France

The contents of the 16 chapters, by several authors, cannot be summarized, but this work does need some comments now that it is completed.

First, it must be admitted that a summary of a work that reports and summarizes 30 years of research into elementary mathematics teaching and learning, in many of their aspects, would appear to be ridiculous. Thus, we are not going to consider this course of action.

However, some comments are needed about the past and present of SEMT, and its foreseeable future.

Although set within the difficulties of a pandemic, SEMT 2021 went ahead, albeit in a hybrid format with about a third of participants on-site. While this meant some difficulties due to the differences in the time zones of participants, staying up late and getting up very early were well worth the effort. The SEMT 2021 research papers and plenary presentations were again of the high standard expected at SEMT. Technical issues were negligible, and the different sessions went smoothly. Charles University Faculty of Education, staff, and students are to be congratulated on their fine work.

Again, the range of presentations offered the participants a window into present research into teaching elementary mathematics across a range of countries and viewpoints. This is, as ever, due to the hard work of the members of the International Programme Committee, the Advisory Board, and the paper reviewers.

Of course, there were comments about not being able to be in Prague for SEMT: an unavoidable issue for many participants. However, it was pointed out that 2023 was more than likely to be in Prague for everyone, and it was only two years away!

This book is also a virtual SEMT, in that in the chapters of the contributors we are able, simply, 'to be' at SEMT over the entire 30 years of its history. Like its recent virtual SEMT, these chapters give the reader some very precise insights into what was at SEMT over the 30 years, and how this fitted into the wider aspects of research into elementary mathematics teaching and learning.

There are participants who have participated at SEMT symposia since the first SEMT in 1991. On the other hand, each symposium attracts several new participants. This diversity guarantees a tradition of collaboration and research into new trends, and results in the high quality of the scientific programme. A confirmation of the high quality of the SEMT symposia is that, since 2005, SEMT proceedings are included in the Web of Science.

In 20 years time, after another ten more SEMT symposia, a group of researchers will, no doubt, wish to add the further history of SEMT, or even produce a 50-year history, using this present work as part of the overall history of changes in elementary mathematics teaching and learning research. Naturally we, the present editors, wish them well in whatever form their endeavour takes.

One thing is certain: there will be further changes in the emphasis and direction of research into elementary mathematics teaching and learning that will come to SEMT. Reading the present volume, it is clear that these future trends are not easily predicted, although some of the later trends, discerned by the present authors, will continue, probably, for some time.

These trends indicate that there is no one, single, solution to many of the issues faced by teachers in the last 30 years. While there have been some strictly national approaches, SEMT presentations have also shown endeavours that have seen many useful and beneficial approaches to addressing these issues in a remarkably innovative and successful manner internationally.

We believe that future International Programme Committee (IPC) members will continue the tradition of selecting themes for SEMT that will draw researchers to present their work in papers, posters, and workshops. Further, the IPC will continue to select plenary speakers who will inspire researchers, both new and experienced, to pursue research in the critical areas of mathematics education in the elementary years that come into existence in the years to come.

The future is not predictable, of course, but we look forward to our 50-year anniversary, with a further 20 years of excellent SEMT symposia to draw upon. We hope that you will join us for part, or all, of this future. And so, as is said as a new ship is launched, may it take all on to new adventures and discoveries, and so, too, we wish future SEMT symposia to take all who support them on to future research adventures and discoveries.

Brian Doig
Jarmila Novotná

Index

For the benefit of digital users, indexed terms that span two pages (e.g., 52–53) may, on occasion, appear on only one of those pages.